Matrix Differential Calculus with Applications in Statistics and Econometrics

WILEY SERIES IN PROBABILITY AND STATISTICS

Established by WALTER A. SHEWHART and SAMUEL S. WILKS

A complete list of the titles in this series appears at the end of this volume

Matrix Differential Calculus with Applications in Statistics and Econometrics

Revised Edition

JAN R. MAGNUS

CentER for Economic Research, Tilburg University

and

HEINZ NEUDECKER

Cesaro, Schagen

JOHN WILEY & SONS

Chichester • New York • Weinheim • Brisbane • Singapore • Toronto

National 01243 779777
International (+44) 1243 779777
e-mail (for orders and customer service enquiries): cs-books@wiley.co.uk
Visit our Home Page on http://www.wiley.co.uk
or http://www.wiley.com

Reprinted December 1999, May 2001, July 2002

Other Wiley Editorial Offices

John Wiley & Sons, Inc., 605 Third Avenue,
New York, NY 10158-0012, USA

Wiley-VCH Verlag GmbH, Pappelallee 3,
D-69469 Weinheim, Germany

Jacaranda Wiley Ltd, 33 Park Road, Milton,
Queensland 4064, Australia

John Wiley & Sons (Asia) Pte Ltd, Clementi Loop #02-01,
Jin Xing Distripark, Singapore 129809

John Wiley & Sons (Canada) Ltd, 22 Worcester Road,
Rexdale, Ontario M9W 1L1, Canada

Library of Congress Cataloging in Publication Data

Magnus, Jan R.
Matrix differential calculus with applications in statistics and econometrics / J.R. Magnus and H. Neudecker. — Rev. ed.
p. cm.
Includes bibliographical references and index.
ISBN 0-471-98632-1 (alk. paper). — ISBN 0-471-98633-X (pbk. : alk. paper)
1. Matrices. 2. Differential calculus. 3. Statistics. 4. Econometrics. I. Neudecker, Heinz. II. Title.
QA188.M345 1999
512.9′434—dc21 98-53556
CIP

British Library Cataloguing in Publication Data

A catalogue record for this book is available from the British Library

ISBN 0-471-98632-1; 0-471-98633-X (pbk)

Contents

Preface

There has been a long-felt need for a book that gives a self-contained and unified treatment of matrix differential calculus, specifically written for econometricians and statisticians. The present book is meant to satisfy this need. It can serve as a textbook for advanced undergraduates and postgraduates in econometrics and as a reference book for practising econometricians. Also mathematical statisticians and psychometricians may find something to their liking in the book.

When used as a textbook it can provide a full semester course. Reasonable proficiency in basic matrix theory is assumed, especially with use of partitioned matrices. The basics of matrix algebra as deemed necessary for a proper understanding of the main subject of the book are summarized in the first of the book's six parts. The book also contains the essentials of multivariable calculus but geared to and often phrased in terms of differentials.

The sequence in which the chapters are read is not of great consequence. It is fully conceivable that practitioners start with Part Three (Differentials: the practice) and dependent on their predilections carry on to any of the applied Parts Five or Six. Those who want a full understanding of the underlying theory should read the whole book, although even then they could go through the necessary matrix algebra when the specific need arises.

Matrix differential calculus as presented in this book is based on differentials, and this sets the book apart from other books in this area. The approach via differentials is in our opinion superior to any other existing approach. Our leading idea is that differentials are more congenial to multivariable functions as they crop up in econometrics, mathematical statistics or psychometrics than derivatives, although from a theoretical point of view the two concepts are equivalent. When there is a specific need for derivatives they will be obtained from differentials.

The book falls into six parts. Part One deals with matrix algebra. It lists—and often also proves—items like the Schur, Jordan and singular-value decompositions, concepts like the Hadamard and Kronecker products, the vec operator, the commutation and duplication matrices, the Moore–Penrose inverse, and results on bordered matrices (and their determinants) and (linearly restricted) quadratic forms.

Part Two, which forms the theoretical heart of the book, is entirely devoted to a thorough development of the theory of differentials. It presents the essentials of calculus but geared to and phrased in terms of differentials. First and second differentials are defined, 'identification' rules for Jacobian and Hessian matrices are given, and chain rules derived. A separate chapter on the theory of (constrained) optimization in terms of differentials concludes this part.

Part Three is the practical core of the book. It contains the rules for working with differentials, lists the differentials of important scalar, vector and matrix functions (*inter alia* eigenvalues, eigenvectors and the Moore–Penrose inverse) and supplies 'identification' tables for Jacobian and Hessian matrices.

Part Four (one chapter on inequalities) owes its existence to our feeling that econometricians should be conversant with inequalities like the Cauchy–Schwarz and Minkowski (and extensions thereof), but should also master a powerful result like Poincaré's separation theorem. This part is to some extent also the case history of a disappointment. When we started writing this book we had the ambition to derive all inequalities by means of matrix differential calculus. After all, every inequality can be rephrased as the solution of an optimization problem. This proved to be an illusion, due to the fact that the Hessian matrix in most cases is singular at the optimum point.

Part Five is entirely devoted to applications of matrix differential calculus to the linear regression model. There is an exhaustive treatment of estimation problems related to the fixed part of the model under various assumptions concerning ranks and (other) constraints. It further has topics relating to the stochastic part of the model, viz. estimation of the error variance and prediction of the error term. There is also a small section on sensitivity analysis. An introductory chapter deals with the necessary statistical preliminaries.

Part Six deals with maximum likelihood estimation, which is of course an ideal source for demonstrating the power of the propagated techniques. In the first of three chapters, several models are being analysed, *inter alia* the multivariate normal distribution, the errors-in-variables model and the nonlinear regression model. There is a discussion of how to deal with symmetry and positive definiteness, and special attention is given to the information matrix. The second chapter in this part deals with simultaneous equations under normality conditions. It investigates both identification and estimation problems, subject to various (non)linear constraints on the parameters. There is a separate treatment of full-information maximum likelihood (FIML) and limited-information maximum likelihood (LIML) with special attention to the derivation of asymptotic variance matrices. The final chapter addresses itself to various psychometric problems, *inter alia* principal components, multimode component analysis, factor analysis and canonical correlation.

All chapters contain many exercises. These are frequently meant to be complementary to the main text. References to equations, theorems and sections are given as follows: equation (1) refers to an equation within the same section; (2.1) refers to an equation in section 2 within the same chapter; and (3.2.1) refers to an equation in section 2 of Chapter 3. Similarly, we refer to theorems and sections within the same chapter by a single serial number, and to theorems and sections in other chapters by double numbers. The notation is mostly standard, except that matrices and vectors are printed in italic, not in bold face. Special symbols are used to denote derivative D and Hessian H. The differential operator is denoted by (roman) d.

A large number of books and papers have been published on the theory and applications of matrix differential calculus. Without attempting to describe their relative virtues and particularities, the interested reader may wish to consult Dwyer and MacPhail (1948), Bodewig (1959), Wilkinson (1965), Dwyer (1967), Neudecker (1967, 1969), Tracy and Dwyer (1969), Tracy and Singh (1972), McDonald and Swaminathan (1973), MacRae (1974), Balestra (1976), Bentler and Lee (1978), Henderson and Searle (1979), Wong and Wong (1979, 1980), Nel (1980), Rogers (1980), Wong (1980, 1985), Graham (1981), McCulloch (1982), Schönemann (1985), Magnus and Neudecker (1985), Pollock (1985), Don (1986), and Kollo (1991). The papers by Henderson and Searle (1979) and Nel (1980) and Rogers' (1980) book contain extensive bibliographies.

The two authors share the responsibility for Parts One, Three, Five and Six, although any new results in Part One are due to Magnus. Parts Two and Four are due to Magnus, although Neudecker contributed some results to Part Four. Magnus is also responsible for the writing and organization of the final text.

We wish to thank our colleagues F. J. H. Don, R. D. H. Heijmans, D. S. G. Pollock and R. Ramer for their critical remarks and contributions. The greatest obligation is owed to Sue Kirkbride at the London School of Economics who patiently and cheerfully typed and retyped the various versions of the book. Partial financial support was provided by the Netherlands Organization for the Advancement of Pure Research (Z. W. O.) and the Suntory Toyota International Centre for Economics and Related Disciplines at the London School of Economics.

Preface to the first revised printing

Since this book first appeared—now almost four years ago—many of our colleagues, students and other readers have pointed out typographical errors and have made suggestions for improving the text. We are particularly grateful to R. D. H. Heijmans, J. F. Kiviet, I. J. Steyn and G. Trenkler. We owe the greatest debt to F. Gerrish, formerly of the School of Mathematics in the Polytechnic, Kingston-upon-Thames, who read Chapters 1-11 with awesome precision and care and made numerous insightful suggestions and constructive remarks. We hope that this printing will continue to trigger comments from our readers.

London/Tilburg/Amsterdam
February 1991

Jan R. Magnus
Heinz Neudecker

Preface to the second revised printing

A further seven years have passed since our first revision in 1991. We are happy to see that our book is still being used by colleagues and students. In this revision we attempted to reach three goals. First, we made a serious attempt to keep the book up-to-date by adding many recent references and new exercises. Secondly, we made numerous small changes throughout the text, improving the clarity of exposition. Finally, we corrected a number of typographical and other errors.

The structure of the book and its philosphy are unchanged. Apart from a large number of small changes, there are two major changes. First, we interchanged Sections 12 and 13 of Chapter 1, since complex numbers need to be discussed before eigenvalues and eigenvectors, and we corrected an error in Theorem 1.7. Secondly, in Chapter 17 on psychometrics, we rewrote Sections 8–10 relating to the Eckart–Young theorem.

We are grateful to Karim Abadir, Paul Bekker, Hamparsum Bozdogan, Michael Browne, Frank Gerrish, Kaddour Hadri, Tõnu Kollo, Shuangzhe Liu, Daan Nel, Albert Satorra, Kazuo Shigemasu, Jos ten Berge, Peter ter Berg, Götz Trenkler, Haruo Yanai and many others for their thoughtful and constructive comments. Of course, we welcome further comments from our readers.

Tilburg/Amsterdam
March 1998

Jan R. Magnus
Heinz Neudecker

Part One—

Matrices

CHAPTER 1

Basic properties of vectors and matrices

1. INTRODUCTION

In this chapter we summarize some of the well-known definitions and theorems of matrix algebra. Most of the theorems will be proved.

2. SETS

A *set* is a collection of objects, called the *elements* (or members) of the set. We write $x \in S$ to mean 'x is an element of S', or 'x belongs to S'. If x does not belong to S we write $x \notin S$. The set that contains no elements is called the *empty set*, denoted $\emptyset$. If a set has at least one element, it is called non-empty.

Sometimes a set can be defined by displaying the elements in braces. For example $A = \{0, 1\}$ or

$$\mathbb{N} = \{1, 2, 3, \dots\}. \tag{1}$$

Notice that A is a finite set (contains a finite number of elements), whereas $\mathbb{N}$ is an infinite set. If P is a property that any element of S has or does not have, then

$$\{x : x \in S, x \text{ satisfies } P\} \tag{2}$$

denotes the set of all the elements of S that have property P.

A set A is called a *subset* of B, written $A \subset B$, whenever every element of A also belongs to B. The notation $A \subset B$ does not rule out the possibility that $A = B$. If $A \subset B$ and $A \neq B$, then we say that A is a *proper subset* of B.

If A and B are two subsets of S, we define

$$A \cup B, \tag{3}$$

the *union* of A and B, as the set of elements of S that belong to A or to B (or to both), and

$$A \cap B, \tag{4}$$

the *intersection* of A and B, as the set of elements of S that belong to both A and B. We say that A and B are (mutually) *disjoint* if they have no common elements, that is, if

$$A \cap B = \varnothing. \tag{5}$$

The *complement* of A relative to B, denoted by $B-A$, is the set $\{x : x \in B, \text{ but } x \notin A\}$. The complement of A (relative to S) is sometimes denoted A^c.

The *Cartesian product* of two sets A and B, written $A \times B$, is the set of all ordered pairs (a, b) such that $a \in A$ and $b \in B$. More generally, the Cartesian product of n sets $A_1, A_2, \ldots, A_n$, written

$$\prod_{i=1}^{n} A_i, \tag{6}$$

is the set of all ordered n-tuples $(a_1, a_2, \ldots, a_n)$ such that $a_i \in A_i$ $(i=1, \ldots, n)$.

The set of (finite) real numbers (the *one-dimensional Euclidean space*) is denoted by $\mathbb{R}$. The *n-dimensional Euclidean space* $\mathbb{R}^n$ is the Cartesian product of n sets equal to $\mathbb{R}$, i.e.

$$\mathbb{R}^n = \mathbb{R} \times \mathbb{R} \times \ldots \times \mathbb{R} \qquad (n \text{ times}). \tag{7}$$

The elements of $\mathbb{R}^n$ are thus the ordered n-tuples $(x_1, x_2, \ldots, x_n)$ of real numbers $x_1, x_2, \ldots, x_n$.

A set S of real numbers is said to be *bounded* if there exists a number M such that $|x| \leqslant M$ for all $x \in S$.

3. MATRICES: ADDITION AND MULTIPLICATION

An $m \times n$ matrix A is a rectangular array of real numbers

$$A = \begin{bmatrix} a_{11} & a_{12} & \ldots & a_{1n} \\ a_{21} & a_{22} & \ldots & a_{2n} \\ \vdots & \vdots & & \vdots \\ a_{m1} & a_{m2} & \ldots & a_{mn} \end{bmatrix}. \tag{1}$$

We sometimes write $A = (a_{ij})$. An $m \times n$ matrix can be regarded as a point in $\mathbb{R}^{m \times n}$. The real numbers a_{ij} are called the *elements* of A.

An $m \times 1$ matrix is a point in $\mathbb{R}^{m \times 1}$ (that is, in $\mathbb{R}^m$) and is called a (column) *vector* of order $m \times 1$. A $1 \times n$ matrix is called a *row vector* (of order $1 \times n$). The elements of a vector are usually called its *components*. Matrices are usually denoted by capital letters, vectors by lower-case letters.

The sum of two matrices A and B of the same order is defined as

$$A + B = (a_{ij}) + (b_{ij}) = (a_{ij} + b_{ij}). \tag{2}$$

The product of a matrix by a scalar λ is

$$\lambda A = A\lambda = (\lambda a_{ij}). \tag{3}$$

The following properties are now easily proved:

$$A+B=B+A \tag{4}$$

$$(A+B)+C=A+(B+C) \tag{5}$$

$$(\lambda+\mu)A=\lambda A+\mu A \tag{6}$$

$$\lambda(A+B)=\lambda A+\lambda B \tag{7}$$

$$\lambda(\mu A)=(\lambda\mu)A. \tag{8}$$

A matrix whose elements are all zero is called a *null matrix* and denoted 0. We have, of course,

$$A+(-1)A=0. \tag{9}$$

If A is an $m \times n$ matrix and B an $n \times p$ matrix (so that A has the same number of columns as B has rows) then we define the product of A and B as

$$AB=\left(\sum_{j=1}^{n} a_{ij}b_{jk}\right). \tag{10}$$

Thus, AB is an $m \times p$ matrix and its ik-th element is $\Sigma_{j=1}^{n} a_{ij}b_{jk}$. The following properties of the matrix product can be established:

$$(AB)C=A(BC) \tag{11}$$

$$A(B+C)=AB+AC \tag{12}$$

$$(A+B)C=AC+BC. \tag{13}$$

These relations hold provided the matrix products exist.

We note that the existence of AB does not imply the existence of BA; and even when both products exist they are not in general equal. (Two matrices A and B for which

$$AB=BA \tag{14}$$

are said to *commute*.) We therefore distinguish between pre-multiplication and post-multiplication: a given $m \times n$ matrix A can be pre-multiplied by a $p \times m$ matrix B to form the product BA; it can also be post-multiplied by an $n \times q$ matrix C to form AC.

4. THE TRANSPOSE OF A MATRIX

The transpose of an $m \times n$ matrix $A=(a_{ij})$ is the $n \times m$ matrix, denoted A', whose ij-th element is a_{ji}.

We have

$$(A')'=A \tag{1}$$

$$(A+B)'=A'+B' \tag{2}$$

$$(AB)'=B'A'. \tag{3}$$

If x is an $n \times 1$ vector then x' is a $1 \times n$ row vector and

$$x'x = \sum_{i=1}^{n} x_i^2. \tag{4}$$

The (Euclidean) *norm* of x is defined as

$$\|x\| = (x'x)^{1/2}. \tag{5}$$

5. SQUARE MATRICES

A matrix is said to be *square* if it has as many rows as it has columns. A square matrix $A = (a_{ij})$ is said to be

symmetric	if $A' = A$,
skew symmetric	if $A' = -A$,
lower triangular	if $a_{ij} = 0$ $(i < j)$,
strictly lower triangular	if $a_{ij} = 0$ $(i \leqslant j)$,
unit lower triangular	if $a_{ij} = 0$ $(i < j)$ and $a_{ii} = 1$ (all i),
upper triangular	if $a_{ij} = 0$ $(i > j)$,
strictly upper triangular	if $a_{ij} = 0$ $(i \geqslant j)$,
unit upper triangular	if $a_{ij} = 0$ $(i > j)$ and $a_{ii} = 1$ (all i),
idempotent	if $A^2 = A$.

For any square $n \times n$ matrix $A = (a_{ij})$ we define dg$A$ or dg(A) as

$$\mathrm{dg}A = \begin{bmatrix} a_{11} & 0 & \dots & 0 \\ 0 & a_{22} & \dots & 0 \\ \vdots & \vdots & & \vdots \\ 0 & 0 & \dots & a_{nn} \end{bmatrix} \tag{1}$$

or, alternatively,

$$\mathrm{dg}A = \mathrm{diag}(a_{11}, a_{22}, \dots, a_{nn}). \tag{2}$$

If $A = \mathrm{dg}(A)$, we say that A is *diagonal*. A particular diagonal matrix is the *identity matrix*,

$$I = \begin{bmatrix} 1 & 0 & \dots & 0 \\ 0 & 1 & \dots & 0 \\ \vdots & \vdots & & \vdots \\ 0 & 0 & \dots & 1 \end{bmatrix} = (\delta_{ij}), \tag{3}$$

where $\delta_{ij} = 1$ if $i = j$ and $\delta_{ij} = 0$ if $i \neq j$ (δ_{ij} is called the Kronecker delta). We have

$$IA = AI = A \tag{4}$$

if A and I have the same order.

A real square matrix A is said to be *orthogonal* if

$$AA' = A'A = I \tag{5}$$

and its columns are *orthonormal*. A rectangular matrix can still have the property that $AA' = I$ or $A'A = I$, but not both. Such a matrix is called *semi-orthogonal*.

Any matrix B satisfying

$$B^2 = A \tag{6}$$

is called a *square root* of A, denoted $A^{1/2}$. Such a matrix need not be unique.

6. LINEAR FORMS AND QUADRATIC FORMS

Let a be an $n \times 1$ vector, A an $n \times n$ matrix and B an $n \times m$ matrix. The expression $a'x$ is called a *linear form* in x, the expression $x'Ax$ a *quadratic form* in x, and the expression $x'By$ a *bilinear form* in x and y. In quadratic forms we may, without loss of generality, assume that A is symmetric, because if not then we can replace A by $(A + A')/2$:

$$x'Ax = x'\left(\frac{A + A'}{2}\right)x. \tag{1}$$

Thus, let A be a symmetric matrix. We say that A is

positive definite	if $x'Ax > 0$ for all $x \neq 0$,
positive semidefinite	if $x'Ax \geqslant 0$ for all x,
negative definite	if $x'Ax < 0$ for all $x \neq 0$,
negative semidefinite	if $x'Ax \leqslant 0$ for all x,
indefinite	if $x'Ax > 0$ for some x and $x'Ax < 0$ for some x.

It is clear that the matrices BB' and $B'B$ are positive semidefinite, and that A is negative (semi)definite if and only if $-A$ is positive (semi)definite. A square null matrix is both positive and negative semidefinite.

The following two theorems are often useful.

Theorem 1

Let $A(m \times n)$, $B(n \times p)$ and $C(n \times p)$ be matrices and let $x(n \times 1)$ be a vector. Then

(a) $$Ax = 0 \leftrightarrow A'Ax = 0,$$

(b) $$AB = 0 \leftrightarrow A'AB = 0,$$

(c) $$A'AB = A'AC \leftrightarrow AB = AC.$$

Proof. (a) Clearly $Ax = 0 \Rightarrow A'Ax = 0$. Conversely, if $A'Ax = 0$, then $(Ax)'(Ax) = x'A'Ax = 0$ and hence $Ax = 0$. (b) This follows from (a). (c) This follows from (b) by substituting $B - C$ for B in (b). □

Theorem 2

Let A be an $m \times n$ matrix, B and C $n \times n$ matrices, B symmetric. Then

(a) $Ax = 0$ for all $n \times 1$ vectors x if and only if $A = 0$,

(b) $x'Bx = 0$ for all $n \times 1$ vectors x if and only if $B = 0$,

(c) $x'Cx = 0$ for all $n \times 1$ vectors x if and only if $C' = -C$.

Proof. The proof is easy and is left to the reader. □

7. THE RANK OF A MATRIX

A set of vectors $x_1, \ldots, x_n$ is said to be *linearly independent* if $\Sigma\alpha_i x_i = 0$ implies all $\alpha_i = 0$. If $x_1, \ldots, x_n$ are not linearly independent, they are said to be linearly dependent.

Let A be an $m \times n$ matrix. The *column rank* of A is the maximum number of linearly independent columns it contains. The *row rank* of A is the maximum number of linearly independent rows it contains. It may be shown that the column rank of A is equal to its row rank. Hence the concept of *rank* is unambiguous. We denote the rank of A by

$$r(A). \tag{1}$$

It is clear that

$$r(A) \leqslant \min(m, n). \tag{2}$$

If $r(A) = m$, we say that A has full row rank. If $r(A) = n$, we say that A has full column rank. If $r(A) = 0$, then A is the null matrix and, conversely, if A is the null matrix, then $r(A) = 0$.

We have the following important results concerning ranks:

$$r(A) = r(A') = r(A'A) = r(AA') \tag{3}$$

$$r(AB) \leqslant \min(r(A), r(B)) \tag{4}$$

$$r(AB) = r(A) \qquad \text{if } B \text{ is square of full rank} \tag{5}$$

$$r(A+B) \leqslant r(A) + r(B) \tag{6}$$

and finally, if A is an $m \times n$ matrix and $Ax = 0$ for some $x \neq 0$, then

$$r(A) \leqslant n - 1. \tag{7}$$

The column space of A ($m \times n$), denoted $\mathcal{M}(A)$, is the set of vectors

$$\mathcal{M}(A) = \{y : y = Ax \text{ for some } x \text{ in } \mathbb{R}^n\}. \tag{8}$$

Thus, $\mathcal{M}(A)$ is the vector space generated by the columns of A. The dimension of this vector space is $r(A)$. We have

$$\mathcal{M}(A)=\mathcal{M}(AA') \tag{9}$$

for any matrix A.

8. THE INVERSE

Let A be a square matrix of order $n \times n$. We say that A is *non-singular* if $r(A)=n$, and that A is *singular* if $r(A)<n$.

If A is non-singular, there exists a non-singular matrix B such that

$$AB=BA=I_n. \tag{1}$$

The matrix B, denoted A^{-1}, is unique and is called the *inverse* of A. We have

$$(A^{-1})'=(A')^{-1} \tag{2}$$

$$(AB)^{-1}=B^{-1}A^{-1} \tag{3}$$

if the inverses exist.

A square matrix P is said to be a *permutation matrix* if each row and each column of P contains a single element 1, and the remaining elements are zero. An $n \times n$ permutation matrix thus contains n ones and $n(n-1)$ zeros. It can be proved that any permutation matrix is non-singular. In fact, it is even true that P is orthogonal, that is,

$$P^{-1}=P' \tag{4}$$

for any permutation matrix P.

9. THE DETERMINANT

Associated with any $n \times n$ matrix A is the determinant $|A|$ defined by

$$|A|=\sum(-1)^{\phi(j_1,\ldots,j_n)}\prod_{i=1}^{n}a_{ij_i} \tag{1}$$

where the summation is taken over all permutations $(j_1, \ldots, j_n)$ of the set of integers $(1, \ldots, n)$, and $\phi(j_1, \ldots, j_n)$ is the number of transpositions required to change $(1, \ldots, n)$ into $(j_1, \ldots, j_n)$. (A transposition consists of interchanging two numbers. It can be shown that the number of transpositions required to transform $(1, \ldots, n)$ into $(j_1, \ldots, j_n)$ is always even or always odd, so that $(-1)^{\phi(j_1,\ldots,j_n)}$ is consistently defined.)

We have

$$|AB|=|A||B| \tag{2}$$

$$|A'|=|A| \tag{3}$$

$$|\alpha A|=\alpha^n|A| \qquad \text{for any scalar } \alpha \tag{4}$$

$$|A^{-1}|=|A|^{-1} \qquad \text{if } A \text{ is non-singular} \tag{5}$$

$$|I_n|=1. \tag{6}$$

A *submatrix* of A is the rectangular array obtained from A by deleting rows and columns. A *minor* is the determinant of a square submatrix of A. The minor of an element a_{ij} is the determinant of the submatrix of A obtained by deleting the i-th row and j-th column. The *cofactor* of a_{ij}, say c_{ij}, is $(-1)^{i+j}$ times the minor of a_{ij}. The matrix $C=(c_{ij})$ is called the *cofactor matrix* of A. The transpose of C is called the *adjoint* of A and will be denoted as $A^{\#}$.

We have

$$|A|=\sum_{j=1}^{n} a_{ij}c_{ij}=\sum_{j=1}^{n} a_{jk}c_{jk} \; (i,k=1,\cdots,n) \tag{7}$$

$$AA^{\#}=A^{\#}A=|A|I \tag{8}$$

$$(AB)^{\#}=B^{\#}A^{\#}. \tag{9}$$

For any square matrix A, a *principal submatrix* of A is obtained by deleting *corresponding* rows and columns. The determinant of a principal submatrix is called a *principal minor.*

Exercises

1. If A is non-singular, show that $A^{\#}=|A|A^{-1}$.
2. Prove that the determinant of a triangular matrix is the product of its diagonal elements.

10. THE TRACE

The *trace* of a square $n \times n$ matrix A, denoted tr A or tr(A), is the sum of its diagonal elements:

$$\operatorname{tr} A=\sum_{i=1}^{n} a_{ii}. \tag{1}$$

We have

$$\operatorname{tr}(A+B)=\operatorname{tr} A+\operatorname{tr} B \tag{2}$$

$$\operatorname{tr}(\lambda A)=\lambda \operatorname{tr} A \qquad \text{if } \lambda \text{ is a scalar} \tag{3}$$

$$\operatorname{tr} A'=\operatorname{tr} A \tag{4}$$

$$\operatorname{tr} AB=\operatorname{tr} BA. \tag{5}$$

We note in (5) that AB and BA, though both square, need not be of the same order.

Corresponding to the vector (Euclidean) norm

$$\|x\|=(x'x)^{1/2} \tag{6}$$

given in (4.5), we now define the matrix (Euclidean) norm as

$$\|A\|=(\operatorname{tr} A'A)^{1/2}. \tag{7}$$

We have

$$\operatorname{tr} A'A \geqslant 0 \tag{8}$$

with equality if and only if $A=0$.

11. PARTITIONED MATRICES

Let A be an $m \times n$ matrix. We can partition A as

$$A=\begin{pmatrix} A_{11} & A_{12} \\ A_{21} & A_{22} \end{pmatrix}, \tag{1}$$

where A_{11} is $m_1 \times n_1$, A_{12} is $m_1 \times n_2$, A_{21} is $m_2 \times n_1$, A_{22} is $m_2 \times n_2$, and $m_1+m_2=m$, $n_1+n_2=n$.

Let B $(m \times n)$ be similarly partitioned into submatrices B_{ij} $(i,j=1,2)$. Then

$$A+B=\begin{pmatrix} A_{11}+B_{11} & A_{12}+B_{12} \\ A_{21}+B_{21} & A_{22}+B_{22} \end{pmatrix}. \tag{2}$$

Now let C $(n \times p)$ be partitioned into submatrices C_{ij} $(i,j=1,2)$ such that C_{11} has n_1 rows (and hence C_{12} too has n_1 rows and C_{21} and C_{22} have n_2 rows). Then we may post-multiply A by C:

$$AC=\begin{pmatrix} A_{11}C_{11}+A_{12}C_{21} & A_{11}C_{12}+A_{12}C_{22} \\ A_{21}C_{11}+A_{22}C_{21} & A_{21}C_{12}+A_{22}C_{22} \end{pmatrix}. \tag{3}$$

The transpose of the matrix A given in (1) is

$$A'=\begin{pmatrix} A'_{11} & A'_{21} \\ A'_{12} & A'_{22} \end{pmatrix}. \tag{4}$$

If the off-diagonal blocks A_{12} and A_{21} are both zero, and A_{11} and A_{22} are square and non-singular, then A is also non-singular and its inverse is

$$A^{-1}=\begin{pmatrix} A_{11}^{-1} & 0 \\ 0 & A_{22}^{-1} \end{pmatrix}. \tag{5}$$

More generally, if A as given in (1) is non-singular and $D=A_{22}-A_{21}A_{11}^{-1}A_{12}$ is also non-singular, then

$$A^{-1}=\begin{pmatrix} A_{11}^{-1}+A_{11}^{-1}A_{12}D^{-1}A_{21}A_{11}^{-1} & -A_{11}^{-1}A_{12}D^{-1} \\ -D^{-1}A_{21}A_{11}^{-1} & D^{-1} \end{pmatrix}. \tag{6}$$

Alternatively, if A is non-singular and $E=A_{11}-A_{12}A_{22}^{-1}A_{21}$ is non-singular, then

$$A^{-1}=\begin{pmatrix} E^{-1} & -E^{-1}A_{12}A_{22}^{-1} \\ -A_{22}^{-1}A_{21}E^{-1} & A_{22}^{-1}+A_{22}^{-1}A_{21}E^{-1}A_{12}A_{22}^{-1} \end{pmatrix}. \tag{7}$$

Of course, if both D and E are non-singular, blocks in (6) and (7) can be interchanged. The results (6) and (7) can be easily extended to a 3×3 matrix partition. We only consider the following symmetric case, where two of the off-diagonal blocks are null matrices.

Theorem 3

If the matrix

$$\begin{bmatrix} A & B & C \\ B' & D & 0 \\ C' & 0 & E \end{bmatrix} \tag{8}$$

is symmetric and non-singular, its inverse is given by

$$\begin{bmatrix} Q^{-1} & -Q^{-1}BD^{-1} & -Q^{-1}CE^{-1} \\ -D^{-1}B'Q^{-1} & D^{-1}+D^{-1}B'Q^{-1}BD^{-1} & D^{-1}B'Q^{-1}CE^{-1} \\ -E^{-1}C'Q^{-1} & E^{-1}C'Q^{-1}BD^{-1} & E^{-1}+E^{-1}C'Q^{-1}CE^{-1} \end{bmatrix} \tag{9}$$

where

$$Q = A - BD^{-1}B' - CE^{-1}C'. \tag{10}$$

Proof. The proof is left to the reader. □

As to the determinants of partitioned matrices we note that

$$\begin{vmatrix} A_{11} & A_{12} \\ 0 & A_{22} \end{vmatrix} = |A_{11}||A_{22}| = \begin{vmatrix} A_{11} & 0 \\ A_{21} & A_{22} \end{vmatrix} \tag{11}$$

if both A_{11} and A_{22} are square matrices.

Exercises

1. Find the determinant and inverse (if it exists) of

$$B = \begin{pmatrix} A & 0 \\ a' & 1 \end{pmatrix}.$$

2. If $|A| \neq 0$, prove that

$$\begin{vmatrix} A & b \\ a' & \alpha \end{vmatrix} = (\alpha - a'A^{-1}b)|A|.$$

3. If $\alpha \neq 0$, prove that

$$\begin{vmatrix} A & b \\ a' & \alpha \end{vmatrix} = \alpha|A - (1/\alpha)ba'|.$$

12. COMPLEX MATRICES

If X and Y are real matrices of the same order, a complex matrix Z can be expressed as

$$Z = X + \mathrm{i}Y. \tag{1}$$

The *complex conjugate* of Z, denoted Z^*, is defined as

$$Z^* = X' - \mathrm{i}Y'. \tag{2}$$

If Z is real, then $Z^* = Z'$. If Z is a scalar, say ζ, we usually write $\bar{\zeta}$ instead of ζ^*.

A square complex matrix Z is said to be *Hermitian* if $Z^* = Z$ (the complex equivalent to a symmetric matrix) and *unitary* if $Z^*Z = I$ (the complex equivalent to an orthogonal matrix).

We shall see in Theorem 4 that the eigenvalues of a real symmetric matrix are real. In general, however, eigenvalues (and hence eigenvectors) are complex. In this book, complex numbers appear only in connection with eigenvalues and eigenvectors of non-symmetric matrices (Chapter 8). A detailed treatment is therefore omitted. Matrices and vectors are assumed to be real, unless it is explicitly specified that they are complex.

13. EIGENVALUES AND EIGENVECTORS

Let A be a square matrix, say $n \times n$. The *eigenvalues* of A are defined as the roots of the *characteristic equation*

$$|\lambda I_n - A| = 0. \tag{1}$$

Equation (1) has n roots, in general complex. Let λ be an eigenvalue of A. Then there exist vectors x and y ($x \neq 0$, $y \neq 0$) such that

$$(\lambda I - A)x = 0, \qquad y'(\lambda I - A) = 0, \tag{2}$$

that is,

$$Ax = \lambda x, \qquad y'A = \lambda y'. \tag{3}$$

The vectors x and y are called a (column) *eigenvector* and *row eigenvector* of A associated with the eigenvalue λ. Eigenvectors are usually normalized in some way to make them unique, for example by $x'x = y'y = 1$ (when x and y are real).

Not all roots of the characteristic equation need to be different. Each root is counted a number of times equal to its multiplicity. When a root (eigenvalue) appears more than once it is called a *multiple eigenvalue*; if it appears only once it is called a *simple eigenvalue*.

Although eigenvalues are in general complex, the eigenvalues of a real symmetric matrix are always real.

Theorem 4

A real symmetric matrix has only real eigenvalues.

Proof. Let λ be an eigenvalue of a real symmetric matrix A and let $x=u+\mathrm{i}v$ be an associated eigenvector. Then

$$A(u+\mathrm{i}v)=\lambda(u+\mathrm{i}v) \tag{4}$$

and hence

$$(u-\mathrm{i}v)'A(u+\mathrm{i}v)=\lambda(u-\mathrm{i}v)'(u+\mathrm{i}v), \tag{5}$$

which leads to

$$u'Au+v'Av=\lambda(u'u+v'v) \tag{6}$$

because of the symmetry of A. This implies that λ is real. □

Let us now prove the following three results, which will be useful to us later.

Theorem 5

If A is an $n \times n$ matrix and G a non-singular $n \times n$ matrix, then A and $G^{-1}AG$ have the same set of eigenvalues (with the same multiplicities).

Proof. From

$$\lambda I_n-G^{-1}AG=G^{-1}(\lambda I_n-A)G \tag{7}$$

we obtain

$$|\lambda I_n-G^{-1}AG|=|G^{-1}||\lambda I_n-A||G|=|\lambda I_n-A| \tag{8}$$

and the result follows. □

Theorem 6

A singular matrix has at least one eigenvalue zero.

Proof. If A is singular then $|A|=0$ and hence $|\lambda I-A|=0$ for $\lambda=0$. □

Theorem 7

An idempotent matrix has only eigenvalues 0 or 1. All eigenvalues of an orthogonal matrix have unit modulus.

Proof. Let A be idempotent. Then $A^2=A$. Thus, if $Ax=\lambda x$, then

$$\lambda x=Ax=A^2x=\lambda Ax=\lambda^2 x \tag{9}$$

and hence $\lambda=\lambda^2$, which implies $\lambda=0$ or $\lambda=1$.

If A is orthogonal, then $A'A = I$. Thus, if $Ax = \lambda x$, then

$$x^*A' = \bar{\lambda}x^*, \tag{10}$$

using the notation of section 12. Hence

$$x^*x = x^*A'Ax = \bar{\lambda}\lambda x^*x. \tag{11}$$

Since $x^*x \neq 0$, we obtain $\bar{\lambda}\lambda = 1$ and hence $|\lambda| = 1$. □

An important theorem regarding positive definite matrices is stated below.

Theorem 8

A symmetric matrix is positive definite (positive semidefinite) if and only if all its eigenvalues are positive (non-negative).

Proof. If A is positive definite and $Ax = \lambda x$, then $x'Ax = \lambda x'x$. Now, $x'Ax > 0$ and $x'x > 0$ imply $\lambda > 0$. The converse will not be proved here. (It follows from Theorem 13.) □

Let us next prove Theorem 9.

Theorem 9

Let A be $m \times n$ and let B be $n \times m (n \geqslant m)$. Then the non-zero eigenvalues of BA and AB are identical, and $|I_m - AB| = |I_n - BA|$.

Proof. Taking determinants on both sides of the equality

$$\begin{pmatrix} I_m - AB & A \\ 0 & I_n \end{pmatrix} \begin{pmatrix} I_m & 0 \\ B & I_n \end{pmatrix} = \begin{pmatrix} I_m & 0 \\ B & I_n \end{pmatrix} \begin{pmatrix} I_m & A \\ 0 & I_n - BA \end{pmatrix}, \tag{12}$$

we obtain

$$|I_m - AB| = |I_n - BA|. \tag{13}$$

Now, let $\lambda \neq 0$. Then

$$\begin{aligned} |\lambda I_n - BA| &= \lambda^n |I_n - B(\lambda^{-1}A)| \\ &= \lambda^n |I_m - (\lambda^{-1}A)B| \\ &= \lambda^{n-m} |\lambda I_m - AB|. \end{aligned} \tag{14}$$

Hence the non-zero eigenvalues of BA are the same as the non-zero eigenvalues of AB, and this is equivalent to the statement in the theorem. □

Without proof we state the following famous result.

Theorem 10 (Cayley–Hamilton)

Let A be an $n \times n$ matrix with eigenvalues $\lambda_1, \ldots, \lambda_n$. Then

$$\prod_{i=1}^{n} (\lambda_i I_n - A) = 0. \tag{15}$$

Finally, we present the following result on eigenvectors.

Theorem 11

Eigenvectors associated with distinct eigenvalues are linearly independent.

Proof. Let $Ax_1 = \lambda_1 x_1$, $Ax_2 = \lambda_2 x_2$ and $\lambda_1 \neq \lambda_2$. Assume that x_1 and x_2 are linearly dependent. Then there is an $\alpha \neq 0$ such that $x_2 = \alpha x_1$ and hence

$$\alpha\lambda_1 x_1 = \alpha A x_1 = A x_2 = \lambda_2 x_2 = \alpha\lambda_2 x_1, \tag{16}$$

that is

$$\alpha(\lambda_1 - \lambda_2) x_1 = 0. \tag{17}$$

Since $\alpha \neq 0$ and $\lambda_1 \neq \lambda_2$, (17) implies $x_1 = 0$, a contradiction. □

Exercise

1. Show that

$$\begin{vmatrix} 0 & I_m \\ I_m & 0 \end{vmatrix} = (-1)^m.$$

14. SCHUR'S DECOMPOSITION THEOREM

In the next few sections we present three decomposition theorems: Schur's theorem, Jordan's theorem and the singular-value decomposition. Each of these theorems will prove useful later in this book. We first state Schur's theorem.

Theorem 12 (Schur decomposition)

Let A be an $n \times n$ matrix. Then there exist a unitary $n \times n$ matrix S (that is, $S^*S = I_n$) and an upper triangular matrix M whose diagonal elements are the eigenvalues of A, such that

$$S^*AS = M. \tag{1}$$

The most important special case of Schur's decomposition theorem is the case where A is symmetric.

Theorem 13

Let A be a real symmetric $n \times n$ matrix. Then there exist an orthogonal $n \times n$ matrix S (that is, $S'S = I_n$) whose columns are eigenvectors of A and a diagonal matrix Λ whose diagonal elements are the eigenvalues of A, such that

$$S'AS = \Lambda. \tag{2}$$

Proof. Using Theorem 12, there exists a unitary matrix $S = R + iT$ with real R and T and an upper triangular matrix M such that $S^*AS = M$. Then,

$$M = S^*AS = (R - iT)'A(R + iT) = (R'AR + T'AT) + i(R'AT - T'AR) \quad (3)$$

and hence, using the symmetry of A,

$$M + M' = 2(R'AR + T'AT). \quad (4)$$

It follows that $M + M'$ is a real matrix and hence, since M is triangular, that M is a real matrix. We thus obtain, from (3),

$$M = R'AR + T'AT. \quad (5)$$

Since A is symmetric, M is symmetric. But, since M is also triangular, M must be diagonal. The columns of S are then eigenvectors of A and, since the diagonal elements of M are real, S can be chosen to be real as well. □

Exercises

1. Let A be a real symmetric $n \times n$ matrix with eigenvalues $\lambda_1 \leqslant \lambda_2 \leqslant \ldots \leqslant \lambda_n$. Use Theorem 13 to prove that
$$\lambda_1 \leqslant \frac{x'Ax}{x'x} \leqslant \lambda_n.$$
2. Hence show that, for any $m \times n$ matrix A,
$$\|Ax\| \leqslant \mu\|x\|,$$
where μ^2 denotes the largest eigenvalue of $A'A$.
3. Let A be an $m \times n$ matrix of rank r. Show that there exists an $n \times (n - r)$ matrix S such that
$$AS = 0, \qquad S'S = I_{n-r}.$$
4. Let A be an $m \times n$ matrix of rank r. Let S be a matrix such that $AS = 0$. Show that $r(S) \leqslant n - r$.

15. THE JORDAN DECOMPOSITION

Schur's theorem tells us that there exists, for every square matrix A, a unitary (possibly orthogonal) matrix S which 'transforms' A into an upper triangular matrix M, whose diagonal elements are the eigenvalues of A.

Jordan's theorem similarly states that there exists a non-singular matrix, say T, which transforms A into an upper triangular matrix M, whose diagonal elements are the eigenvalues of A. The difference between the two decomposition theorems is that in Jordan's theorem less structure is put on the matrix T (non-singular, but not necessarily unitary) and more structure on the matrix M.

Theorem 14 (Jordan decomposition)

Let A be an $n\times n$ matrix and denote by $J_k(\lambda)$ a $k\times k$ matrix of the form

$$J_k(\lambda)=\begin{bmatrix}\lambda & 1 & 0 & \dots & 0\\ 0 & \lambda & 1 & \dots & 0\\ \vdots & \vdots & \vdots & & \vdots\\ 0 & 0 & 0 & \dots & 1\\ 0 & 0 & 0 & \dots & \lambda\end{bmatrix} \tag{1}$$

where $J_1(\lambda)=\lambda$. Then there exists a non-singular $n\times n$ matrix T such that

$$T^{-1}AT=\begin{bmatrix}J_{k_1}(\lambda_1) & 0 & \dots & 0\\ 0 & J_{k_2}(\lambda_2) & \dots & 0\\ \vdots & \vdots & \ddots & \vdots\\ 0 & 0 & \dots & J_{k_r}(\lambda_r)\end{bmatrix} \tag{2}$$

with $k_1+k_2+\dots+k_r=n$. The λ_i are the eigenvalues of A, not necessarily distinct.

The most important special case of Theorem 14 is Theorem 15.

Theorem 15

Let A be an $n\times n$ matrix with distinct eigenvalues. Then there exist a non-singular $n\times n$ matrix T and a diagonal $n\times n$ matrix Λ whose diagonal elements are the eigenvalues of A, such that

$$T^{-1}AT=\Lambda. \tag{3}$$

Proof. Immediate from Theorem 14 (or Theorem 11). □

Exercise

1. Show that $(\lambda I_k-J_k(\lambda))^k=0$, and use this fact to prove Theorem 10.
2. Show that Theorem 15 remains valid when A is complex.

16. THE SINGULAR-VALUE DECOMPOSITION

The third important decomposition theorem is the singular-value decomposition.

Theorem 16 (singular-value decomposition)

Let A be a real $m\times n$ matrix with $r(A)=r>0$. Then there exist an $m\times r$ matrix S such that $S'S=I_r$, an $n\times r$ matrix T such that $T'T=I_r$ and an $r\times r$ diagonal matrix Λ with positive diagonal elements, such that

$$A=S\Lambda^{1/2}T'. \tag{1}$$

Proof. Since AA' is a real $m \times m$ symmetric (in fact, positive semidefinite) matrix of rank r (by (7.3)), its non-zero eigenvalues are all positive (Theorem 8). From Theorem 13 there exists an orthogonal $m \times m$ matrix $(S:S_*)$ such that

$$AA'S = S\Lambda, \qquad AA'S_* = 0, \qquad SS' + S_*S_*' = I_m, \tag{2}$$

where Λ is an $r \times r$ diagonal matrix having these r positive eigenvalues as diagonal elements. Define $T = A'S\Lambda^{-1/2}$. Then we see that

$$A'AT = T\Lambda, \qquad T'T = I_r. \tag{3}$$

Thus, since (2) implies $A'S_* = 0$ by Theorem 1(b), we have

$$A = (SS' + S_*S_*')A = SS'A = S\Lambda^{1/2}(A'S\Lambda^{-1/2})' = S\Lambda^{1/2}T'. \tag{4}$$

□

We see from (2) and (3) that the semi-orthogonal matrices S and T satisfy

$$AA'S = S\Lambda, A'AT = T\Lambda. \tag{5}$$

Hence, Λ contains the r non-zero eigenvalues of AA' (and of $A'A$) and S (by construction) and T contain corresponding eigenvectors. A common mistake in applying the singular-value decomposition is to find S, T and Λ from (5). This is incorrect because, given S, T is not unique! The correct procedure is to find S and Λ from $AA'S = S\Lambda$ and then define $T = A'S\Lambda^{-1/2}$. Alternatively we can find T and Λ from $A'AT = T\Lambda$ and define $S = AT\Lambda^{-1/2}$.

17. FURTHER RESULTS CONCERNING EIGENVALUES

Let us now prove the following theorems, all of which concern eigenvalues.

Theorem 17

Let A be a square $n \times n$ matrix with eigenvalues $\lambda_1, \ldots, \lambda_n$. Then

$$\operatorname{tr} A = \sum_{i=1}^{n} \lambda_i \tag{1}$$

$$|A| = \prod_{i=1}^{n} \lambda_i. \tag{2}$$

Proof. We write, using Theorem 12, $S^*AS = M$. Then

$$\operatorname{tr} A = \operatorname{tr} SMS^* = \operatorname{tr} MS^*S = \operatorname{tr} M = \sum_i \lambda_i \tag{3}$$

and

$$|A| = |SMS^*| = |S|\,|M|\,|S^*| = |M| = \prod_i \lambda_i. \tag{4}$$

□

Theorem 18

If A has r non-zero eigenvalues, then $r(A) \geq r$.

Proof. We write again, by Theorem 12, $S^*AS=M$. We partition

$$M=\begin{pmatrix} M_1 & M_2 \\ 0 & M_3 \end{pmatrix}, \tag{5}$$

where M_1 is a non-singular upper triangular $r\times r$ matrix and M_3 is *strictly* upper triangular. Since $r(A)=r(M)\geqslant r(M_1)=r$, the result follows. □

The following example shows that it is indeed possible that $r(A)>r$. Let

$$A=\begin{pmatrix} 1 & -1 \\ 1 & -1 \end{pmatrix}. \tag{6}$$

Then $r(A)=1$ and both eigenvalues of A are zero.

Theorem 19

Let A be an $n\times n$ matrix. If λ is a simple eigenvalue of A, then $r(\lambda I-A)=n-1$. Conversely, if $r(\lambda I-A)=n-1$, then λ is an eigenvalue of A, but not necessarily a simple eigenvalue.

Proof. Let $\lambda_1,\ldots,\lambda_n$ be the eigenvalues of A. Then $B=\lambda I-A$ has eigenvalues $\lambda-\lambda_i$ $(i=1,\ldots,n)$ and since λ is a simple eigenvalue of A, B has a simple eigenvalue zero. Hence $r(B)\leqslant n-1$. Also, since B has $n-1$ non-zero eigenvalues, $r(B)\geqslant n-1$ (Theorem 18). Hence $r(B)=n-1$. Conversely, if $r(B)=n-1$, then B has at least one zero eigenvalue and hence $\lambda=\lambda_i$ for at least one i. □

Corollary

An $n\times n$ matrix with a simple zero eigenvalue has rank $n-1$.

Theorem 20

If A is symmetric and has r non-zero eigenvalues, then $r(A)=r$.

Proof. Using Theorem 13, we have $S'AS=\Lambda$ and hence

$$r(A)=r(S\Lambda S')=r(\Lambda)=r. \tag{7}$$

□

Theorem 21

If A is an idempotent matrix with r eigenvalues equal to one, then $r(A)=\operatorname{tr} A=r$.

Proof. By Theorem 12, $S^*AS=M$ (upper triangular), where

$$M=\begin{pmatrix} M_1 & M_2 \\ 0 & M_3 \end{pmatrix} \tag{8}$$

with M_1 a *unit* upper triangular $r\times r$ matrix and M_3 a *strictly* upper triangular matrix. Since A is idempotent, so is M and hence

$$\begin{pmatrix} M_1^2 & M_1M_2+M_2M_3 \\ 0 & M_3^2 \end{pmatrix} = \begin{pmatrix} M_1 & M_2 \\ 0 & M_3 \end{pmatrix}. \tag{9}$$

This implies that M_1 is idempotent; it is non-singular, hence $M_1 = I_r$ (see exercise 1). Also, M_3 is idempotent and all its eigenvalues are zero, hence $M_3 = 0$ (see exercise 2), so that

$$M = \begin{pmatrix} I_r & M_2 \\ 0 & 0 \end{pmatrix}. \tag{10}$$

Hence,

$$r(A) = r(M) = r(I_r : M_2) = r. \tag{11}$$

Also, by Theorem 17,

$$\operatorname{tr} A = (\text{sum of eigenvalues of } A) = r. \tag{12}$$

□

We note that in Theorem 21 the matrix A is not required to be symmetric. If A is idempotent *and* symmetric, then it is positive semidefinite. Since its eigenvalues are only 0 and 1, it then follows from Theorem 13 that A can be written as

$$A = GG', \qquad G'G = I_r \tag{13}$$

where r denotes the rank of A.

Exercises

1. The only non-singular idempotent matrix is the identity matrix.
2. The only idempotent matrix whose eigenvalues are all zero is the null matrix.
3. If A is a positive semidefinite $n \times n$ matrix with $r(A) = r$, then there exists an $n \times r$ matrix G such that

$$A = GG', \qquad G'G = \Lambda \tag{14}$$

where Λ is an $r \times r$ diagonal matrix containing the positive eigenvalues of A.

18. POSITIVE (SEMI)DEFINITE MATRICES

Positive (semi)definite matrices were introduced in section 6. We have already seen that AA' and $A'A$ are both positive semidefinite and that the eigenvalues of a positive (semi)definite matrix are all positive (non-negative) (Theorem 8). We now present some more properties of positive (semi)definite matrices.

Theorem 22

Let A be positive definite and B positive semidefinite. Then

$$|A+B| \geqslant |A| \tag{1}$$

with equality if and only if $B=0$.

Proof. Let Λ be a positive definite diagonal matrix such that

$$S'AS = \Lambda, \qquad S'S = I. \tag{2}$$

Then, $SS' = I$ and

$$A+B=S\Lambda^{1/2}(I+\Lambda^{-1/2}S'BS\Lambda^{-1/2})\Lambda^{1/2}S' \tag{3}$$

and hence, using (9.2),

$$\begin{aligned}|A+B| &= |S\Lambda^{1/2}|\,|I+\Lambda^{-1/2}S'BS\Lambda^{-1/2}|\,|\Lambda^{1/2}S'| \\ &= |S\Lambda^{1/2}\Lambda^{1/2}S'|\,|I+\Lambda^{-1/2}S'BS\Lambda^{-1/2}| \\ &= |A|\,|I+\Lambda^{-1/2}S'BS\Lambda^{-1/2}|. \end{aligned} \tag{4}$$

If $B=0$ then $|A+B|=|A|$. If $B\neq 0$, then the matrix $\Lambda^{-1/2}S'BS\Lambda^{-1/2}$ will be positive semidefinite with at least one positive eigenvalue. Hence $|I+\Lambda^{-1/2}S'BS\Lambda^{-1/2}|>1$ and $|A+B|>|A|$. □

Theorem 23

Let A be positive definite and B symmetric of the same order. Then there exist a non-singular matrix P and a diagonal matrix Λ such that

$$A=PP', \qquad B=P\Lambda P'. \tag{5}$$

Proof. Let $C=A^{-1/2}BA^{-1/2}$. Since C is symmetric, there exist by Theorem 13 an orthogonal matrix S and a diagonal matrix Λ such that

$$S'CS=\Lambda, \quad S'S=I. \tag{6}$$

Now define

$$P=A^{1/2}S. \tag{7}$$

Then,

$$PP'=A^{1/2}SS'A^{1/2}=A^{1/2}A^{1/2}=A \tag{8}$$

and

$$P\Lambda P'=A^{1/2}S\Lambda S'A^{1/2}=A^{1/2}CA^{1/2}=A^{1/2}A^{-1/2}BA^{-1/2}A^{1/2}=B. \tag{9}$$

(If B is positive semidefinite, so is Λ.) □

For two symmetric matrices A and B we shall write $A\geqslant B$ (or $B\leqslant A$) if $A-B$ is positive semidefinite, and $A>B$ (or $B<A$) if $A-B$ is positive definite.

Theorem 24

Let A and B be positive definite $n\times n$ matrices. Then $A>B$ if and only if $B^{-1}>A^{-1}$.

Proof. By Theorem 23 there exist a non-singular matrix P and a positive definite diagonal matrix $\Lambda=\text{diag}(\lambda_1,\ldots,\lambda_n)$ such that

$$A=PP', \qquad B=P\Lambda P'. \tag{10}$$

Then

$$A-B=P(I-\Lambda)P', \qquad B^{-1}-A^{-1}=P'^{-1}(\Lambda^{-1}-I)P^{-1}. \tag{11}$$

If $A-B$ is positive definite, then $I-\Lambda$ is positive definite and hence $0<\lambda_i<1$ $(i=1,\ldots,n)$. This implies that $\Lambda^{-1}-I$ is positive definite and hence that $B^{-1}-A^{-1}$ is positive definite. □

Theorem 25

Let A and B be positive definite matrices such that $A-B$ is positive semidefinite. Then $|A|\geqslant|B|$ with equality if and only if $A=B$.

Proof. Let $C=A-B$. Then B is positive definite and C is positive semidefinite. Thus, by Theorem 22, $|B+C|\geqslant|B|$ with equality if and only if $C=0$, that is, $|A|\geqslant|B|$ with equality if and only if $A=B$. □

A useful special case of Theorem 25 is Theorem 26.

Theorem 26

Let A be positive definite with $|A|=1$. If also $I-A$ is positive semidefinite, then $A=I$.

Proof. This follows immediately from Theorem 25. □

19. THREE FURTHER RESULTS FOR POSITIVE DEFINITE MATRICES

Let us now prove Theorem 27.

Theorem 27

Let A be a positive definite $n\times n$ matrix, and let B be the $(n+1)\times(n+1)$ matrix

$$B=\begin{pmatrix} A & b \\ b' & \alpha \end{pmatrix}. \tag{1}$$

Then, (i)

$$|B|\leqslant\alpha|A| \tag{2}$$

with equality if and only if $b=0$; and (ii) B is positive definite if and only if $|B|>0$.

Proof. Define the $(n+1)\times(n+1)$ matrix

$$P=\begin{pmatrix} I_n & -A^{-1}b \\ 0' & 1 \end{pmatrix}. \tag{3}$$

Then

$$P'BP=\begin{pmatrix} A & 0 \\ 0' & \alpha-b'A^{-1}b \end{pmatrix}, \tag{4}$$

so that

$$|B|=|P'BP|=|A|(\alpha-b'A^{-1}b). \tag{5}$$

(Compare exercise 11.2.) Statement (i) of the theorem is an immediate consequence of (5).

To prove (ii) we note that $|B|>0$ iff $\alpha-b'A^{-1}b>0$ (from (5)), which is the case iff $P'BP$ is positive definite (from (4)). This in turn is true iff B is positive definite. □

An immediate consequence of Theorem 27, proved by induction, is the following.

Theorem 28

If $A=(a_{ij})$ is a positive definite $n \times n$ matrix, then

$$|A| \leqslant \prod_{i=1}^{n} a_{ii} \tag{6}$$

with equality if and only if A is diagonal.

Another consequence of Theorem 27 is Theorem 29.

Theorem 29

A symmetric $n \times n$ matrix A is positive definite if and only if all principal minors $|A_k|$ $(k=1, \ldots, n)$ are positive.

Note. The $k \times k$ matrix A_k is obtained from A by deleting the last $n-k$ rows and columns of A. Notice that $A_n = A$.

Proof. Let $E_k = (I_k : 0)$ be a $k \times n$ matrix, so that $A_k = E_k A E_k'$. Let y be an arbitrary $k \times 1$ vector, $y \neq 0$. Then

$$y' A_k y = (E_k' y)' A (E_k' y) > 0$$

since $E_k' y \neq 0$ and A is positive definite. Hence A_k is positive definite, and, in particular, $|A_k| > 0$.

The converse follows by repeated application of Theorem 27 (ii). □

Exercises

1. If A is positive definite show that the matrix

$$\begin{pmatrix} A & b \\ b' & b'A^{-1}b \end{pmatrix}$$

is positive semidefinite and singular, and find the eigenvector associated with the zero eigenvalue.

2. Hence show that, for positive definite A,

$$x'Ax - 2b'x \geqslant -b'A^{-1}b$$

for every x, with equality if and only if $x = A^{-1}b$.

20. A USEFUL RESULT

If A is a positive definite $n \times n$ matrix, then, in accordance with Theorem 28,

$$|A| = \prod_{i=1}^{n} a_{ii} \tag{1}$$

if and only if A is diagonal. If A is merely symmetric, then condition (1), while obviously necessary, is no longer sufficient for the diagonality of A. For example, the matrix

$$A=\begin{bmatrix}2 & 3 & 3\\ 3 & 2 & 3\\ 3 & 3 & 2\end{bmatrix} \tag{2}$$

has determinant $|A|=8$ (its eigenvalues are -1, -1 and 8), thus satisfying (1), but A is not diagonal.

Theorem 30 gives a necessary *and* sufficient condition for the diagonality of a symmetric matrix.

Theorem 30

A real symmetric matrix is diagonal if and only if its eigenvalues and its diagonal elements coincide.

Proof. Let $A = (a_{ij})$ be a symmetric $n \times n$ matrix. The 'only if' part of the theorem is trivial. To prove the 'if' part, assume that $\lambda_i(A) = a_{ii}$, $i = 1, \ldots, n$, and consider the matrix

$$B=A+kI, \tag{3}$$

where $k>0$ is such that B is positive definite. Then

$$\lambda_i(B)=\lambda_i(A)+k=a_{ii}+k=b_{ii} \qquad (i=1, \ldots, n), \tag{4}$$

and hence

$$|B|=\prod_{i=1}^{n} \lambda_i(B)=\prod_{i=1}^{n} b_{ii}. \tag{5}$$

It then follows from Theorem 28 that B is diagonal, and hence that A is diagonal. □

MISCELLANEOUS EXERCISES

1. If A and B are square matrices such that $AB=0$, $A\neq 0$, $B\neq 0$, then prove that $|A|=|B|=0$.
2. If x and y are vectors of the same order, prove that $x'y=\operatorname{tr} yx'$.
3. Let P and Q be square matrices and $|Q|\neq 0$. Show that

$$\begin{vmatrix}P & R\\ S & Q\end{vmatrix}=|Q|\,|P-RQ^{-1}S|.$$

4. Show that $(I-AB)^{-1}=I+A(I-BA)^{-1}B$, if the inverses exist.
5. Show that

$$(\alpha I-A)^{-1}-(\beta I-A)^{-1}=(\beta-\alpha)(\beta I-A)^{-1}(\alpha I-A)^{-1}.$$

6. If A is positive definite, show that $A+A^{-1}-2I$ is positive semidefinite.
7. For any symmetric matrices A and B, show that $AB-BA$ is skew symmetric.
8. Prove that the eigenvalues λ_i of $(A+B)^{-1}A$, where A is positive semidefinite and B positive definite, satisfy $0\leqslant\lambda_i<1$.

9. Let x and y be $n \times 1$ vectors. Prove that xy' has $n-1$ zero eigenvalues and one eigenvalue $x'y$.
10. Show that $|I+xy'|=1+x'y$.
11. Let $\mu = 1 + x'y$. If $\mu \neq 0$, show that $(I + xy')^{-1} = I - (1/\mu)xy'$.
12. Show that $(I+AA')^{-1}A = A(I+A'A)^{-1}$.
13. Show that $A(A'A)^{1/2} = (AA')^{1/2}A$.
14. (Monotonicity of the entropic complexity). Let A_n be a positive definite $n \times n$ matrix and define

$$\varphi(n) = \frac{n}{2}\log \operatorname{tr}(A_n/n) - \frac{1}{2}\log |A_n|.$$

Let A_{n+1} be a positive definite $(n+1) \times (n+1)$ matrix such that

$$A_{n+1} = \begin{pmatrix} A_n & a_n \\ a_n' & \alpha_n \end{pmatrix}.$$

Then,

$$\varphi(n+1) \geqslant \varphi(n)$$

with equality if and only if

$$a_n = 0, \quad \alpha_n = \operatorname{tr} A_n/n$$

(Bozdogan 1990, 1994).

15. Let A be positive definite, $X'X = I$, and $B = XX'A - AXX'$. Show that

$$|X'AX||X'A^{-1}X| = |A+B|/|A|$$

(Bloomfield and Watson 1975).

BIBLIOGRAPHICAL NOTES

§1. There are many excellent introductory texts on matrix algebra. We mention in particular Hadley (1961), Bellman (1970) and Rao (1973, chap. 1). More advanced are Gantmacher (1959) and Mirsky (1961).
§8. For a proof that each permutation matrix is orthogonal and some examples, see Marcus and Minc (1964, sec. 4.8.2).
§9. Aitken (1939, chap. 5) contains a useful discussion of adjoint matrices.
§12. See Bellman (1970, chap. 11, theorem 8). The Cayley–Hamilton theorem is quite easily proved using the Jordan decomposition (Theorem 14).
§13. We are grateful to Abadir and Hadri (1996) for pointing out an error in the previous edition of this book.
§14. Schur's theorem is proved in Bellman (1970, chap. 11, th. 4).
§15. Jordan's theorem is not usually proved in introductory texts. For a full proof see Gantmacher (1959, vol. I, p. 201).

CHAPTER 2

Kronecker products, the vec operator and the Moore–Penrose inverse

1. INTRODUCTION

This chapter develops some matrix tools that will prove useful to us later. The first of these is the Kronecker product, which transforms two matrices $A = (a_{ij})$ and $B = (b_{st})$ into a matrix $C = (a_{ij}b_{st})$. The vec operator transforms a matrix into a vector by stacking its columns one underneath the other. We shall see that the Kronecker product and the vec operator are intimately connected. Finally we discuss the Moore–Penrose inverse, which generalizes the concept of the inverse of a non-singular matrix to singular square matrices and rectangular matrices.

2. THE KRONECKER PRODUCT

Let A be an $m \times n$ matrix and B a $p \times q$ matrix. The $mp \times nq$ matrix defined by

$$\begin{bmatrix} a_{11}B & \dots & a_{1n}B \\ \vdots & & \vdots \\ a_{m1}B & \dots & a_{mn}B \end{bmatrix} \tag{1}$$

is called the *Kronecker product* of A and B and written $A \otimes B$.

Observe that, while the matrix product AB only exists if the number of columns in A equals the number of rows in B or if either A or B is a scalar, the Kronecker product $A \otimes B$ is defined for any pair of matrices A and B.

The following three properties justify the name Kronecker *product*:

$$A \otimes B \otimes C = (A \otimes B) \otimes C = A \otimes (B \otimes C) \tag{2}$$

$$(A+B) \otimes (C+D) = A \otimes C + A \otimes D + B \otimes C + B \otimes D \tag{3}$$

if $A+B$ and $C+D$ exist, and

$$(A \otimes B)(C \otimes D) = AC \otimes BD \tag{4}$$

if AC and BD exist.

If α is a scalar, then

$$\alpha \otimes A = \alpha A = A\alpha = A \otimes \alpha. \tag{5}$$

(This property can be used, for example, to prove that $(A \otimes b)B = (AB) \otimes b$, by writing $B = B \otimes 1$.) Another useful property concerns two column vectors a and b (not necessarily of the same order):

$$a' \otimes b = ba' = b \otimes a'. \tag{6}$$

The transpose of a Kronecker product is

$$(A \otimes B)' = A' \otimes B'. \tag{7}$$

If A and B are square matrices (not necessarily of the same order), then

$$\operatorname{tr}(A \otimes B) = (\operatorname{tr} A)(\operatorname{tr} B). \tag{8}$$

If A and B are non-singular, then

$$(A \otimes B)^{-1} = A^{-1} \otimes B^{-1}. \tag{9}$$

Exercises

1. Prove properties (2)–(9) above.
2. If A is a partitioned matrix,

$$A = \begin{pmatrix} A_{11} & A_{12} \\ A_{21} & A_{22} \end{pmatrix},$$

then $A \otimes B$ takes the form

$$A \otimes B = \begin{pmatrix} A_{11} \otimes B & A_{12} \otimes B \\ A_{21} \otimes B & A_{22} \otimes B \end{pmatrix}.$$

3. EIGENVALUES OF A KRONECKER PRODUCT

Let us now demonstrate the following result.

Theorem 1

Let A be an $m \times m$ matrix with eigenvalues $\lambda_1, \lambda_2, \ldots, \lambda_m$, and let B be a $p \times p$ matrix with eigenvalues $\mu_1, \mu_2, \ldots, \mu_p$. Then the mp eigenvalues of $A \otimes B$ are $\lambda_i \mu_j$ $(i = 1, \ldots, m; j = 1, \ldots, p)$.

Proof. By Schur's theorem (Theorem 1.12) there exist non-singular (in fact, unitary) matrices S and T such that

$$S^{-1}AS = L, \qquad T^{-1}BT = M, \tag{1}$$

where L and M are upper triangular matrices whose diagonal elements are the eigenvalues of A and B respectively. Thus,

$$(S^{-1} \otimes T^{-1})(A \otimes B)(S \otimes T) = L \otimes M. \tag{2}$$

Since $S^{-1} \otimes T^{-1}$ is the inverse of $S \otimes T$, it follows from Theorem 1.5 that $A \otimes B$ and $(S^{-1} \otimes T^{-1})(A \otimes B)(S \otimes T)$ have the same set of eigenvalues, and hence that $A \otimes B$ and $L \otimes M$ have the same set of eigenvalues. But $L \otimes M$ is an upper triangular matrix by virtue of the fact that L and M are upper triangular; its eigenvalues are therefore its diagonal elements $\lambda_i \mu_j$. This concludes the proof. □

Remark. If x is an eigenvector of A and y an eigenvector of B, then $x \otimes y$ is clearly an eigenvector of $A \otimes B$. It is not generally true, however, that every eigenvector of $A \otimes B$ is the Kronecker product of an eigenvector of A and an eigenvector of B. For example, let

$$A = B = \begin{pmatrix} 0 & 1 \\ 0 & 0 \end{pmatrix}, \qquad e_1 = \begin{pmatrix} 1 \\ 0 \end{pmatrix}, \qquad e_2 = \begin{pmatrix} 0 \\ 1 \end{pmatrix}. \tag{3}$$

Both eigenvalues of A (and B) are zero and the only eigenvector is e_1. The four eigenvalues of $A \otimes B$ are all zero (in concordance with Theorem 1), but the eigenvectors of $A \otimes B$ are not just $e_1 \otimes e_1$, but also $e_1 \otimes e_2$ and $e_2 \otimes e_1$.

Theorem 1 has several important corollaries. First, if A and B are positive (semi)definite, then $A \otimes B$ is positive (semi)definite. Secondly, since the determinant of $A \otimes B$ is equal to the product of its eigenvalues, we obtain

$$|A \otimes B| = |A|^p |B|^m, \tag{4}$$

where A is an $m \times m$ matrix and B is a $p \times p$ matrix. Thirdly, we can obtain the rank of $A \otimes B$ from Theorem 1 as follows. The rank of $A \otimes B$ is equal to the rank of $AA' \otimes BB'$. The rank of the latter (symmetric, in fact, positive semidefinite) matrix equals the number of non-zero (in this case, positive) eigenvalues it possesses. According to Theorem 1, the eigenvalues of $AA' \otimes BB'$ are $\lambda_i \mu_j$, where λ_i are the eigenvalues of AA' and μ_j are the eigenvalues of BB'. Now, $\lambda_i \mu_j$ is non-zero if and only if both λ_i and μ_j are non-zero. Hence, the number of non-zero eigenvalues of $AA' \otimes BB'$ is the product of the number of non-zero eigenvalues of AA' and the number of non-zero eigenvalues of BB'. Thus the rank of $A \otimes B$ is

$$r(A \otimes B) = r(A) r(B). \tag{5}$$

Exercise

1. Show that $A \otimes B$ is non-singular if and only if A and B are non-singular, and relate this result to (2.9).

4. THE VEC OPERATOR

Let A be an $m \times n$ matrix and a_j its j-th column; then vec A is the $mn \times 1$ vector

$$\text{vec } A = \begin{bmatrix} a_1 \\ a_2 \\ \vdots \\ a_n \end{bmatrix}. \tag{1}$$

Thus the vec operator transforms a matrix into a vector by stacking the columns of the matrix one underneath the other. Notice that vec A is defined for *any* matrix A, not just for square matrices. Also notice that vec $A =$ vec B does not imply $A = B$, unless A and B are matrices of the same order.

A very simple but often useful property is

$$\text{vec } a' = \text{vec } a = a \tag{2}$$

for any column vector a. The basic connection between the vec operator and the Kronecker product is

$$\text{vec } ab' = b \otimes a \tag{3}$$

for any two column vectors a and b (not necessarily of the same order). This follows because the j-th column of ab' is $b_j a$. Stacking the columns of ab' thus yields $b \otimes a$.

The basic connection between the vec operator and the trace is

$$(\text{vec } A)' \text{ vec } B = \text{tr } A'B, \tag{4}$$

where A and B are matrices of the same order. This is easy to verify since both the left side and the right side of equation (4) are equal to

$$\sum_i \sum_j a_{ij} b_{ij}.$$

Let us now generalize the basic properties of (3) and (4). The generalization of (3) is the following well-known result.

Theorem 2

Let A, B and C be three matrices such that the matrix product ABC is defined. Then,

$$\text{vec } ABC = (C' \otimes A) \text{ vec } B. \tag{5}$$

Proof. Assume that B has q columns denoted $b_1, b_2, \ldots, b_q$. Similarly let $e_1, e_2, \ldots, e_q$ denote the columns of the $q \times q$ identity matrix I_q, so that

$$B = \sum_{j=1}^{q} b_j e_j'.$$

Then, using (3),

$$\operatorname{vec} ABC = \operatorname{vec} \sum_{j=1}^{q} Ab_j e_j' C = \sum_{j=1}^{q} \operatorname{vec} (Ab_j)(C'e_j)'$$

$$= \sum_{j=1}^{q} (C'e_j \otimes Ab_j) = (C' \otimes A) \sum_{j=1}^{q} (e_j \otimes b_j)$$

$$= (C' \otimes A) \sum_{j=1}^{q} \operatorname{vec} b_j e_j' = (C' \otimes A) \operatorname{vec} B. \qquad (6)$$

□

One special case of Theorem 2 is

$$\operatorname{vec} AB = (B' \otimes I_m) \operatorname{vec} A = (B' \otimes A) \operatorname{vec} I_n = (I_q \otimes A) \operatorname{vec} B, \qquad (7)$$

where A is an $m \times n$ matrix and B is an $n \times q$ matrix. Another special case arises when the matrix C in (5) is replaced by a vector. Then we obtain, using (2),

$$ABd = (d' \otimes A) \operatorname{vec} B = (A \otimes d') \operatorname{vec} B', \qquad (8)$$

where d is a $q \times 1$ vector.

The equality (4) can be generalized as follows.

Theorem 3

Let A, B, C and D be four matrices such that the matrix product $ABCD$ is defined and square. Then,

$$\operatorname{tr} ABCD = (\operatorname{vec} D')'(C' \otimes A) \operatorname{vec} B = (\operatorname{vec} D)'(A \otimes C') \operatorname{vec} B'. \qquad (9)$$

Proof. We have, using (4) and (5),

$$\operatorname{tr} ABCD = \operatorname{tr} D(ABC) = (\operatorname{vec} D')' \operatorname{vec} ABC$$

$$= (\operatorname{vec} D')'(C' \otimes A) \operatorname{vec} B. \qquad (10)$$

The second equality is proved in precisely the same way starting from $\operatorname{tr} ABCD = \operatorname{tr} D'(C'B'A')$. □

Exercises

1. For any $m \times n$ matrix A, prove that

$$\operatorname{vec} A = (I_n \otimes A) \operatorname{vec} I_n = (A' \otimes I_m) \operatorname{vec} I_m.$$

2. If A, B and V are square matrices of the same order and $V = V'$, prove that

$$(\operatorname{vec} V)'(A \otimes B) \operatorname{vec} V = (\operatorname{vec} V)'(B \otimes A) \operatorname{vec} V.$$

5. THE MOORE–PENROSE (MP) INVERSE

The inverse of a matrix is defined when the matrix is square and non-singular. For many purposes it is useful to generalize the concept of invertibility to singular matrices and, indeed, to non-square matrices. One such generalization that is particularly useful because of its uniqueness is the *Moore–Penrose inverse* (*MP inverse*).

Definition

An $n \times m$ matrix X is the MP inverse of a real $m \times n$ matrix A if

$$AXA = A \tag{1}$$

$$XAX = X \tag{2}$$

$$(AX)' = AX \tag{3}$$

$$(XA)' = XA. \tag{4}$$

We shall denote the MP inverse of A as A^+.

Exercises

1. What is the MP inverse of a non-singular matrix?
2. What is the MP inverse of a scalar?
3. What is the MP inverse of a null matrix?

6. EXISTENCE AND UNIQUENESS OF THE MP INVERSE

Let us now demonstrate the following theorem.

Theorem 4

For each A, A^+ exists and is unique.

Proof (uniqueness). Assume that two matrices B and C both satisfy the four defining conditions. Then,

$$AB = (AB)' = B'A' = B'(ACA)' = B'A'C'A' = (AB)'(AC)' = ABAC = AC. \tag{1}$$

Similarly,

$$BA = (BA)' = A'B' = (ACA)'B' = A'C'A'B' = (CA)'(BA)' = CABA = CA. \tag{2}$$

Hence,

$$B = BAB = BAC = CAC = C. \tag{3}$$

Proof (existence). Let A be an $m \times n$ matrix with $r(A) = r$. If $r = 0$, then $A = 0$ and $A^+ = 0$ satisfies the four defining equations. Assume therefore $r > 0$. According to Theorem 1.16 there exist semi-orthogonal matrices S and T and a positive definite diagonal $r \times r$ matrix Λ such that

$$A = S\Lambda^{1/2}T', \qquad S'S = T'T = I_r. \tag{4}$$

Now define

$$B = T\Lambda^{-1/2}S'. \tag{5}$$

Then,

$$ABA = S\Lambda^{1/2}T'T\Lambda^{-1/2}S'S\Lambda^{1/2}T' = S\Lambda^{1/2}T' = A \tag{6}$$

$$BAB = T\Lambda^{-1/2}S'S\Lambda^{1/2}T'T\Lambda^{-1/2}S' = T\Lambda^{-1/2}S' = B \tag{7}$$

$$AB = S\Lambda^{1/2}T'T\Lambda^{-1/2}S' = SS' \quad \text{is symmetric} \tag{8}$$

$$BA = T\Lambda^{-1/2}S'S\Lambda^{1/2}T' = TT' \quad \text{is symmetric.} \tag{9}$$

Hence B is the unique MP inverse of A. □

7. SOME PROPERTIES OF THE MP INVERSE

Having established that for any matrix A there exists one, and only one, MP inverse A^+, let us now derive some of its properties.

Theorem 5

(i) $A^+ = A^{-1}$ for non-singular A,
(ii) $(A^+)^+ = A$,
(iii) $(A')^+ = (A^+)'$,
(iv) $A^+ = A$ if A is symmetric and idempotent,
(v) AA^+ and A^+A are idempotent,
(vi) A, A^+, AA^+ and A^+A have the same rank,
(vii) $A'AA^+ = A' = A^+AA'$,
(viii) $A'A^{+\prime}A^+ = A^+ = A^+A^{+\prime}A'$,
(ix) $(A'A)^+ = A^+A^{+\prime}$, $(AA')^+ = A^{+\prime}A^+$,
(x) $A(A'A)^+A'A = A = AA'(AA')^+A$,
(xi) $A^+ = (A'A)^+A' = A'(AA')^+$,
(xii) $A^+ = (A'A)^{-1}A'$ if A has full column rank,
(xiii) $A^+ = A'(AA')^{-1}$ if A has full row rank,
(xiv) $A = 0 \leftrightarrow A^+ = 0$,
(xv) $AB = 0 \leftrightarrow B^+A^+ = 0$,
(xvi) $A^+B = 0 \leftrightarrow A'B = 0$,
(xvii) $(A \otimes B)^+ = A^+ \otimes B^+$.

Proof. (i)–(v), (xiv) and (xvii) are established by direct substitution in the defining equations. To prove (vi), notice that each A, A^+, AA^+ and A^+A can be obtained from the others by pre- and post-multiplication by suitable matrices. Thus their ranks must all be equal. (vii) and (viii) follow from the symmetry of AA^+ and A^+A. (ix) is established by substitution in the defining equations using (vii) and (viii). (x) follows from (ix) and (vii); (xi) follows from (ix) and (viii); (xii) and (xiii) follow from (xi) and (i). To prove (xv), note that $B^+A^+ = (B'B)^+B'A'(AA')^+$, using (xi). Finally, to prove (xvi) we use (xi) and (x) and write $A^+B = 0 \leftrightarrow (A'A)^+A'B = 0 \leftrightarrow A'A(A'A)^+A'B = 0 \leftrightarrow A'B = 0$. □

Exercises

1. Determine a^+, where a is a column vector.
2. If $r(A)=1$, show that $A^+=(\operatorname{tr} AA')^{-1}A'$.
3. Show that
$$(AA^+)^+=AA^+ \qquad \text{and} \qquad (A^+A)^+=A^+A.$$
4. If A is block diagonal, then A^+ is also block diagonal. For example,
$$A=\begin{pmatrix} A_1 & 0 \\ 0 & A_2 \end{pmatrix} \quad \text{iff} \quad A^+=\begin{pmatrix} A_1^+ & 0 \\ 0 & A_2^+ \end{pmatrix}.$$
5. Show that the converse of (iv) does not hold. [*Hint*: Consider $A=-I$.]
6. Let A be an $m \times n$ matrix. If A has full row rank, show that $AA^+=I_m$; if A has full column rank, show that $A^+A=I_n$.
7. If A is symmetric, then A^+ is also symmetric and $AA^+=A^+A$.
8. Show that $(AT')^+=TA^+$ for any matrix T satisfying $T'T=I$.
9. Prove the results of Theorem 5 using the singular-value decomposition.
10. If $|A| \neq 0$ then $(AB)^+ = B^+(ABB^+)^+$.

8. FURTHER PROPERTIES

In this section we discuss some further properties of the Moore–Penrose inverse. We first prove Theorem 6, which is related to Theorem 1.1.

Theorem 6

$A'AB=A'C \leftrightarrow AB=AA^+C$.

Proof. If $AB=AA^+C$, then
$$A'AB=A'AA^+C=A'C, \tag{1}$$
using Theorem 5(vii). Conversely, if $A'AB=A'C$, then
$$AA^+C=A(A'A)^+A'C=A(A'A)^+A'AB=AB, \tag{2}$$
using Theorem 5(xi) and (x). □

Next, let us prove Theorem 7.

Theorem 7

If $|BB'| \neq 0$, then $(AB)(AB)^+=AA^+$.

Proof. Since $|BB'| \neq 0$, B has full row rank and $BB^+ = I$ (exercise 7.6). Then, using $A' = A'AA^+$,

$$\begin{aligned} AB(AB)^+ &= (AB)^{+\prime}(AB)' = (AB)^{+\prime}B'A' = (AB)^{+\prime}B'A'AA^+ \\ &= (AB)^{+\prime}(AB)'AA^+ = AB(AB)^+AA^+ = AB(AB)^+ABB^+A^+ \\ &= ABB^+A^+ = AA^+. \end{aligned} \tag{3}$$

□

To complete this section we present the following two theorems on idempotent matrices.

Theorem 8

Let $A = A' = A^2$ and $AB = B$. Then $A - BB^+$ is symmetric idempotent with rank $r(A) - r(B)$. In particular, if $r(A) = r(B)$, then $A = BB^+$.

Proof. Let $C = A - BB^+$. Then $C = C'$, $CB = 0$ and $C^2 = C$. Hence C is idempotent. Its rank is

$$r(C) = \operatorname{tr} C = \operatorname{tr} A - \operatorname{tr} BB^+ = r(A) - r(B). \tag{4}$$

Clearly, if $r(A) = r(B)$, then $C = 0$. □

Theorem 9

Let A be a symmetric idempotent $n \times n$ matrix and let $AB = 0$. If $r(A) + r(B) = n$, then $A = I_n - BB^+$.

Proof. Let $C = I_n - A$. Then C is symmetric idempotent and $CB = B$. Further $r(C) = n - r(A) = r(B)$. Hence, by Theorem 8, $C = BB^+$, that is, $A = I_n - BB^+$. □

Exercises

1. Show that
$$X'V^{-1}X(X'V^{-1}X)^+X' = X'$$
for any positive definite matrix V.
2. Hence show that if $\mathcal{M}(R') \subset \mathcal{M}(X')$, then
$$R(X'V^{-1}X)^+R'(R(X'V^{-1}X)^+R')^+R = R$$
for any positive definite matrix V.
3. Let V be a positive semidefinite $n \times n$ matrix of rank r. Let Λ be an $r \times r$ diagonal matrix with positive diagonal elements and let S be a semi-orthogonal $n \times r$ matrix such that
$$VS = S\Lambda, \qquad S'S = I_r.$$
Then
$$V = S\Lambda S', \qquad V^+ = S\Lambda^{-1}S'.$$

4. Show that the condition, in Theorem 7, that BB' is non-singular is not necessary. [*Hint*: Take $B=A^+$.]
5. Prove Theorem 6 using the singular-value decomposition.
6. Show that $ABB^+(ABB^+)^+ = AB(AB)^+$.

9. THE SOLUTION OF LINEAR EQUATION SYSTEMS

An important property of the Moore–Penrose inverse is that it enables us to find *explicit* solutions of a system of linear equations. We shall first prove Theorem 10.

Theorem 10

The general solution of the homogeneous equation $Ax=0$ is

$$x=(I-A^+A)q, \tag{1}$$

where q is an arbitrary vector of appropriate order.

Proof. Clearly, $x=(I-A^+A)q$ is a solution of $Ax=0$. Also, any arbitrary solution x of the equation $Ax=0$ satisfies

$$x=(I-A^+A)x, \tag{2}$$

which shows the existence of a vector q (namely x) such that $x=(I-A^+A)q$. □

The solution of $Ax=0$ is *unique* if, and only if, A has full column rank, since this means that $A'A$ is non-singular and hence that $A^+A=I$. The unique solution is, of course, $x=0$. If the solution is not unique, then there exist an infinite number of solutions given by (1).

The homogeneous equation $Ax=0$ always has at least one solution, namely $x=0$. The inhomogeneous equation

$$Ax=b \tag{3}$$

does not necessarily have any solution for x. If there exists at least one solution, we say that the vector equation (3) is *consistent*.

Theorem 11

Let A be a given $m\times n$ matrix and b a given $m\times 1$ vector. The following four statements are equivalent:

(a) the vector equation $Ax=b$ has a solution for x,
(b) $b\in\mathcal{M}(A)$,
(c) $r(A:b)=r(A)$,
(d) $AA^+b=b$.

Proof. It is easy to show that (a), (b) and (c) are equivalent. Let us show that (a) and (d) are equivalent, too. Suppose $Ax=b$ is consistent. Then there exists an $\bar{x}$ such that $A\bar{x}=b$. Hence, $b=A\bar{x}=AA^+A\bar{x}=AA^+b$. Now suppose that $AA^+b=b$ and let $\tilde{x}=A^+b$. Then $A\tilde{x}=AA^+b=b$. □

Having established conditions for the existence of a solution of the inhomogeneous vector equation $Ax=b$, we now proceed to give the general solution.

Theorem 12

A necessary and sufficient condition for the vector equation $Ax=b$ to have a solution is that

$$AA^+b=b, \tag{4}$$

in which case the general solution is

$$x=A^+b+(I-A^+A)q, \tag{5}$$

where q is an arbitrary vector of appropriate order.

Proof. That (4) is necessary and sufficient for the consistency of $Ax=b$ follows from Theorem 11. Let us show that the general solution is given by (5). Assume $AA^+b=b$ and define

$$x^\circ=x-A^+b. \tag{6}$$

Then, by Theorem 10,

$$\begin{aligned} &Ax=b \leftrightarrow Ax=AA^+b \leftrightarrow A(x-A^+b)=0 \\ &\quad\leftrightarrow Ax^\circ=0 \leftrightarrow x^\circ=(I-A^+A)q \leftrightarrow x=A^+b+(I-A^+A)q. \end{aligned} \tag{7}$$

□

The system $Ax = b$ is consistent *for every b* if and only if A has full *row* rank (since $AA^+ = I$ in that case). If the system is consistent, its solution is *unique* if and only if A has full *column* rank. Clearly if A has full row rank *and* full column rank then A is non-singular and the unique solution is $A^{-1}b$.

We now apply Theorem 12 to the matrix equation $AXB=C$. This yields the following theorem.

Theorem 13

A necessary and sufficient condition for the matrix equation $AXB=C$ to have a solution is that

$$AA^+CB^+B=C, \tag{8}$$

in which case the general solution is

$$X=A^+CB^+ +Q-A^+AQBB^+, \tag{9}$$

where Q is an arbitrary matrix of appropriate order.

Proof. Write the matrix equation $AXB=C$ as a vector equation $(B'\otimes A)\operatorname{vec} X = \operatorname{vec} C$, and apply Theorem 12, remembering that $(B'\otimes A)^+ = B^{+\prime}\otimes A^+$. □

Exercises

1. The matrix equation $AXB = C$ is consistent for every C if and only if A has full row rank and B has full column rank.
2. The solution of $AXB=C$, if it exists, is unique if and only if A has full column rank and B has full row rank.
3. The general solution of $AX=0$ is $X=(I-A^+A)Q$.
4. The general solution of $XA=0$ is $X=Q(I-AA^+)$.

MISCELLANEOUS EXERCISES

1. (Alternative proof of the uniqueness of the MP inverse.) Let B and C be two MP inverses of A. Let $Z=C-B$, and show that
   (i) $AZA=0$,
   (ii) $Z=ZAZ+BAZ+ZAB$,
   (iii) $(AZ)'=AZ$,
   (iv) $(ZA)'=ZA$.
   Now show that (i) and (iii) imply $AZ=0$ and that (i) and (iv) imply $ZA=0$. [*Hint*: If $P=P'$ and $P^2=0$, then $P=0$.] Conclude that $Z=0$.
2. Any matrix X that satisfies $AXA=A$ is called a *generalized inverse* of A and denoted A^-. Show that A^- exists and that
$$A^- = A^+ + Q - A^+AQAA^+, \qquad Q \text{ arbitrary.}$$
3. Show that A^-A is idempotent, but not, in general, symmetric. However, if A^-A is symmetric, then $A^-A=A^+A$ and hence *unique*. A similar result holds, of course, for AA^-.
4. Show that $A(A'A)^-A' = A(A'A)^+A'$ and hence is symmetric and idempotent.
5. Show that a necessary and sufficient condition for the equation $Ax=b$ to have a solution is that $AA^-b=b$, in which case the general solution is $x=A^-b+(I-A^-A)q$ where q is an arbitrary vector of appropriate order. (Compare Theorem 12.)
6. Show that $(AB)^+ = B^+A^+$ if A has full column rank and B has full row rank.
7. Show that $(A'A)^2B=A'A$ if and only if $A^+=B'A'$.
8. If A and B are positive semidefinite and $AB=BA$, show that $(B^{1/2}A^+B^{1/2})^+ = B^{+1/2}AB^{+1/2}$ (Liu 1995).
9. Let b be an $n\times 1$ vector with only positive elements $b_1,\ldots,b_n$. Let $B = \operatorname{dg}(b_1,\ldots,b_n)$ and $M = I_n - (1/n)\imath\imath'$, where $\imath$ denotes the $n\times 1$ sum-vector $(1,1,\ldots,1)'$. Then, $(B-bb')^+ = MB^{-1}M$ (Tanabe and Sagae 1992, Neudecker 1995).
10. If A and B are positive semidefinite, then $A\otimes A - B\otimes B$ is positive semidefinite if and only if $A-B$ is positive semidefinite (Neudecker and Satorra 1993).

11. If A and B are positive semidefinite, show that $\operatorname{tr} AB \geqslant 0$ (see also Exercise 11.5.1).
12. Let A be a symmetric $m \times m$ matrix, B an $m \times n$ matrix, $C = AB$ and $M = I_m - CC^+$. Prove that

$$(AC)^+ = C^+A^+(I_m - (MA^+)^+MA^+)$$

(Abdullah, Neudecker and Liu 1992).
13. Let A, B and $A - B$ be positive semidefinite matrices. Necessary and sufficient for $B^+ - A^+$ to be positive semidefinite is $r(A) = r(B)$ (Milliken and Akdeniz 1977, Neudecker 1989b).
14. For complex matrices we replace the transpose sign (′) by the complex conjugate sign (*) in the definition and the properties of the MP inverse. Show that these properties, thus amended, remain valid for complex matrices.

BIBLIOGRAPHICAL NOTES

§2–§3. See MacDuffee (1933, pp. 81–4) for some early references on the Kronecker product. The original interest in the Kronecker product focused on the determinantal result (3.4).
§4. The 'vec' notation was introduced by Koopmans *et al.* (1950). Theorem 2 is due to Roth (1934).
§5–§8. The Moore–Penrose inverse was introduced by Moore (1920, 1935) and rediscovered by Penrose (1955). There is a huge literature on generalized inverses, of which the Moore–Penrose inverse is one example. The interested reader may wish to consult Rao and Mitra (1971), Pringle and Rayner (1971), Boullion and Odell (1971), or Ben-Israel and Greville (1974).
§9. The results in this section are due to Penrose (1956).

CHAPTER 3

Miscellaneous matrix results

1. INTRODUCTION

In this final chapter of Part One we shall discuss some more specialized topics which will be applied later in this book. These include some further results on adjoint matrices (sections 2 and 3), Hadamard products (section 6), the commutation and the duplication matrix (sections 7–10) and some results on the bordered Gramian matrix with applications to the solution of certain matrix equations (sections 13 and 14).

2. THE ADJOINT MATRIX

We recall from section 1.9 that the cofactor c_{ij} of the element a_{ij} of any square matrix A is $(-1)^{i+j}$ times the determinant of the submatrix obtained from A by deleting row i and column j. The matrix $C=(c_{ij})$ is called the *cofactor matrix* of A. The transpose of C is called the *adjoint* matrix of A and we use the notation

$$A^{\#}=C'. \tag{1}$$

We also recall the following two properties:

$$AA^{\#}=A^{\#}A=|A|I \tag{2}$$

$$(AB)^{\#}=B^{\#}A^{\#}. \tag{3}$$

Let us now prove some further properties of the adjoint matrix.

Theorem 1

Let A be an $n\times n$ matrix ($n\geqslant 2$), and let $A^{\#}$ be the adjoint matrix of A. Then

(a) if $r(A)=n$, then

$$A^{\#}=|A|A^{-1}, \tag{4}$$

(b) if $r(A)=n-1$, then

$$A^{\#}=(-1)^{k+1}\mu(A)\frac{xy'}{y'(A^{k-1})^{+}x} \tag{5}$$

where k denotes the multiplicity of the zero eigenvalue of $A (1 \leqslant k \leqslant n)$, $\mu(A)$ is the product of the $n-k$ non-zero eigenvalues of A (if $k=n$, we put $\mu(A)=1$), and x and y are $n \times 1$ vectors satisfying $Ax = A'y = 0$, and

(c) if $r(A) \leqslant n-2$, then

$$A^{\#} = 0. \tag{6}$$

Before giving the proof of Theorem 1 we formulate the following two important corollaries.

Theorem 2

Let A be an $n \times n$ matrix ($n \geqslant 2$). Then

$$r(A^{\#}) = \begin{cases} n & \text{if } r(A) = n \\ 1 & \text{if } r(A) = n-1 \\ 0 & \text{if } r(A) \leqslant n-2. \end{cases} \tag{7}$$

Theorem 3

Let A be an $n \times n$ matrix ($n \geqslant 2$) possessing a simple eigenvalue 0. Then $r(A) = n - 1$, and

$$A^{\#} = \mu(A) \frac{xy'}{y'x} \tag{8}$$

where $\mu(A)$ is the product of the $n-1$ non-zero eigenvalues of A, and x and y satisfy $Ax = A'y = 0$.

A direct proof of Theorem 3 is given in the miscellaneous exercises to Chapter 8.

Exercises

1. Why is $y'x \neq 0$ in (8)?
2. Show that $y'x = 0$ in (5) if $k \geqslant 2$.
3. Let A be an $n \times n$ matrix. Show that
   (i) $|A^{\#}| = |A|^{n-1}$ $(n \geqslant 2)$,
   (ii) $(\alpha A)^{\#} = \alpha^{n-1} A^{\#}$ $(n \geqslant 2)$,
   (iii) $(A^{\#})^{\#} = |A|^{n-2} A$ $(n \geqslant 3)$.

3. PROOF OF THEOREM 1

If $r(A) = n$, the result follows immediately from (2.2). To prove that $A^{\#} = 0$ if $r(A) \leqslant n-2$, we express the cofactor c_{ij} as

$$c_{ij} = (-1)^{i+j} |E_i' A E_j|, \tag{1}$$

where E_j is the $n\times(n-1)$ matrix obtained from I_n by deleting column j. Now, $E_i'AE_j$ is an $(n-1)\times(n-1)$ matrix whose rank satisfies

$$r(E_i'AE_j)\leqslant r(A)\leqslant n-2. \tag{2}$$

It follows that $E_i'AE_j$ is singular and hence that $c_{ij}=0$. Since this holds for arbitrary i and j, we have $C=0$ and thus $A^{\#}=0$.

Finally assume $r(A)=n-1$. Let $\lambda_1,\lambda_2,\ldots,\lambda_n$ be the eigenvalues of A, and assume

$$\lambda_1=\lambda_2=\ldots=\lambda_k=0, \tag{3}$$

while the remaining $n-k$ eigenvalues are non-zero. By Jordan's decomposition theorem (Theorem 1.14), there exists a non-singular matrix T such that

$$T^{-1}AT=J, \tag{4}$$

where

$$J=\begin{pmatrix} J_1 & 0 \\ 0 & J_2 \end{pmatrix}. \tag{5}$$

Here J_1 is the $k\times k$ matrix

$$J_1=\begin{bmatrix} 0 & 1 & 0 & \ldots & 0 \\ 0 & 0 & 1 & \ldots & 0 \\ \vdots & \vdots & \vdots & & \vdots \\ 0 & 0 & 0 & \ldots & 1 \\ 0 & 0 & 0 & \ldots & 0 \end{bmatrix} \tag{6}$$

and J_2 is the $(n-k)\times(n-k)$ matrix

$$J_2=\begin{bmatrix} \lambda_{k+1} & \delta_{k+1} & 0 & \ldots & 0 \\ 0 & \lambda_{k+2} & \delta_{k+2} & \ldots & 0 \\ \vdots & \vdots & \vdots & & \vdots \\ 0 & 0 & 0 & \ldots & \delta_{n-1} \\ 0 & 0 & 0 & \ldots & \lambda_n \end{bmatrix} \tag{7}$$

where $\delta_j(k+1\leqslant j\leqslant n-1)$ can take the values zero or one only.

It is easy to see that every cofactor of J vanishes, with the exception of the cofactor of the element in the $(k,1)$ position. Hence

$$J^{\#}=(-1)^{k+1}\mu(A)e_1e_k', \tag{8}$$

where e_1 and e_k are the first and k-th unit vectors of order $n\times 1$, and

$$\mu(A)=\prod_{j=k+1}^{n}\lambda_j.$$

Using (2.3), (4) and (8), we obtain

$$\begin{aligned} A^{\#} &= (TJT^{-1})^{\#} = (T^{-1})^{\#} J^{\#} T^{\#} \\ &= TJ^{\#}T^{-1} = (-1)^{k+1}\mu(A)(Te_1)(e_k'T^{-1}). \end{aligned} \tag{9}$$

From (5)–(7) we have $Je_1 = 0$ and $e_k'J = 0'$. Hence, using (4),

$$ATe_1 = 0 \qquad \text{and} \qquad e_k'T^{-1}A = 0'. \tag{10}$$

Further, since $r(A) = n-1$, the vectors x and y satisfying $Ax = A'y = 0$ are unique up to a factor of proportionality. Hence

$$x = \alpha Te_1 \qquad \text{and} \qquad y' = \beta e_k'T^{-1} \tag{11}$$

for some real α and β. Now,

$$A^{k-1}Te_k = TJ^{k-1}T^{-1}Te_k = TJ^{k-1}e_k = Te_1, \tag{12}$$

and

$$e_1'T^{-1}A^{k-1} = e_1'T^{-1}TJ^{k-1}T^{-1} = e_1'J^{k-1}T^{-1} = e_k'T^{-1}. \tag{13}$$

It follows that

$$\begin{aligned} y'(A^{k-1})^{+}x &= \alpha\beta e_k'T^{-1}(A^{k-1})^{+}Te_1 \\ &= \alpha\beta e_1'T^{-1}A^{k-1}(A^{k-1})^{+}A^{k-1}Te_k \\ &= \alpha\beta e_1'T^{-1}A^{k-1}Te_k = \alpha\beta e_1'J^{k-1}e_k = \alpha\beta. \end{aligned} \tag{14}$$

Hence, from (11) and (14),

$$\frac{xy'}{y'(A^{k-1})^{+}x} = (Te_1)(e_k'T^{-1}). \tag{15}$$

Inserting (15) in (9) concludes the proof. □

4. TWO RESULTS CONCERNING BORDERED DETERMINANTS

The adjoint matrix also appears in the evaluation of the determinant of a bordered matrix, as the following theorem demonstrates.

Theorem 4

Let A be an $n \times n$ matrix, and let x and y be $n \times 1$ vectors. Then

$$\begin{vmatrix} A & x \\ y' & 0 \end{vmatrix} = -y'A^{\#}x. \tag{1}$$

Proof. Let A_i be the $(n-1) \times n$ matrix obtained from A by deleting row i, and let A_{ij} be the $(n-1) \times (n-1)$ matrix obtained from A by deleting row i and column j.

Then, using (1.9.7),

$$\begin{vmatrix} A & x \\ y' & 0 \end{vmatrix} = \sum_i x_i(-1)^{n+i+1} \begin{vmatrix} A_i \\ y' \end{vmatrix} = \sum_{i,j} x_i(-1)^{n+i+1} y_j(-1)^{n+j}|A_{ij}|$$

$$= -\sum_{i,j}(-1)^{i+j}x_i y_j|A_{ij}| = -\sum_{i,j} x_i y_j A^{\#}_{ji} = -y'A^{\#}x. \tag{2}$$

□

As one of many special cases of Theorem 4, we mention Theorem 5.

Theorem 5

Let A be a symmetric $n \times n$ matrix ($n \geqslant 2$) of rank $r(A) = n - 1$. Let u be an $n \times 1$ eigenvector of A associated with the (simple) zero eigenvalue, so that $Au = 0$. Then,

$$\begin{vmatrix} A & u \\ u' & \alpha \end{vmatrix} = -\left(\prod_{i=1}^{n-1} \lambda_i\right) u'u, \tag{3}$$

where $\lambda_1, \ldots, \lambda_{n-1}$ are the non-zero eigenvalues of A.

Proof. Without loss of generality we may take $\alpha = 0$. The result then follows immediately from Theorems 3 and 4. □

Exercise

1. Prove that $|A + \alpha s s'| = |A| + \alpha s' A^{\#} s$ (Rao and Bhimasankaram 1992).

5. THE MATRIX EQUATION $AX = 0$

In this section we shall be interested in finding the general solutions of the matrix equation $AX = 0$, where A is an $n \times n$ matrix with rank $n - 1$.

Theorem 6

Let A be an $n \times n$ matrix (possibly complex) with rank $n - 1$. Let u and v be eigenvectors of A associated with the eigenvalue zero (not necessarily simple), such that

$$Au = 0, \qquad v^*A = 0'. \tag{1}$$

The general solution of the equation

$$AX = 0 \tag{2}$$

is

$$X = uq' \tag{3}$$

where q is an arbitrary vector of appropriate order.

Moreover, the general solution of the equations

$$AX=0, \qquad XA=0 \tag{4}$$

is

$$X=\mu uv^* \tag{5}$$

where μ is an arbitrary scalar.

Proof. If $AX = 0$, then it follows from the complex analogue of exercise 1.14.4 that $X = 0$ or $r(X) = 1$. Since $Au = 0$ and $r(X) \leqslant 1$, each column of X must be a multiple of u, that is

$$X=uq' \tag{6}$$

for some vector q of appropriate order. Similarly, if $XA=0$, then

$$X=pv^* \tag{7}$$

for some vector p of appropriate order. If $AX=XA=0$, we obtain by combining (6) and (7),

$$X=\mu uv^* \tag{8}$$

for some scalar μ. □

6. THE HADAMARD PRODUCT

If $A=(a_{ij})$ and $B=(b_{ij})$ are matrices of the same order, say $m \times n$, then we define the Hadamard product of A and B as

$$A \odot B=(a_{ij}b_{ij}). \tag{1}$$

Thus, the Hadamard product $A \odot B$ is also an $m \times n$ matrix and its ij-th element is $a_{ij}b_{ij}$.

The following properties are immediate consequences of the definition:

$$A \odot B=B \odot A \tag{2}$$

$$(A \odot B)'=A' \odot B' \tag{3}$$

$$(A \odot B) \odot C=A \odot (B \odot C) \tag{4}$$

so that the brackets in (4) can be deleted without ambiguity. Further

$$(A+B) \odot (C+D)=A \odot C+A \odot D+B \odot C+B \odot D \tag{5}$$

$$A \odot 0=0 \tag{6}$$

$$A \odot I=\mathrm{dg}\, A \tag{7}$$

$$A \odot J=A=J \odot A \tag{8}$$

where J is a matrix consisting of ones only.

The following two theorems are of importance.

Theorem 7

Let A, B and C be $m \times n$ matrices, let $\imath = (1, 1, \ldots, 1)'$ be an $n \times 1$ vector and let $\Gamma = \operatorname{diag}(\gamma_1, \gamma_2, \ldots, \gamma_m)$ with $\gamma_i = \sum_{j=1}^{n} a_{ij}$. Then

(a) $$\operatorname{tr} A'(B \odot C) = \operatorname{tr}(A' \odot B')C \tag{9}$$

(b) $$\imath' A'(B \odot C)\imath = \operatorname{tr} B'\Gamma C. \tag{10}$$

Proof. To prove (a) we note that $A'(B \odot C)$ and $(A' \odot B')C$ have the same diagonal elements, namely

$$[A'(B \odot C)]_{ii} = \sum_h a_{hi} b_{hi} c_{hi} = [(A' \odot B')C]_{ii}. \tag{11}$$

To prove (b) we write

$$\imath' A'(B \odot C)\imath = \sum_{i,j,h} a_{hi} b_{hj} c_{hj} = \sum_{j,h} \gamma_h b_{hj} c_{hj} = \operatorname{tr} B'\Gamma C. \tag{12}$$

□

Theorem 8

Let A and B be square $n \times n$ matrices and $\imath = (1, 1, \ldots, 1)'$ be an $n \times 1$ vector. Further let M be a diagonal $n \times n$ matrix and m an $n \times 1$ vector such that

$$M = \operatorname{diag}(\mu_1, \mu_2, \ldots, \mu_n), \qquad m = M\imath. \tag{13}$$

Then

(a) $$\operatorname{tr} AMB'M = m'(A \odot B)m \tag{14}$$

(b) $$\operatorname{tr} AB' = \imath'(A \odot B)\imath \tag{15}$$

(c) $$MA \odot B'M = M(A \odot B')M. \tag{16}$$

Proof. To prove (a) we write

$$\operatorname{tr} AMB'M = \sum_i (AMB'M)_{ii} = \sum_{i,j} \mu_i \mu_j a_{ij} b_{ij} = m'(A \odot B)m. \tag{17}$$

Taking $M = I_n$, we obtain (b) as a special case of (a). Finally, to prove (c) we have

$$\begin{aligned}(MA \odot B'M)_{ij} &= (MA)_{ij}(B'M)_{ij} = (\mu_i a_{ij})(\mu_j b_{ji}) \\ &= \mu_i \mu_j (A \odot B')_{ij} = (M(A \odot B')M)_{ij}.\end{aligned} \tag{18}$$

□

7. THE COMMUTATION MATRIX K_{mn}

Let A be an $m \times n$ matrix. The vectors vec A and vec A' clearly contain the same mn components, but in a different order. Hence there exists a unique $mn \times mn$

permutation matrix which transforms vec A into vec A'. This matrix is called the *commutation matrix* and is denoted K_{mn} or $K_{m,n}$. (If $m = n$, we often write K_n instead of K_{nn}.) Thus

$$K_{mn} \operatorname{vec} A = \operatorname{vec} A'. \tag{1}$$

Since K_{mn} is a permutation matrix it is orthogonal, i.e. $K'_{mn} = K^{-1}_{mn}$, see (1.8.4). Also, pre-multiplying (1) by K_{nm} gives $K_{nm}K_{mn}$ vec $A =$ vec A so that $K_{nm}K_{mn} = I_{mn}$. Hence,

$$K'_{mn} = K^{-1}_{mn} = K_{nm}. \tag{2}$$

Further, using (2.4.2),

$$K_{n1} = K_{1n} = I_n. \tag{3}$$

The key property of the commutation matrix (and the one from which it derives its name) enables us to interchange ('commute') the two matrices of a Kronecker product.

Theorem 9

Let A be an $m \times n$ matrix, B a $p \times q$ matrix and b a $p \times 1$ vector. Then

(a) $$K_{pm}(A \otimes B) = (B \otimes A)K_{qn} \tag{4}$$

(b) $$K_{pm}(A \otimes B)K_{nq} = B \otimes A \tag{5}$$

(c) $$K_{pm}(A \otimes b) = b \otimes A \tag{6}$$

(d) $$K_{mp}(b \otimes A) = A \otimes b. \tag{7}$$

Proof. Let X be an arbitrary $q \times n$ matrix. Then, by repeated application of (1) and Theorem 2.2,

$$K_{pm}(A \otimes B) \operatorname{vec} X = K_{pm} \operatorname{vec} BXA'$$

$$= \operatorname{vec} AX'B' = (B \otimes A) \operatorname{vec} X' = (B \otimes A)K_{qn} \operatorname{vec} X. \tag{8}$$

Since X is arbitrary, (a) follows. The remaining results are immediate consequences of (a). □

An important application of the commutation matrix is that it allows us to transform the vec of a Kronecker product into the Kronecker product of the vecs, a crucial property in the differentiation of Kronecker products.

Theorem 10

Let A be an $m \times n$ matrix and B a $p \times q$ matrix. Then

$$\operatorname{vec}(A \otimes B) = (I_n \otimes K_{qm} \otimes I_p)(\operatorname{vec} A \otimes \operatorname{vec} B). \tag{9}$$

Proof. Let a_i $(i = 1, \ldots, n)$ and b_j $(j = 1, \ldots, q)$ denote the columns of A and B, respectively. Also, let e_i $(i = 1, \ldots, n)$ and u_j $(j = 1, \ldots, q)$ denote the columns of

I_n and I_q, respectively. Then we can write A and B as

$$A=\sum_{i=1}^{n} a_i e_i', \qquad B=\sum_{j=1}^{q} b_j u_j', \tag{10}$$

and we obtain

$$\begin{aligned}
\operatorname{vec}(A\otimes B) &= \sum_{i=1}^{n}\sum_{j=1}^{q} \operatorname{vec}(a_i e_i' \otimes b_j u_j') \\
&= \sum_{i,j} \operatorname{vec}(a_i\otimes b_j)(e_i\otimes u_j)' = \sum_{i,j}(e_i\otimes u_j\otimes a_i\otimes b_j) \\
&= \sum_{i,j}(I_n\otimes K_{qm}\otimes I_p)(e_i\otimes a_i\otimes u_j\otimes b_j) \\
&= (I_n\otimes K_{qm}\otimes I_p)\left[\left(\sum_i \operatorname{vec} a_i e_i'\right)\otimes\left(\sum_j \operatorname{vec} b_j u_j'\right)\right] \\
&= (I_n\otimes K_{qm}\otimes I_p)(\operatorname{vec} A\otimes \operatorname{vec} B).
\end{aligned} \tag{11}$$

□

Closely related to the matrix K_n is the matrix $\frac{1}{2}(I_{n^2}+K_n)$. Some properties of this matrix follow below.

Theorem 11

Let $N_n=\frac{1}{2}(I_{n^2}+K_n)$. Then

(a) $$N_n=N_n'=N_n^2 \tag{12}$$

(b) $$r(N_n)=\operatorname{tr} N_n=\tfrac{1}{2}n(n+1) \tag{13}$$

(c) $$N_n K_n = N_n = K_n N_n. \tag{14}$$

Proof. The proof is easy and is left to the reader. □

Exercise

1. Let A $(m\times n)$ and B $(p\times q)$ be two matrices. Show that

$$\operatorname{vec}(A\otimes B)=(I_n\otimes G)\operatorname{vec} A=(H\otimes I_p)\operatorname{vec} B,$$

where

$$G=(K_{qm}\otimes I_p)(I_m\otimes \operatorname{vec} B), \qquad H=(I_n\otimes K_{qm})(\operatorname{vec} A\otimes I_q).$$

8. THE DUPLICATION MATRIX D_n

Let A be a square $n\times n$ matrix. Then $\operatorname{v}(A)$ will denote the $\frac{1}{2}n(n+1)\times 1$ vector that is obtained from vec A by eliminating all supradiagonal elements of A. For example, if $n=3$,

$$\text{vec}\,A=(a_{11}\;a_{21}\;a_{31}\;a_{12}\;a_{22}\;a_{32}\;a_{13}\;a_{23}\;a_{33})', \tag{1}$$

and

$$\text{v}(A)=(a_{11}\;a_{21}\;a_{31}\;a_{22}\;a_{32}\;a_{33})'. \tag{2}$$

In this way, for symmetric A, $\text{v}(A)$ contains only the generically distinct elements of A. Since the elements of vec A are those of $\text{v}(A)$ with some repetitions, there exists a unique $n^2 \times \frac{1}{2}n(n+1)$ matrix which transforms, for symmetric A, $\text{v}(A)$ into vec A. This matrix is called the *duplication matrix* and is denoted D_n. Thus,

$$D_n\text{v}(A)=\text{vec}\,A \qquad (A=A'). \tag{3}$$

Let $A=A'$ and $D_n\text{v}(A)=0$. Then vec $A=0$, and so $\text{v}(A)=0$. Since the symmetry of A does not restrict $\text{v}(A)$, it follows that the columns of D_n are linearly independent. Hence D_n has full column rank $\frac{1}{2}n(n+1)$, $D_n'D_n$ is non-singular, and D_n^+, the Moore–Penrose inverse of D_n, equals

$$D_n^+=(D_n'D_n)^{-1}D_n'. \tag{4}$$

Since D_n has full column rank, $\text{v}(A)$ can be uniquely solved from (3) and we have

$$\text{v}(A)=D_n^+\,\text{vec}\,A \qquad (A=A'). \tag{5}$$

Some further properties of D_n are easily derived from its definition (3).

Theorem 12

(a)
$$K_nD_n=D_n \tag{6}$$

(b)
$$D_nD_n^+=\tfrac{1}{2}(I_{n^2}+K_n) \tag{7}$$

(c)
$$D_nD_n^+(b\otimes A)=\tfrac{1}{2}(b\otimes A+A\otimes b) \tag{8}$$

for any $n\times 1$ vector b and $n\times n$ matrix A.

Proof. Let X be a symmetric $n\times n$ matrix. Then

$$K_nD_n\text{v}(X)=K_n\,\text{vec}\,X=\text{vec}\,X=D_n\text{v}(X). \tag{9}$$

Since the symmetry of X does not restrict $\text{v}(X)$, we obtain (a). To prove (b), let $N_n=\frac{1}{2}(I_{n^2}+K_n)$. Then, from (a), $N_nD_n=D_n$. Now, N_n is symmetric idempotent with $r(N_n)=r(D_n)=\frac{1}{2}n(n+1)$ (Theorem 11(b)). Then, by Theorem 2.8, $N_n=D_nD_n^+$. Finally, (c) follows from (b) and the fact that $K_n(b\otimes A)=A\otimes b$. □

Much of the interest in the duplication matrix is due to the importance of the matrices $D_n^+(A\otimes A)D_n$ and $D_n'(A\otimes A)D_n$, some of whose properties follow below.

Theorem 13

Let A be an $n\times n$ matrix. Then

(a)
$$D_nD_n^+(A\otimes A)D_n=(A\otimes A)D_n \tag{10}$$

(b)
$$D_nD_n^+(A\otimes A)D_n^{+\prime}=(A\otimes A)D_n^{+\prime}, \tag{11}$$

and if A is non-singular,

(c)
$$(D_n^+(A\otimes A)D_n)^{-1}=D_n^+(A^{-1}\otimes A^{-1})D_n \tag{12}$$

(d)
$$(D_n'(A\otimes A)D_n)^{-1}=D_n^+(A^{-1}\otimes A^{-1})D_n^{+\prime}. \tag{13}$$

Proof. Let $N_n=\frac{1}{2}(I+K_n)$. Then, since

$$D_nD_n^+=N_n,\qquad N_n(A\otimes A)=(A\otimes A)N_n, \tag{14}$$

$$N_nD_n=D_n,\qquad N_nD_n^{+\prime}=D_n^{+\prime}, \tag{15}$$

we obtain (a) and (b). To prove (c) we write

$$\begin{aligned}D_n^+(A\otimes A)D_nD_n^+(A^{-1}\otimes A^{-1})D_n&=D_n^+(A\otimes A)N_n(A^{-1}\otimes A^{-1})D_n\\&=D_n^+(A\otimes A)(A^{-1}\otimes A^{-1})N_nD_n=D_n^+D_n=I_{\frac{1}{2}n(n+1)}.\end{aligned} \tag{16}$$

Finally, to prove (d), we use (c) and $D_n^+=(D_n'D_n)^{-1}D_n'$ and write

$$\begin{aligned}(D_n'(A\otimes A)D_n)^{-1}&=(D_n'D_nD_n^+(A\otimes A)D_n)^{-1}=(D_n^+(A\otimes A)D_n)^{-1}(D_n'D_n)^{-1}\\&=D_n^+(A^{-1}\otimes A^{-1})D_n(D_n'D_n)^{-1}=D_n^+(A^{-1}\otimes A^{-1})D_n^{+\prime}.\end{aligned} \tag{17}$$

□

Finally, we state, without proof, two further properties of the duplication matrix which we shall need later.

Theorem 14

Let A be an $n\times n$ matrix. Then

(a)
$$D_n'\operatorname{vec}A=\mathrm{v}(A+A'-\mathrm{dg}\,A) \tag{18}$$

(b)
$$|D_n^+(A\otimes A)D_n^{+\prime}|=2^{-\frac{1}{2}n(n-1)}|A|^{n+1}. \tag{19}$$

9. RELATIONSHIP BETWEEN D_{n+1} AND D_n, I

Let A_1 be a symmetric $(n+1)\times(n+1)$ matrix. We wish to express $D_{n+1}'(A_1\otimes A_1)D_{n+1}$ and $D_{n+1}^+(A_1\otimes A_1)D_{n+1}^{+\prime}$ as partitioned matrices. In particular, we wish to know whether $D_n'(A\otimes A)D_n$ is a submatrix of $D_{n+1}'(A_1\otimes A_1)D_{n+1}$ and $D_n^+(A\otimes A)D_n^{+\prime}$ is a submatrix of $D_{n+1}^+(A_1\otimes A_1)D_{n+1}^{+\prime}$ when A is the appropriate submatrix of A_1. The next theorem answers a slightly more general question in the affirmative.

Theorem 15

Let

$$A_1=\begin{pmatrix}\alpha & a'\\ a & A\end{pmatrix},\qquad B_1=\begin{pmatrix}\beta & b'\\ b & B\end{pmatrix}$$

where A and B are symmetric $n \times n$ matrices, a and b are $n \times 1$ vectors and α and β are scalars. Then

(i) $D'_{n+1}(A_1 \otimes B_1)D_{n+1} =$

$$\begin{bmatrix} \alpha\beta & \alpha b' + \beta a' & (a' \otimes b')D_n \\ \alpha b + \beta a & \alpha B + \beta A + ab' + ba' & (a' \otimes B + b' \otimes A)D_n \\ D'_n(a \otimes b) & D'_n(a \otimes B + b \otimes A) & D'_n(A \otimes B)D_n \end{bmatrix}$$

(ii) $D^+_{n+1}(A_1 \otimes B_1)D^{+\prime}_{n+1} =$

$$\begin{bmatrix} \alpha\beta & \tfrac{1}{2}(\alpha b' + \beta a') & (a' \otimes b')D^{+\prime}_n \\ \tfrac{1}{2}(\alpha b + \beta a) & \tfrac{1}{4}(\alpha B + \beta A + ab' + ba') & \tfrac{1}{2}(a' \otimes B + b' \otimes A)D^{+\prime}_n \\ D^+_n(a \otimes b) & \tfrac{1}{2}D^+_n(a \otimes B + b \otimes A) & D^+_n(A \otimes B)D^{+\prime}_n \end{bmatrix}.$$

In particular,

(iii) $$D'_{n+1}D_{n+1} = \begin{bmatrix} 1 & 0 & 0 \\ 0 & 2I_n & 0 \\ 0 & 0 & D'_nD_n \end{bmatrix}$$

(iv) $$D^+_{n+1}D^{+\prime}_{n+1} = (D'_{n+1}D_{n+1})^{-1} = \begin{bmatrix} 1 & 0 & 0 \\ 0 & \tfrac{1}{2}I_n & 0 \\ 0 & 0 & (D'_nD_n)^{-1} \end{bmatrix}.$$

Proof. Let X_1 be an arbitrary symmetric $(n+1) \times (n+1)$ matrix partitioned conformably with A_1 and B_1 as

$$X_1 = \begin{pmatrix} \xi & x' \\ x & X \end{pmatrix}. \tag{1}$$

Then,

$$\begin{aligned} \operatorname{tr} A_1X_1B_1X_1 &= (\operatorname{vec} X_1)'(A_1 \otimes B_1)(\operatorname{vec} X_1) \\ &= (\mathrm{v}(X_1))'D'_{n+1}(A_1 \otimes B_1)D_{n+1}\mathrm{v}(X_1) \end{aligned} \tag{2}$$

and also

$$\begin{aligned} \operatorname{tr} A_1X_1B_1X_1 &= \alpha\beta\xi^2 + 2\xi(\alpha b'x + \beta a'x) + \alpha x'Bx + \beta x'Ax + 2(a'x)(b'x) \\ &\quad + 2\xi a'Xb + 2(x'BXa + x'AXb) + \operatorname{tr} AXBX \end{aligned}$$

$$= \begin{bmatrix} \xi \\ x \\ \mathrm{v}(X) \end{bmatrix}' \begin{bmatrix} \alpha\beta & \alpha b' + \beta a' & (a' \otimes b')D_n \\ \alpha b + \beta a & \alpha B + \beta A + ab' + ba' & (a' \otimes B + b' \otimes A)D_n \\ D'_n(a \otimes b) & D'_n(a \otimes B + b \otimes A) & D'_n(A \otimes B)D_n \end{bmatrix} \begin{bmatrix} \xi \\ x \\ \mathrm{v}(X) \end{bmatrix}. \tag{3}$$

The first result now follows from (2) and (3), the symmetry of all matrices concerned and the fact that

$$(\mathrm{v}(X_1))' = (\xi, x', (\mathrm{v}(X))'). \tag{4}$$

By letting $A_1 = B_1 = I_{n+1}$, we obtain (iii) as a special case of (i).

(iv) follows from (iii). Pre- and post-multiplying (i) by $(D'_{n+1}D_{n+1})^{-1}$ as given in (iv) yields (ii). □

10. RELATIONSHIP BETWEEN D_{n+1} and D_n, II

Related to Theorem 15 is the following result.

Theorem 16

Let

$$A_1 = \begin{pmatrix} \alpha & b' \\ a & A \end{pmatrix}, \tag{1}$$

where A is an $n \times n$ matrix (not necessarily symmetric), a and b are $n \times 1$ vectors and α is a scalar. Then

$$D'_{n+1} \operatorname{vec} A_1 = \begin{bmatrix} \alpha \\ a+b \\ D'_n \operatorname{vec} A \end{bmatrix}, \qquad D^+_{n+1} \operatorname{vec} A_1 = \begin{bmatrix} \alpha \\ \frac{1}{2}(a+b) \\ D^+_n \operatorname{vec} A \end{bmatrix}. \tag{2}$$

Proof. We have, using Theorem 14(a),

$$D'_{n+1} \operatorname{vec} A_1 = \mathrm{v}(A_1 + A'_1 - \mathrm{dg}\, A_1)$$

$$= \begin{bmatrix} \alpha \\ a \\ \mathrm{v}(A) \end{bmatrix} + \begin{bmatrix} \alpha \\ b \\ \mathrm{v}(A') \end{bmatrix} - \begin{bmatrix} \alpha \\ 0 \\ \mathrm{v}(\mathrm{dg}\, A) \end{bmatrix} = \begin{bmatrix} \alpha \\ a+b \\ D'_n \operatorname{vec} A \end{bmatrix}. \tag{3}$$

Also, using Theorem 15(iv),

$$D^+_{n+1} \operatorname{vec} A_1 = \begin{bmatrix} 1 & 0 & 0 \\ 0 & \frac{1}{2}I_n & 0 \\ 0 & 0 & (D'_nD_n)^{-1} \end{bmatrix} \begin{bmatrix} \alpha \\ a+b \\ D'_n \operatorname{vec} A \end{bmatrix}$$

$$= \begin{bmatrix} \alpha \\ \frac{1}{2}(a+b) \\ D^+_n \operatorname{vec} A \end{bmatrix}. \tag{4}$$

□

As a corollary of Theorem 16 we obtain Theorem 17.

Theorem 17

Let A be an $n\times p$ matrix and b a $p\times 1$ vector. Then

$$D'_{n+1}\begin{bmatrix} b' \\ A \\ 0_1 \end{bmatrix} = \begin{bmatrix} b' \\ A \\ 0_2 \end{bmatrix}, \qquad D^+_{n+1}\begin{bmatrix} b' \\ A \\ 0_1 \end{bmatrix} = \begin{bmatrix} b' \\ \tfrac{1}{2}A \\ 0_2 \end{bmatrix}, \tag{5}$$

where 0_1 and 0_2 denote null matrices of orders $n(n+1)\times p$ and $\frac{1}{2}n(n+1)\times p$ respectively.

Proof. Let β_i be the i-th component of b and let a_i be the i-th column of A $(i=1,\ldots,p)$. Define the $(n+1)\times(n+1)$ matrices

$$C_i=\begin{pmatrix} \beta_i & 0' \\ a_i & 0 \end{pmatrix} \qquad (i=1,\ldots,p). \tag{6}$$

Then, using Theorem 16,

$$\operatorname{vec} C_i=\begin{bmatrix} \beta_i \\ a_i \\ 0 \end{bmatrix}, \qquad D'_{n+1}\operatorname{vec} C_i=\begin{bmatrix} \beta_i \\ a_i \\ 0 \end{bmatrix}, \qquad D^+_{n+1}\operatorname{vec} C_i=\begin{bmatrix} \beta_i \\ \tfrac{1}{2}a_i \\ 0 \end{bmatrix} \tag{7}$$

for $i=1,\ldots,p$, and the result follows from the fact that

$$\begin{bmatrix} b' \\ A \\ 0_1 \end{bmatrix} =(\operatorname{vec} C_1, \operatorname{vec} C_2, \ldots, \operatorname{vec} C_p). \tag{8}$$

□

11. CONDITIONS FOR A QUADRATIC FORM TO BE POSITIVE (NEGATIVE) SUBJECT TO LINEAR CONSTRAINTS

Many optimization problems take the form

$$\text{minimize} \quad x'Ax \tag{1}$$

$$\text{subject to} \quad Bx=0, \tag{2}$$

and, as we shall see later (Theorem 7.12), this problem also arises when we try to establish second-order conditions for Lagrange minimization (maximization). The following theorem is then of importance.

Theorem 18

Let A be a symmetric $n\times n$ matrix and B an $m\times n$ matrix with full row rank m. Let A_{rr} denote the $r\times r$ matrix in the top left corner of A, and B_r the $m\times r$ matrix whose columns are the first r columns of B ($r = 1, \ldots, n$). Assume that $|B_m| \neq 0$. Define the $(m + r) \times (m + r)$ matrices

$$\Delta_r = \begin{pmatrix} 0 & B_r \\ B'_r & A_{rr} \end{pmatrix} \qquad (r = 1, 2, \ldots, n), \tag{3}$$

and let $\Gamma = \{x \in \mathbb{R}^n : x \neq 0, Bx = 0\}$. Then

(i) $x'Ax > 0$ for all $x \in \Gamma$ if only if

$$(-1)^m |\Delta_r| > 0 \qquad (r = m+1, \ldots, n); \tag{4}$$

(ii) $x'Ax < 0$ for all $x \in \Gamma$ if and only if

$$(-1)^r |\Delta_r| > 0 \qquad (r = m+1, \ldots, n). \tag{5}$$

Proof. We partition B and x conformably as

$$B = (B_m : B_*), \qquad x = (x'_1, x'_2)' \tag{6}$$

where B_* is an $m \times (n - m)$ matrix and $x_1 \in \mathbb{R}^m, x_2 \in \mathbb{R}^{n-m}$. The constraint $Bx = 0$ can then be written as

$$B_m x_1 + B_* x_2 = 0, \tag{7}$$

that is,

$$x_1 + B_m^{-1} B_* x_2 = 0, \tag{8}$$

or equivalently,

$$x = Qx_2, \qquad Q = \begin{pmatrix} -B_m^{-1} B_* \\ I_{n-m} \end{pmatrix}. \tag{9}$$

Hence we can write the constraint set Γ as

$$\Gamma = \{x \in \mathbb{R}^n : x = Qy, \ y \neq 0, \ y \in \mathbb{R}^{n-m}\}, \tag{10}$$

and we see that $x'Ax > 0$ (<0) for all $x \in \Gamma$ if and only if the $(n-m) \times (n-m)$ matrix $Q'AQ$ is positive definite (negative definite).

Next we investigate the signs of the $n-m$ principal minors of $Q'AQ$. For $k = 1, 2, \ldots, n-m$, let E_k be the $k \times (n-m)$ selection matrix

$$E_k = (I_k : 0) \tag{11}$$

and let C_k be the $k \times k$ matrix in the top left corner of $Q'AQ$. Then

$$C_k = E_k Q'AQE'_k. \tag{12}$$

We partition $B_* = (B_{*1} : B_{*2})$, where B_{*1} is an $m \times k$ matrix and B_{*2} an $m \times (n-m-k)$ matrix, and define the $(m+k) \times k$ matrix

$$Q_k = \begin{pmatrix} -B_m^{-1} B_{*1} \\ I_k \end{pmatrix}. \tag{13}$$

We then have

$$QE_k' = \begin{bmatrix} -B_m^{-1}B_{*1} & -B_m^{-1}B_{*2} \\ I_k & 0 \\ 0 & I_{n-m-k} \end{bmatrix} \begin{pmatrix} I_k \\ 0 \end{pmatrix} = \begin{bmatrix} -B_m^{-1}B_{*1} \\ I_k \\ 0 \end{bmatrix} = \begin{pmatrix} Q_k \\ 0 \end{pmatrix} \tag{14}$$

and hence

$$C_k = (Q_k':0)\begin{pmatrix} A_{m+k,\,m+k} & * \\ * & * \end{pmatrix}\begin{pmatrix} Q_k \\ 0 \end{pmatrix} = Q_k' A_{m+k,\,m+k} Q_k, \tag{15}$$

where $*$'s indicate matrices the precise form of which is of no relevance. Now, let T_k be the non-singular $(m+k)\times(m+k)$ matrix

$$T_k = \begin{pmatrix} B_m & B_{*1} \\ 0 & I_k \end{pmatrix}. \tag{16}$$

Its inverse is

$$T_k^{-1} = \begin{pmatrix} B_m^{-1} & -B_m^{-1}B_{*1} \\ 0 & I_k \end{pmatrix} \tag{17}$$

and one verifies easily that

$$B_{m+k}T_k^{-1} = (B_m:B_{*1})\begin{pmatrix} B_m^{-1} & -B_m^{-1}B_{*1} \\ 0 & I_k \end{pmatrix} = (I_m:0). \tag{18}$$

Hence,

$$\begin{pmatrix} I_m & 0 \\ 0 & T_k^{-1\prime} \end{pmatrix}\begin{pmatrix} 0 & B_{m+k} \\ B_{m+k}' & A_{m+k,\,m+k} \end{pmatrix}\begin{pmatrix} I_m & 0 \\ 0 & T_k^{-1} \end{pmatrix} = \begin{pmatrix} 0 & B_{m+k}T_k^{-1} \\ T_k^{-1\prime}B_{m+k}' & T_k^{-1\prime}A_{m+k,\,m+k}T_k^{-1} \end{pmatrix}$$

$$= \begin{bmatrix} 0 & I_m & 0 \\ I_m & * & * \\ 0 & * & C_k \end{bmatrix} = \begin{bmatrix} I_m & 0 & 0 \\ * & I_m & * \\ * & 0 & C_k \end{bmatrix} \begin{bmatrix} 0 & I_m & 0 \\ I_m & 0 & 0 \\ 0 & 0 & I_k \end{bmatrix}. \tag{19}$$

Taking determinants on both sides of (19) we obtain

$$|T_k^{-1}|^2 |\Delta_{m+k}| = (-1)^m |C_k| \tag{20}$$

(see exercise 1), and hence

$$(-1)^m |\Delta_{m+k}| = |T_k|^2 |C_k| \qquad (k=1,\ldots,n-m). \tag{21}$$

Thus, $x'Ax>0$ for all $x\in\Gamma$ iff $Q'AQ$ is positive definite iff $|C_k|>0$ $(k=1,\ldots,n-m)$ iff $(-1)^m|\Delta_{m+k}|>0$ $(k=1,\ldots,n-m)$.

Similarly, $x'Ax<0$ for all $x\in\Gamma$ iff $Q'AQ$ is negative definite iff $(-1)^k|C_k|>0$ $(k=1,\ldots,n-m)$ iff $(-1)^{m+k}|\Delta_{m+k}|>0$ $(k=1,\ldots,n-m)$. □

12. NECESSARY AND SUFFICIENT CONDITIONS FOR $r(A:B)=r(A)+r(B)$

Let us now prove Theorem 19.

Theorem 19

Let A and B be two matrices with the same number of rows. Then the following seven statements are equivalent.

(i) $\mathcal{M}(A)\cap\mathcal{M}(B)=\{0\}$,
(ii) $r(AA'+BB')=r(A)+r(B)$,
(iii) $A'(AA'+BB')^+A$ is idempotent,
(iv) $A'(AA'+BB')^+A=A^+A$,
(v) $B'(AA'+BB')^+B$ is idempotent,
(vi) $B'(AA'+BB')^+B=B^+B$,
(vii) $A'(AA'+BB')^+B=0$.

Proof. (ii)→(i): Since $r(AA'+BB')=r(A:B)$, (ii) implies $r(A:B)=r(A)+r(B)$. Hence the linear space spanned by the columns of A and the linear space spanned by the columns of B are disjoint, that is, $\mathcal{M}(A)\cap\mathcal{M}(B)=\{0\}$.

(i)→(iii): We shall show that (i) implies that the eigenvalues of the matrix $(AA' + BB')^+AA'$ are either zero or one. Then by Theorem 1.9 the same is true for the symmetric matrix $A'(AA' + BB')^+A$, thus proving its idempotency. Let λ be an eigenvalue of $(AA' + BB')^+AA'$, and x a corresponding eigenvector, so that

$$(AA'+BB')^+AA'x=\lambda x. \tag{1}$$

Since

$$(AA'+BB')(AA'+BB')^+A=A, \tag{2}$$

we have

$$\begin{aligned} AA'x&=(AA'+BB')(AA'+BB')^+AA'x\\ &=\lambda(AA'+BB')x, \end{aligned} \tag{3}$$

and hence

$$(1-\lambda)AA'x=\lambda BB'x. \tag{4}$$

Now, since $\mathcal{M}(AA')\cap\mathcal{M}(BB')=\{0\}$, (4) implies

$$(1-\lambda)AA'x=0. \tag{5}$$

Thus, $AA'x=0$ implies $\lambda=0$ by (1) and $AA'x\neq 0$ implies $\lambda=1$ by (5). Hence $\lambda=0$ or $\lambda=1$.

(iii)→(vii): If (iii) holds, then

$$\begin{aligned}A'(AA'+BB')^+A &= A'(AA'+BB')^+AA'(AA'+BB')^+A \\ &= A'(AA'+BB')^+(AA'+BB')(AA'+BB')^+A \\ &\quad - A'(AA'+BB')^+BB'(AA'+BB')^+A \\ &= A'(AA'+BB')^+A - A'(AA'+BB')^+BB'(AA'+BB')^+A. \qquad (6)\end{aligned}$$

Hence

$$A'(AA'+BB')^+BB'(AA'+BB')^+A = 0, \qquad (7)$$

which implies (vii).

(v)→(vii): This is proved similarly.

(vii)→(iv): If (vii) holds, then, using (2),

$$A = (AA'+BB')(AA'+BB')^+A = AA'(AA'+BB')^+A. \qquad (8)$$

Pre-multiplication with A^+ gives (iv).

(vii)→(vi): This is proved similarly.

(iv)→(iii) and (vi)→(v): Trivial.

(vii)→(ii): We know already that (vii) implies (iv) and (vi). Hence

$$\begin{pmatrix} A' \\ B' \end{pmatrix}(AA'+BB')^+(A:B) = \begin{bmatrix} A^+A & 0 \\ 0 & B^+B \end{bmatrix}. \qquad (9)$$

The rank of the matrix on the left side of (9) is $r(A:B)$; the rank of the matrix on the right side is $r(A^+A)+r(B^+B)$. It follows that

$$r(AA'+BB') = r(A:B) = r(A^+A)+r(B^+B) = r(A)+r(B). \qquad (10)$$

This completes the proof. □

13. THE BORDERED GRAMIAN MATRIX

Let A be a positive semidefinite $n \times n$ matrix and B an $n \times k$ matrix. The symmetric $(n+k)\times(n+k)$ matrix

$$Z = \begin{pmatrix} A & B \\ B' & 0 \end{pmatrix}, \qquad (1)$$

called a bordered Gramian matrix, is of great interest in optimization theory. We first prove Theorem 20.

Theorem 20

Let $N = A + BB'$ and $C = B'N^+B$. Then

(i) $\mathcal{M}(A) \subset \mathcal{M}(N)$, $\mathcal{M}(B) \subset \mathcal{M}(N)$, $\mathcal{M}(B') = \mathcal{M}(C)$,

(ii) $NN^+A = A$, $NN^+B = B$,
(iii) $C^+C = B^+B$, $r(C) = r(B)$.

Proof. Let $A = TT'$ and recall from (1.7.9) that $\mathcal{M}(Q) = \mathcal{M}(QQ')$ for any Q. Then

$$\mathcal{M}(A) = \mathcal{M}(T) \subset \mathcal{M}(T:B) = \mathcal{M}(TT' + BB') = \mathcal{M}(N), \tag{2}$$

and similarly $\mathcal{M}(B) \subset \mathcal{M}(N)$. Hence $NN^+A = A$ and $NN^+B = B$. Next, let $N^+ = FF'$ and define $G = B'F$. Then $C = GG'$. Using (ii) and the fact that $G'(GG')(GG')^+ = G'$ for any G, we obtain

$$B(I - CC^+) = NN^+B(I - CC^+) = NFG'(I - GG'(GG')^+) = 0, \tag{3}$$

and hence $\mathcal{M}(B') \subset \mathcal{M}(C)$. Since obviously $\mathcal{M}(C) \subset \mathcal{M}(B')$, we find that $\mathcal{M}(B') = \mathcal{M}(C)$.

Finally, to prove (iii), we note that $\mathcal{M}(B') = \mathcal{M}(C)$ implies that the ranks of B' and C must be equal and hence that $r(B) = r(C)$; we also have

$$(B'B^{+\prime})C = (B'B^{+\prime})(B'N^+B) = B'N^+B = C. \tag{4}$$

As $B'B^{+\prime}$ is symmetric idempotent and $r(B'B^{+\prime}) = r(B') = r(C)$, it follows (by Theorem 2.8) that $B'B^{+\prime} = CC^+$ and hence that $B^+B = C^{+\prime}C' = C^+C$ (exercise 2.7.7). This concludes the proof. □

Next we obtain the Moore–Penrose inverse of Z.

Theorem 21

The Moore–Penrose inverse of Z is

$$Z^+ = \begin{pmatrix} D & E \\ E' & -F \end{pmatrix} \tag{5}$$

where

$$D = N^+ - N^+BC^+B'N^+ \tag{6}$$

$$E = N^+BC^+ \tag{7}$$

$$F = C^+ - CC^+, \tag{8}$$

and

$$N = A + BB', \qquad C = B'N^+B. \tag{9}$$

Moreover,

$$ZZ^+ = Z^+Z = \begin{pmatrix} NN^+ & 0 \\ 0 & CC^+ \end{pmatrix}. \tag{10}$$

Proof. Let G be defined by

$$G = \begin{pmatrix} N^+ - N^+BC^+B'N^+ & N^+BC^+ \\ C^+B'N^+ & -C^+ + CC^+ \end{pmatrix}. \tag{11}$$

Then

$$ZG = \begin{pmatrix} AN^+ - AN^+BC^+B'N^+ + BC^+B'N^+ & AN^+BC^+ - BC^+ + BCC^+ \\ B'N^+ - B'N^+BC^+B'N^+ & B'N^+BC^+ \end{pmatrix}$$

$$= \begin{pmatrix} NN^+ & 0 \\ 0 & CC^+ \end{pmatrix}, \tag{12}$$

which we obtain by replacing A by $N - BB'$, and using the definition of C and the results $NN^+B = B$ and $CC^+B' = B'$ (see Theorem 20). Since Z and G are both symmetric and ZG is also symmetric by (12), it follows that $ZG = GZ$ and so GZ is also symmetric. To show that $ZGZ = Z$ and $GZG = G$ is straightforward. This concludes the proof. □

In the special case where $\mathcal{M}(B) \subset \mathcal{M}(A)$, the results can be simplified. This case is worth stating as a separate theorem.

Theorem 22

In the special case where $\mathcal{M}(B) \subset \mathcal{M}(A)$, we have

$$AA^+B = B, \qquad \Gamma\Gamma^+ = B^+B \tag{13}$$

where $\Gamma = B'A^+B$. Furthermore,

$$Z^+ = \begin{pmatrix} A^+ - A^+B\Gamma^+B'A^+ & A^+B\Gamma^+ \\ \Gamma^+B'A^+ & -\Gamma^+ \end{pmatrix} \tag{14}$$

and

$$ZZ^+ = Z^+Z = \begin{pmatrix} AA^+ & 0 \\ 0 & B^+B \end{pmatrix}. \tag{15}$$

Proof. We could prove the theorem as a special case of the previous results. Below, however, we present a simple direct proof. The first statement of (13) follows from $\mathcal{M}(B) \subset \mathcal{M}(A)$. To prove the second statement of (13) we write $A = TT'$ with $|T'T| \neq 0$ and $B = TS$, so that

$$\Gamma = B'A^+B = S'T'(TT')^+TS = S'S. \tag{16}$$

Then, using Theorem 2.7,

$$B^+B = (TS)^+(TS) = S^+S = (S'S)^+S'S = \Gamma^+\Gamma = \Gamma\Gamma^+. \tag{17}$$

As a consequence we also have $\Gamma\Gamma^+B' = B'$. Now, let G be defined by

$$G = \begin{pmatrix} A^+ - A^+B\Gamma^+B'A^+ & A^+B\Gamma^+ \\ \Gamma^+B'A^+ & -\Gamma^+ \end{pmatrix}. \tag{18}$$

Then,

$$ZG = \begin{pmatrix} AA^+ - AA^+B\Gamma^+B'A^+ + B\Gamma^+B'A^+ & AA^+B\Gamma^+ - B\Gamma^+ \\ B'A^+ - B'A^+B\Gamma^+B'A^+ & B'A^+B\Gamma^+ \end{pmatrix}$$

$$= \begin{pmatrix} AA^+ & 0 \\ 0 & \Gamma\Gamma^+ \end{pmatrix} = \begin{pmatrix} AA^+ & 0 \\ 0 & B^+B \end{pmatrix}, \tag{19}$$

using the facts $AA^+B = B$, $\Gamma\Gamma^+B' = B'$ and $\Gamma\Gamma^+ = B^+B$. To show that $G = Z^+$ is then straightforward. □

14. THE EQUATIONS $X_1A + X_2B' = G_1$, $X_1B = G_2$

The two matrix equations in X_1 and X_2,

$$X_1A + X_2B' = G_1 \tag{1}$$

$$X_1B = G_2, \tag{2}$$

where A is positive semidefinite, can be written equivalently as

$$\begin{pmatrix} A & B \\ B' & 0 \end{pmatrix}\begin{pmatrix} X_1' \\ X_2' \end{pmatrix} = \begin{pmatrix} G_1' \\ G_2' \end{pmatrix}. \tag{3}$$

The properties of the matrix Z studied in the previous section enable us to solve these equations.

Theorem 23

The matrix equation in X_1 and X_2,

$$\begin{pmatrix} A & B \\ B' & 0 \end{pmatrix}\begin{pmatrix} X_1' \\ X_2' \end{pmatrix} = \begin{pmatrix} G_1' \\ G_2' \end{pmatrix}, \tag{4}$$

where A, B, G_1 and G_2 are given matrices (of appropriate orders) and A is positive semidefinite, has a solution if and only if

$$\mathcal{M}(G_1') \subset \mathcal{M}(A:B) \quad \text{and} \quad \mathcal{M}(G_2') \subset \mathcal{M}(B') \tag{5}$$

in which case the general solution is

$$X_1 = G_1(N^+ - N^+BC^+B'N^+) + G_2C^+B'N^+ + Q_1(I - NN^+) \tag{6}$$

and

$$X_2 = G_1N^+BC^+ + G_2(I - C^+) + Q_2(I - B^+B) \tag{7}$$

where

$$N = A + BB', \qquad C = B'N^+B \tag{8}$$

and Q_1 and Q_2 are arbitrary matrices of appropriate orders.

Moreover, if $\mathcal{M}(B) \subset \mathcal{M}(A)$, then we may take $N = A$.

Proof. Let $X = (X_1 : X_2)$, $G = (G_1 : G_2)$ and

$$Z = \begin{pmatrix} A & B \\ B' & 0 \end{pmatrix}. \tag{9}$$

Then equation (4) can be written as

$$ZX' = G'. \tag{10}$$

A solution of (10) exists if and only if

$$ZZ^+G' = G', \tag{11}$$

and if a solution exists it takes the form

$$X' = Z^+G' + (I - Z^+Z)Q' \tag{12}$$

where Q is an arbitrary matrix of appropriate order (Theorem 2.13).

Now, (11) is equivalent, by Theorem 21, to the two equations

$$NN^+G_1' = G_1', \qquad CC^+G_2' = G_2'. \tag{13}$$

The equations (13) in their turn are equivalent to

$$\mathcal{M}(G_1') \subset \mathcal{M}(N) = \mathcal{M}(A:B) \tag{14}$$

and

$$\mathcal{M}(G_2') \subset \mathcal{M}(C) = \mathcal{M}(B'), \tag{15}$$

using Theorems 2.11 and 20. This proves (5).

Using (12) and the expression for Z^+ in Theorem 21, we obtain the general solutions

$$X_1' = (N^+ - N^+BC^+B'N^+)G_1' + N^+BC^+G_2' + (I - NN^+)Q_1' \tag{16}$$

and

$$\begin{aligned} X_2' &= C^+B'N^+G_1' + (CC^+ - C^+)G_2' + (I - CC^+)P' \\ &= C^+B'N^+G_1' + (I - C^+)G_2' + (I - CC^+)(P' - G_2') \\ &= C^+B'N^+G_1' + (I - C^+)G_2' + (I - B^+B)Q_2', \end{aligned} \tag{17}$$

using Theorem 20 (iii) and letting $Q = (Q_1 : P)$ and $Q_2 = P - G_2$.

The special case where $\mathcal{M}(B) \subset \mathcal{M}(A)$ follows from Theorem 22. □

An important special case of Theorem 23 arises when we take $G_1 = 0$.

Theorem 24

The matrix equation in X_1 and X_2,

$$\begin{pmatrix} A & B \\ B' & 0 \end{pmatrix} \begin{pmatrix} X_1' \\ X_2' \end{pmatrix} = \begin{pmatrix} 0 \\ G' \end{pmatrix}, \tag{18}$$

where A, B and G are given matrices (of appropriate orders) and A is positive semidefinite, has a solution if and only if

$$\mathcal{M}(G') \subset \mathcal{M}(B') \tag{19}$$

in which case the general solution for X_1 is

$$X_1 = G(B'N^+B)^+B'N^+ + Q(I - NN^+) \tag{20}$$

where $N = A + BB'$ and Q is arbitrary (of appropriate order).

Moreover, if $\mathcal{M}(B) \subset \mathcal{M}(A)$, then the general solution can be written as

$$X_1 = G(B'A^+B)^+B'A^+ + Q(I - AA^+). \tag{21}$$

Proof. This follows immediately from Theorem 23. □

Exercise

1. Give the general solution for X_2 in Theorem 24.

MISCELLANEOUS EXERCISES

1. $D_n' = D_n^+(I_{n^2} + K_n - \mathrm{dg}K_n) = D_n^+(2I_{n^2} - \mathrm{dg}K_n)$.
2. $D_n^+ = \frac{1}{2}D_n'(I_{n^2} + \mathrm{dg}K_n)$.
3. $D_nD_n' = I_{n^2} + K_n - \mathrm{dg}K_n$.
4. Let e_i denote a unit vector of order m, that is, e_i has unity in its ith position and zeros elsewhere. Let u_j be a unit vector of order n. Define the $m^2 \times m$ and $n^2 \times n$ matrices

$$W_m = (\mathrm{vec}\, e_1e_1', \ldots, \mathrm{vec}\, e_me_m'), \quad W_n = (\mathrm{vec}\, u_1u_1', \ldots, \mathrm{vec}\, u_nu_n').$$

Let A and B be $m \times n$ matrices. Prove that

$$A \odot B = W_m'(A \otimes B)W_n.$$

BIBLIOGRAPHICAL NOTES

§2. A good discussion on adjoint matrices can be found in Aitken (1939, chap. 5). Theorem 1(b) appears to be new.
§6. For a review of the properties of the Hadamard product, see Styan (1973). Browne (1974) was the first to present the relation between the Hadamard and Kronecker products (square case). Faliva (1983) and Liu (1995) treated the rectangular case. See also Neudecker, Liu and Polasek (1995) and Neudecker, Polasek and Liu (1995) for a survey and applications.
§7. The commutation matrix was systematically studied by Magnus and Neudecker (1979). See also Magnus and Neudecker (1986). Theorem 10 is due to Neudecker and Wansbeek (1983). The matrix N_n was introduced by Browne (1974). For a rigorous and extensive treatment see Magnus (1988).
§8. See Browne (1974) and Magnus and Neudecker (1980, 1986) for further properties of the duplication matrix. Theorem 14 follows from equations (60), (62) and (64) in Magnus and Neudecker (1986). A systematic treatment of linear structures (of which symmetry is one example) is given in Magnus (1988).
§9–§10. See Holly and Magnus (1988).
§11. See also Debreu (1952), Black and Morimoto (1968) and Farebrother (1977).
§12. See also Chipman (1964).
§13. See Pringle and Rayner (1971, chap. 3), Rao (1973, sec. 41.1) and Magnus (1990).

Part Two—

Differentials: the theory

CHAPTER 4

Mathematical preliminaries

1. INTRODUCTION

Chapters 4–7, which constitute Part Two of this monograph, consist of two principal parts. The first part discusses *differentials*; the second part deals with *extremum problems*.

The use of differentials in both applied and theoretical work is widespread, but a satisfactory treatment of differentials is not so widespread in textbooks on economics and mathematics for economists. Indeed, some authors still claim that dx and dy stand for 'infinitesimally small changes in x and y'. The purpose therefore of Chapters 5 and 6 is to provide a systematic theoretical discussion of differentials.

We begin, however, by reviewing some basic concepts which will be used throughout.

2. INTERIOR POINTS AND ACCUMULATION POINTS

Let c be a point in $\mathbb{R}^n$ and r a positive number. The set of all points x in $\mathbb{R}^n$ whose distance from c is less than r is called an *n-ball* of radius r and centre c, and is denoted by $B(c)$ or $B(c; r)$. Thus,

$$B(c;\ r) = \{x : x \in \mathbb{R}^n,\ \|x - c\| < r\}. \tag{1}$$

An n-ball $B(c)$ is sometimes called a *neighbourhood* of c. The two words are used interchangeably.

Let S be a subset of $\mathbb{R}^n$, and assume that $c \in S$ and $x \in \mathbb{R}^n$, not necessarily in S. Then

(a) if there is an n-ball $B(c)$, all of whose points belong to S, then c is called an *interior point* of S;

(b) if *every* n-ball $B(x)$ contains at least one point of S distinct from x, then x is called an *accumulation point* of S;

(c) if $c \in S$ is not an accumulation point of S, then c is called an *isolated point* of S;
(d) if every n-ball $B(x)$ contains at least one point of S and at least one point of $\mathbb{R}^n - S$, then x is called a *boundary point* of S.

We further define:

(e) the *interior* of S, denoted $\mathring{S}$, as the set of all interior points of S;
(f) the *derived set* of S, denoted S', as the set of all accumulation points of S;
(g) the *closure* of S, denoted $\bar{S}$, as $S \cup S'$ (that is, to obtain $\bar{S}$, we adjoin all accumulation points of S to S);
(h) the *boundary* of S, denoted ∂S, as the set of all boundary points of S.

Theorem 1

Let S be a subset of $\mathbb{R}^n$. If $x \in \mathbb{R}^n$ is an accumulation point of S, then every n-ball $B(x)$ contains infinitely many points of S.

Proof. Suppose there is an n-ball $B(x)$ which contains only a finite number of points of S distinct from x, say $a_1, a_2, \ldots, a_p$. Let

$$r = \min_{1 \leq i \leq p} \|x - a_i\|. \tag{2}$$

Then $r > 0$, and the n-ball $B(x; r)$ contains no point of S distinct from x. This contradiction completes the proof. □

Exercises

1. Show that x is a boundary point of a set S in $\mathbb{R}^n$ if, and only if, x is a boundary point of $\mathbb{R}^n - S$.
2. Show that x is a boundary point of a set S in $\mathbb{R}^n$ if, and only if,
   (a) $x \in S$ and x is an accumulation point of $\mathbb{R}^n - S$, or
   (b) $x \notin S$ and x is an accumulation point of S.

3. OPEN AND CLOSED SETS

A set S in $\mathbb{R}^n$ is said to be

(a) *open*, if all its points are interior points;
(b) *closed*, if it contains all its accumulation points;
(c) *bounded*, if there is a real number $r > 0$ and a point c in $\mathbb{R}^n$ such that S lies entirely within the n-ball $B(c; r)$; and
(d) *compact*, if it is closed and bounded.

For example, let A be an interval in $\mathbb{R}$, that is, a set with the property that, if $a \in A$, $b \in A$ and $a < b$, then $a < c < b$ implies $c \in A$. For $a < b \in \mathbb{R}$ the *open intervals in* $\mathbb{R}$ are

$$(a, b), \qquad (a, \infty), \qquad (-\infty, b), \qquad \mathbb{R}; \tag{1}$$

the *closed intervals* are

$$[a, b], \qquad [a, \infty), \qquad (-\infty, b], \qquad \mathbb{R}; \tag{2}$$

the *bounded intervals* are

$$(a, b), \qquad [a, b], \qquad (a, b], \qquad [a, b); \tag{3}$$

and the only type of *compact interval* is

$$[a, b]. \tag{4}$$

This example shows that a set can be both open *and* closed. In fact, the *only* sets in $\mathbb{R}^n$ which are both open and closed are $\varnothing$ and $\mathbb{R}^n$. It is also possible that a set is neither open nor closed as the 'half-open' interval $(a, b]$ shows.

It is clear that S is open if and only if $S=\mathring{S}$, and that S is closed if and only if $S=\bar{S}$. An important example of an open set is the n-ball.

Theorem 2

Every n-ball is an open set in $\mathbb{R}^n$.

Proof. Let $B(c; r)$ be a given n-ball with radius r and centre c, and let x be an arbitrary point of $B(c; r)$. We have to prove that x is an interior point of $B(c; r)$, i.e. that there exists a $\delta>0$ such that $B(x; \delta) \subset B(c; r)$. Now, let

$$\delta = r - \|x-c\|. \tag{5}$$

Then $\delta>0$, and, for any $y \in B(x; \delta)$,

$$\|y-c\| \leqslant \|y-x\| + \|x-c\| < \delta + r - \delta = r, \tag{6}$$

so that $y \in B(c; r)$. Thus $B(x; \delta) \subset B(c; r)$, and x is an interior point of $B(c; r)$.

□

The next theorem characterizes a closed set as the complement of an open set.

Theorem 3

A set S in $\mathbb{R}^n$ is closed if, and only if, its complement $\mathbb{R}^n - S$ is open.

Proof. Assume first that S is closed. Let $x \in \mathbb{R}^n - S$. Then $x \notin S$ and, since S contains all its accumulation points, x is not an accumulation point of S. Hence there exists an n-ball $B(x)$ which does not intersect S, i.e. $B(x) \subset \mathbb{R}^n - S$. It follows that x is an interior point of $\mathbb{R}^n - S$, and hence that $\mathbb{R}^n - S$ is open.

To prove the converse, assume that $\mathbb{R}^n - S$ is open. Let $x \in \mathbb{R}^n$ be an accumulation point of S. We must show that $x \in S$. Assume that $x \notin S$. Then $x \in \mathbb{R}^n - S$ and, since every point of $\mathbb{R}^n - S$ is an interior point, there exists an n-ball $B(x) \subset \mathbb{R}^n - S$. Hence $B(x)$ contains no points of S thereby contradicting the fact that x is an accumulation point of S. It follows that $x \in S$, and hence that S is closed. □

The next two theorems show how to construct further open and closed sets from given ones.

Theorem 4

The union of any collection of open sets is open, and the intersection of a finite collection of open sets is open.

Proof. Let F be a collection of open sets and let S denote their union,

$$S=\bigcup_{A\in F} A.$$

Assume $x\in S$. Then there is at least one set of F, say A, such that $x\in A$. Since A is open, x is an interior point of A, and hence of S. It follows that S is open.

Next let F be a *finite* collection of open sets, $F=\{A_1, A_2, \ldots, A_k\}$ and let

$$T=\bigcap_{j=1}^{k} A_j.$$

Assume $x\in T$. (If T is empty, there is nothing to prove.) Then x belongs to every set in F. Since each set in F is open, there exist k n-balls $B(x; r_j)\subset A_j$, $j=1, \ldots, k$. Let

$$r=\min_{1\leqslant j\leqslant k} r_j.$$

Then $x\in B(x; r)\subset T$. Hence x is an interior point of T. It follows that T is open. □

Note. The intersection of an infinite collection of open sets need not be open. For example,

$$\bigcap_{n\in\mathbb{N}}\left(-\frac{1}{n}, \frac{1}{n}\right)=\{0\}. \tag{7}$$

Theorem 5

The union of a finite collection of closed sets is closed, and the intersection of any collection of closed sets is closed.

Proof. Let F be a finite collection of closed sets, $F=\{A_1, A_2, \ldots, A_k\}$, and let

$$S=\bigcup_{j=1}^{k} A_j.$$

Then

$$\mathbb{R}^n-S=\bigcap_{j=1}^{k} (\mathbb{R}^n-A_j).$$

Since each A_j is closed, $\mathbb{R}^n - A_j$ is open (Theorem 3), and, by Theorem 4, so is their (finite) intersection

$$\bigcap_{j=1}^{k}(\mathbb{R}^n - A_j).$$

Hence $\mathbb{R}^n - \bar{S}$ is open, and $\bar{S}$ is closed. The second statement is proved similarly. □

Finally, we present the following simple relation between open and closed sets.

Theorem 6

If A is open and B is closed, then $A-B$ is open and $B-A$ is closed.

Proof. It is easy to see that $A - B = A \cap (\mathbb{R}^n - B)$, the intersection of two open sets. Hence, by Theorem 4, $A - B$ is open. Similarly, since $B - A = B \cap (\mathbb{R}^n - A)$, the intersection of two closed sets, it is closed by Theorem 5. □

4. THE BOLZANO–WEIERSTRASS THEOREM

Theorem 1 implies that a set cannot have an accumulation point unless it contains infinitely many points to begin with. The converse, however, is not true. For example, $\mathbb{N}$ is an infinite set without accumulation points. We shall now show that infinite sets which are *bounded* always have an accumulation point.

Theorem 7 (Bolzano–Weierstrass)

Every bounded infinite subset of $\mathbb{R}^n$ has an accumulation point in $\mathbb{R}^n$.

Proof. Let us prove the theorem for $n=1$. The case $n>1$ is proved similarly. Since S is bounded, it lies in some interval $[-a, a]$. Since S contains infinitely many points, either $[-a, 0]$ or $[0, a]$ (or both) contains infinitely many points of S. Call this interval $[a_1, b_1]$. Bisect $[a_1, b_1]$ and obtain an interval $[a_2, b_2]$ containing infinitely many points of S. Continuing this process we find a countable sequence of intervals $[a_n, b_n]$, $n=1, 2, \ldots$. The intersection

$$\bigcap_{n=1}^{\infty} [a_n, b_n]$$

of these intervals is a set consisting of only one point, say c (which may or may not belong to S). We shall show that c is an accumulation point of S. Let $\varepsilon>0$, and consider the neighbourhood $(c-\varepsilon, c+\varepsilon)$ of c. Then we can find an $n_0=n_0(\varepsilon)$ such that $[a_{n_0}, b_{n_0}] \subset (c-\varepsilon, c+\varepsilon)$. Since $[a_{n_0}, b_{n_0}]$ contains infinitely many points of S, so does $(c-\varepsilon, c+\varepsilon)$. Hence c is an accumulation point of S. □

5. FUNCTIONS

Let S and T be two sets. If with each element $x \in S$ there is associated exactly one element $y \in T$, denoted $f(x)$, then f is said to be a *function* from S to T. We write

$$f: S \to T, \tag{1}$$

and say that f is defined on S with values in T. The set S is called the *domain* of f; the set of all values of f, i.e.

$$\{y: y = f(x),\ x \in S\} \tag{2}$$

is called the *range* of f, and is a subset of T.

A function $\phi: S \to \mathbb{R}$ defined on a set S with values in $\mathbb{R}$ is called *real-valued*. A function $f: S \to \mathbb{R}^m (m > 1)$ whose values are points in $\mathbb{R}^m$ is called a *vector function*.

A real-valued function $\phi: S \to \mathbb{R}$, $S \subset \mathbb{R}$, is said to be *increasing* on S if for every pair of points x and y in S,

$$\phi(x) \leqslant \phi(y) \qquad \text{whenever } x < y. \tag{3}$$

We say that ϕ is *strictly increasing* on S if

$$\phi(x) < \phi(y) \qquad \text{whenever } x < y. \tag{4}$$

(Strictly) decreasing functions are similarly defined. A function is called (strictly) monotonic on S if it is either (strictly) increasing or (strictly) decreasing on S.

A vector function $f: S \to \mathbb{R}^m$, $S \subset \mathbb{R}^n$ is said to be bounded if there is a real number M such that

$$\|f(x)\| \leqslant M \qquad \text{for all } x \text{ in } S. \tag{5}$$

A function $f : \mathbb{R}^n \to \mathbb{R}^m$ is said to be *affine* if there exist an $m \times n$ matrix A and an $m \times 1$ vector b such that $f(x) = Ax + b$ for every x in $\mathbb{R}^n$. If $b = 0$, the function f is said to be *linear*.

6. THE LIMIT OF A FUNCTION

Definition

Let $f : S \to \mathbb{R}^m$ be defined on a set S in $\mathbb{R}^n$ with values in $\mathbb{R}^m$. Let c be an accumulation point of S. Suppose there exists a point b in $\mathbb{R}^m$ with the property that for every $\varepsilon > 0$ there is a $\delta > 0$ such that

$$\|f(x) - b\| < \varepsilon \tag{1}$$

for all points x in S, $x \neq c$, for which

$$\|x - c\| < \delta. \tag{2}$$

Then we say that the limit of $f(x)$ is b, as x tends to c, and we write

$$\lim_{x \to c} f(x) = b. \tag{3}$$

Note. The requirement that c is an accumulation point of S guarantees that there will be points $x \neq c$ in S sufficiently close to c. However, c need not be a point of S. Moreover, even if $c \in S$, we may have

$$f(c) \neq \lim_{x \to c} f(x).$$

We have the following rules for calculating with limits of vector functions.

Theorem 8

Let f and g be two vector functions defined on $S \subset \mathbb{R}^n$ with values in $\mathbb{R}^m$. Let c be an accumulation point of S, and assume that

$$\lim_{x \to c} f(x) = a, \qquad \lim_{x \to c} g(x) = b. \tag{4}$$

Then,

(a) $\lim_{x \to c} (f+g)(x) = a+b$,

(b) $\lim_{x \to c} (\lambda f)(x) = \lambda a$ for every scalar λ,

(c) $\lim_{x \to c} f(x)'g(x) = a'b$,

(d) $\lim_{x \to c} \|f(x)\| = \|a\|$.

Proof. The proof is left as an exercise to the reader. □

Exercises

1. Let $\phi : \mathbb{R} \to \mathbb{R}$ be defined by $\phi(x) = x$ if $x \neq 0$, $\phi(0) = 1$. Show that $\phi(x) \to 0$ as $x \to 0$.
2. Let $\phi : \mathbb{R} - \{0\} \to \mathbb{R}$ be defined by $\phi(x) = x \sin(1/x)$ if $x \neq 0$. Show that $\phi(x) \to 0$ as $x \to 0$.

7. CONTINUOUS FUNCTIONS AND COMPACTNESS

Let $\phi : S \to \mathbb{R}$ be a real-valued function defined on a set S in $\mathbb{R}^n$. Let c be a point of S. Then we say that ϕ is *continuous at* c if for every $\varepsilon > 0$ there is a $\delta > 0$ such that

$$|\phi(c+u) - \phi(c)| < \varepsilon \tag{1}$$

for all points $c+u$ in S for which $\|u\| < \delta$. If ϕ is continuous at every point of S, we say that ϕ is *continuous on* S.

Continuity is discussed in more detail in section 5.2. Here we only prove the following important theorem.

Theorem 9

Let $\phi : S \to \mathbb{R}$ be a real-valued function defined on a compact set S in $\mathbb{R}^n$. If ϕ is continuous on S, then ϕ is bounded on S.

Proof. Suppose that ϕ is not bounded on S. Then there exists, for every $k \in \mathbb{N}$, an $x_k \in S$ such that $|\phi(x_k)| \geqslant k$. The set

$$A = \{x_1, x_2, \ldots\} \tag{2}$$

contains infinitely many points, and $A \subset S$. Since S is a bounded set, so is A. Hence, by the Bolzano–Weierstrass theorem (Theorem 7), A has an accumulation point, say x_0. Then x_0 is also an accumulation point of S and hence $x_0 \in S$, since S is closed.

Now choose an integer p such that

$$p > 1 + |\phi(x_0)|, \tag{3}$$

and define the set $A_p \subset A$ by

$$A_p = \{x_p, x_{p+1}, \ldots\}, \tag{4}$$

so that

$$|\phi(x)| \geqslant p \qquad \text{for all } x \in A_p. \tag{5}$$

Since ϕ is continuous at x_0, there exists an n-ball $B(x_0)$ such that

$$|\phi(x) - \phi(x_0)| < 1 \qquad \text{for all } x \in S \cap B(x_0). \tag{6}$$

In particular,

$$|\phi(x) - \phi(x_0)| < 1 \qquad \text{for all } x \in A_p \cap B(x_0). \tag{7}$$

The set $A_p \cap B(x_0)$ is not empty. (In fact, it contains infinitely many points, because $A \cap B(x_0)$ contains infinitely many points, see Theorem 1.) For any $x \in A_p \cap B(x_0)$ we have

$$|\phi(x)| < 1 + |\phi(x_0)| < p, \tag{8}$$

using (7) and (3), and also, from (5),

$$|\phi(x)| \geqslant p. \tag{9}$$

This contradiction shows that ϕ must be bounded on S. □

8. CONVEX SETS

Definition 1

A subset S of $\mathbb{R}^n$ is called a *convex set* if, for every pair of points x and y in S and every real θ satisfying $0 < \theta < 1$, we have

$$\theta x + (1 - \theta) y \in S. \tag{1}$$

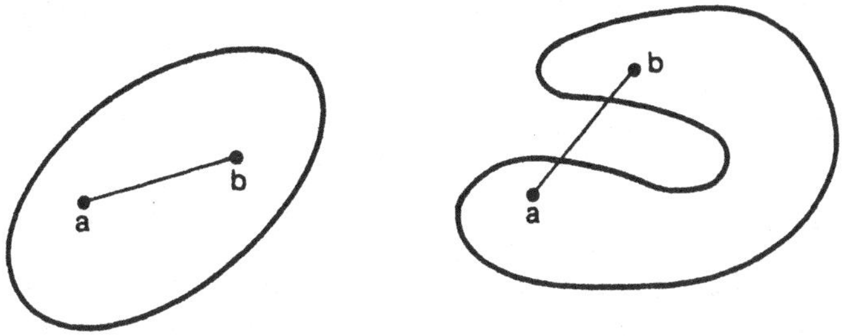

Figure 1 Convex and non-convex sets in $\mathbb{R}^2$

In other words, S is convex if the line segment joining any two points of S lies entirely inside S (see Figure 1).

Convex sets need not be closed, open, or compact. A single point and the whole space $\mathbb{R}^n$ are trivial examples of convex sets. Another example of a convex set is the n-ball.

Theorem 10

Every n-ball in $\mathbb{R}^n$ is convex.

Proof. Let $B(c; r)$ be an n-ball with radius $r > 0$ and centre c. Let x and y be points in $B(c; r)$ and let $\theta \in (0, 1)$. Then

$$\begin{aligned}\|\theta x + (1-\theta)y - c\| &= \|\theta(x-c) + (1-\theta)(y-c)\| \\ &\leqslant \theta\|x-c\| + (1-\theta)\|y-c\| < \theta r + (1-\theta)r = r. \quad (2)\end{aligned}$$

Hence the point $\theta x + (1-\theta)y$ lies in $B(c; r)$. □

Another important property of convex sets is the following.

Theorem 11

The intersection of any collection of convex sets is convex.

Proof. Let F be a collection of convex sets and let S denote their intersection,

$$S = \bigcap_{A \in F} A.$$

Assume x and $y \in S$. (If S is empty, or consists of only one point, there is nothing to prove.) Then x and y belong to every set in F. Since each set in F is convex, the point $\theta x+(1-\theta)y$, $\theta\in(0, 1)$, also belongs to every set in F, and hence to S. It follows that S is convex. □

Note. The union of convex sets is usually not convex.

Definition 2

Let $x_1, x_2, \ldots, x_k$ be k points in $\mathbb{R}^n$. A point $x \in \mathbb{R}^n$ is called a *convex combination* of these points if there exist k real numbers $\lambda_1, \lambda_2, \ldots, \lambda_k$ such that

$$x=\sum_{i=1}^{k} \lambda_i x_i, \qquad \lambda_i \geqslant 0(i = 1,\ldots, k), \qquad \sum_{i=1}^{k} \lambda_i=1. \tag{3}$$

Theorem 12

Let S be a convex set in $\mathbb{R}^n$. Then every convex combination of a finite number of points in S lies in S.

Proof (by induction). The theorem is clearly true for each pair of points in S. Suppose it is true for all collections of k points in S. Let $x_1, \ldots, x_{k+1}$ be $k+1$ arbitrary points in S, and let $\lambda_1, \ldots, \lambda_{k+1}$ be arbitrary real numbers satisfying $\lambda_i \geqslant 0$ $(i=1, \ldots, k+1)$ and

$$\sum_{i=1}^{k+1} \lambda_i=1.$$

Define

$$x=\sum_{i=1}^{k+1} \lambda_i x_i.$$

Assume that $\lambda_{k+1} \neq 1$. (If $\lambda_{k+1}=1$, then $x=x_{k+1}\in S$.) Then we can write x as

$$x=\lambda_0 y+\lambda_{k+1}x_{k+1} \tag{4}$$

with

$$\lambda_0=\sum_{i=1}^{k} \lambda_i, \qquad y=\sum_{i=1}^{k} (\lambda_i/\lambda_0)x_i. \tag{5}$$

By the induction hypothesis, y lies in S. Hence, by the definition of a convex set, $x\in S$. □

Exercises

1. Consider a set S in $\mathbb{R}^n$ with the property that for any pair of points x and y in S, their midpoint $\frac{1}{2}(x+y)$ also belongs to S. Show, by means of a counter-example, that S need not be convex.
2. Show, by means of a counter-example, that the union of two convex sets need not be convex.
3. Let S be a convex set in $\mathbb{R}^n$. Show that $\bar{S}$ and $\mathring{S}$ are convex.

9. CONVEX AND CONCAVE FUNCTIONS

Let $\phi : S \to \mathbb{R}$ be a real-valued function defined on a *convex* set S in $\mathbb{R}^n$. Then

(a) ϕ is said to be *convex* on S, if

$$\phi(\theta x+(1-\theta)y) \leqslant \theta\phi(x)+(1-\theta)\phi(y) \tag{1}$$

for every pair of points x, y in S and every $\theta \in (0, 1)$ (see Figure 2);

(b) ϕ is said to be *strictly convex* on S, if

$$\phi(\theta x+(1-\theta)y) < \theta\phi(x)+(1-\theta)\phi(y) \tag{2}$$

for every pair of points x, y in S, $x \neq y$, and every $\theta \in (0, 1)$;

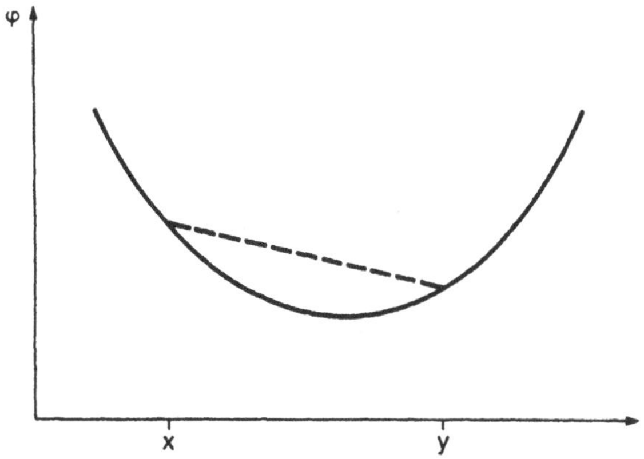

***Figure* 2** A convex function

(c) ϕ is said to be *(strictly) concave* if $\psi \equiv -\phi$ is (strictly) convex.

Note. It is essential in the definition that S is a convex set, since we require that $\theta x+(1-\theta)y \in S$ if $x, y \in S$.

It is clear that a strictly convex (concave) function is convex (concave).

Examples of strictly convex functions in one dimension are $\phi(x)=x^2$ and $\phi(x)=e^x$ $(x>0)$; the function $\phi(x)=\log x$ $(x>0)$ is strictly concave. These functions are continuous (and even differentiable) on their respective domains. That these properties are not necessary is shown by the functions

$$\phi(x)=\begin{cases} x^2, & \text{if } x>0 \\ 1\ , & \text{if } x=0 \end{cases} \tag{3}$$

(strictly convex on $[0, \infty)$ but discontinuous at the boundary point $x=0$) and

$$\phi(x)=|x| \tag{4}$$

(convex on $\mathbb{R}$ but not differentiable at the interior point $x = 0$). Thus, a convex function may have a discontinuity at a boundary point, and may not be differentiable at an interior point. However, *every convex (and concave) function is continuous on its interior.*

The following three theorems give further properties of convex functions.

Theorem 13

An affine function is convex as well as concave, but not strictly so.

Proof. Since ϕ is an affine function, we have

$$\phi(x) = \alpha + a'x \tag{5}$$

for some scalar α and vector a. Hence

$$\phi(\theta x + (1-\theta)y) = \theta\phi(x) + (1-\theta)\phi(y) \tag{6}$$

for every $\theta \in (0, 1)$. □

Theorem 14

Let ϕ and ψ be two convex functions on a convex set S in $\mathbb{R}^n$. Then

$$\alpha\phi + \beta\psi \tag{7}$$

is convex (concave) on S, if $\alpha \geqslant 0$ ($\leqslant 0$) and $\beta \geqslant 0$ ($\leqslant 0$).

Moreover, if ϕ is convex and ψ strictly convex on S, then $\alpha\phi + \beta\psi$ is strictly convex (concave) on S if $\alpha \geqslant 0$ ($\leqslant 0$) and $\beta > 0$ (< 0).

Proof. The proof is a direct consequence of the definition and is left to the reader. □

Theorem 15

Every increasing convex (concave) function of a convex (concave function is convex (concave). Every strictly increasing convex (concave) function of a strictly convex (concave) function is strictly convex (concave).

Proof. Let ϕ be a convex function defined on a convex set S in $\mathbb{R}^n$, let ψ be an increasing convex function of one variable defined on the range of ϕ and let $\eta(x) = \psi[\phi(x)]$. Then

$$\begin{aligned} \eta(\theta x + (1-\theta)y) &= \psi[\phi(\theta x + (1-\theta)y)] \leqslant \psi[\theta\phi(x) + (1-\theta)\phi(y)] \\ &\leqslant \theta\psi[\phi(x)] + (1-\theta)\psi[\phi(y)] = \theta\eta(x) + (1-\theta)\eta(y), \end{aligned} \tag{8}$$

for every $x, y \in S$ and $\theta \in (0, 1)$. (The first inequality follows from the convexity of ϕ and the fact that ψ is increasing; the second inequality follows from the convexity of ψ.) Hence η is convex. The other statements are proved similarly. □

Exercises

1. Show that $\phi(x)=\log x$ is strictly concave and $\phi(x)=x\log x$ is strictly convex on $(0, \infty)$. (Compare exercise 7.8.1.)
2. Show that the quadratic form $x'Ax$ $(A=A')$ is convex iff A is positive semidefinite, and concave iff A is negative semidefinite.
3. Show that the norm
$$\phi(x)=\|x\|=(x_1^2+x_2^2+\ldots+x_n^2)^{1/2}$$
is convex.
4. An increasing function of a convex function is not necessarily convex. Give an example.
5. Prove the following statements by providing an example.
   (a) A strictly increasing, convex function of a convex function is convex, but not necessarily strictly so.
   (b) An increasing convex function of a strictly convex function is convex, but not necessarily strictly so.
   (c) An increasing, strictly convex function of a convex function is convex, but not necessarily strictly so.
6. Show that $\phi(X) = \operatorname{tr} X$ is both convex and concave on $\mathbb{R}^{n\times n}$.
7. If ϕ is convex on $S \subset \mathbb{R}$, $x_i \in S(i = 1,\ldots,n)$, $\alpha_i \geqslant 0(i = 1,\ldots,n)$, and
$$\sum_{i=1}^{n} \alpha_i=1,$$
then
$$\phi\left(\sum_{i=1}^{n} \alpha_i x_i\right)\leqslant \sum_{i=1}^{n} \alpha_i\phi(x_i).$$

BIBLIOGRAPHICAL NOTES

§1. For a list of frequently occurring errors in the economic literature concerning problems of maxima and minima, see Sydsaeter (1974). A careful development of mathematical analysis at the intermediate level is given in Rudin (1964) and Apostol (1974). More advanced, but highly recommended, is Dieudonné (1969).
§9. The fact that convex and concave functions are continuous on their interior is discussed, for example, in Luenberger (1969, sec. 7.9) and Fleming (1977, th. 3.5).

CHAPTER 5

Differentials and differentiability

1. INTRODUCTION

Let us consider a function $f : S \to \mathbb{R}^m$, defined on a set S in $\mathbb{R}^n$ with values in $\mathbb{R}^m$. If $m=1$, the function is called *real-valued* (and we shall use ϕ instead of f to emphasize this); if $m \geq 2$, f is called a *vector function*. Examples of vector functions are

$$f(x)=\begin{pmatrix} x^2 \\ x^3 \end{pmatrix}, \qquad f(x,y)=\begin{bmatrix} xy \\ x \\ y \end{bmatrix}, \qquad f(x,y,z)=\begin{pmatrix} x+y+z \\ x^2+y^2+z^2 \end{pmatrix}. \qquad (1)$$

Note that m may be larger or smaller than n or equal to n. In the first example $n=1$, $m=2$, in the second example $n=2$, $m=3$, and in the third example $n=3$, $m=2$.

In this chapter, we extend the *one-dimensional* theory of differential calculus (concerning real-valued functions $\phi : \mathbb{R} \to \mathbb{R}$) to functions from $\mathbb{R}^n$ to $\mathbb{R}^m$. The extension from real-valued functions of one variable to real-valued functions of several variables is far more significant than the extension from real-valued functions to vector functions. Indeed, for most purposes a vector function can be viewed as a vector of m real-valued functions. Yet, as we shall see shortly, there are good reasons to study vector functions rather than merely real-valued functions.

Throughout this chapter, and, indeed, throughout this book, we shall emphasize the fundamental idea of a *differential* rather than that of a derivative.

2. CONTINUITY

We first review the concept of continuity. Intuitively a function f is continuous at a point c if $f(x)$ can be made arbitrarily close to $f(c)$ by taking x sufficiently close to c; in other words, if points close to c are mapped by f into points close to $f(c)$.

Definition

Let $f : S \to \mathbb{R}^m$ be a function defined on a set S in $\mathbb{R}^n$ with values in $\mathbb{R}^m$. Let c be a point of S. Then we say that f is *continuous at* c if for every $\varepsilon > 0$ there exists a $\delta > 0$ such that

$$\|f(c+u)-f(c)\| < \varepsilon \tag{1}$$

for all points $c+u$ in S for which $\|u\| < \delta$. If f is continuous at every point of S, we say f is *continuous on* S.

The definition is a straightforward generalization of the definition in section 4.7 concerning continuity of real-valued functions ($m=1$). Note that f has to be defined at the point c in order to be continuous at c. Some authors require that c is an accumulation point of S, but this is not assumed here. If c is an isolated point of S (a point of S which is not an accumulation point of S), then every f defined at c will be continuous at c because for sufficiently small δ there is only one point $c+u$ in S satisfying $\|u\| < \delta$, namely the point c itself; then

$$\|f(c+u)-f(c)\| = 0 < \varepsilon. \tag{2}$$

If c is an accumulation point of S, the definition of continuity implies that

$$\lim_{u \to 0} f(c+u) = f(c). \tag{3}$$

Geometrical intuition seems to show that if $f : S \to \mathbb{R}^m$ is continuous at c, it must also be continuous near c. This intuition is wrong for two reasons. First, the point c may be an isolated point of S, in which case there exists a neighbourhood of c where f is not even defined. Secondly, even if c is an accumulation point of S, it may be that every neighbourhood of c contains points of S at which f is not continuous. For example, the real-valued function $\phi : \mathbb{R} \to \mathbb{R}$ defined by

$$\phi(x) = \begin{cases} x & (x \text{ rational}), \\ 0 & (x \text{ irrational}), \end{cases} \tag{4}$$

is continuous at $x=0$, but at no other point.

If $f : S \to \mathbb{R}^m$, the formula

$$f(x) = (f_1(x), \ldots, f_m(x))' \tag{5}$$

defines m real-valued functions $f_i : S \to \mathbb{R}$ ($i = 1, \ldots, m$). These functions are called the *component functions* of f and we write

$$f = (f_1, f_2, \ldots, f_m)'. \tag{6}$$

Theorem 1

Let S be a subset of $\mathbb{R}^n$. A function $f : S \to \mathbb{R}^m$ is continuous at a point c in S if and only if each of its component functions is continuous at c.

If c is an accumulation point of a set S in $\mathbb{R}^n$ and $f: S \to \mathbb{R}^m$ is continuous at c, then we can write (3) as

$$f(c+u)=f(c)+R_c(u), \tag{7}$$

where

$$\lim_{u\to 0} R_c(u)=0. \tag{8}$$

We may call equation (7) the *Taylor formula of order zero*. It says that continuity at an accumulation point of S and 'zero-order approximation' (approximation of $f(c+u)$ by a polynomial of degree zero, that is a constant) are equivalent properties. In the next section we discuss the equivalence of differentiability and first-order (that is linear) approximation.

Exercises

1. Prove Theorem 1.
2. Let S be a set in $\mathbb{R}^n$. If $f: S \to \mathbb{R}^m$ and $g: S \to \mathbb{R}^m$ are continuous on S, then so is the function $f+g: S \to \mathbb{R}^m$.
3. Let S be a set in $\mathbb{R}^n$ and T a set in $\mathbb{R}^m$. Suppose that $g: S \to \mathbb{R}^m$ and $f: T \to \mathbb{R}^p$ are continuous on S and T, respectively, and that $g(x) \in T$ when $x \in S$. Then the composite function $h: S \to \mathbb{R}^p$ defined by $h(x)=f(g(x))$ is continuous on S.
4. Let S be a set in $\mathbb{R}^n$. If the real-valued functions $\phi: S \to \mathbb{R}$, $\psi: S \to \mathbb{R}$ and $\chi: S \to \mathbb{R}-\{0\}$ are continuous on S, then so are the real-valued functions $\phi\psi: S \to \mathbb{R}$ and $\phi/\chi: S \to \mathbb{R}$.
5. Let $\phi: (0, 1) \to \mathbb{R}$ be defined by

$$\phi(x)=\begin{cases} 1/q & (x \text{ rational}, x=p/q) \\ 0 & (x \text{ irrational}), \end{cases}$$

where $p, q \in \mathbb{N}$ have no common factor. Show that ϕ is continuous at every irrational point, and discontinuous at every rational point.

3. DIFFERENTIABILITY AND LINEAR APPROXIMATION

In the one-dimensional case, the equation

$$\lim_{u\to 0} \frac{\phi(c+u)-\phi(c)}{u}=\phi'(c) \tag{1}$$

defining the derivative at c is equivalent to the equation

$$\phi(c+u)=\phi(c)+u\phi'(c)+r_c(u), \tag{2}$$

where the remainder $r_c(u)$ is of *smaller order* than u as $u \to 0$, that is

$$\lim_{u\to 0} \frac{r_c(u)}{u}=0. \tag{3}$$

Equation (2) is called the *first-order Taylor formula*. If for the moment we think of the point c as fixed and the increment u as variable, then the increment of the function, that is the quantity $\phi(c+u)-\phi(c)$, consists of two terms, namely, a part $u\phi'(c)$ which is proportional to u, and an 'error' which can be made as small as we please relative to u by making u itself small enough. Thus the smaller the interval about the point c which we consider, the more accurately is the function $\phi(c + u)$—which is a function of u—represented by its affine part $\phi(c) + u\phi'(c)$. We now define the expression

$$\mathrm{d}\phi(c; u) = u\phi'(c) \tag{4}$$

as the *(first) differential of* ϕ *at* c *with increment* u.

The notation $\mathrm{d}\phi(c; u)$ rather than $\mathrm{d}\phi(c, u)$ emphasizes the different roles of c and u. The first point, c, must be a point where $\phi'(c)$ exists, whereas the second point, u, is an arbitrary point in $\mathbb{R}$.

Although the concept of differential is as a rule only used when u is small, there is in principle no need to restrict u in any way. In particular, the differential $\mathrm{d}\phi(c; u)$ is a number *which has nothing to do with infinitely small quantities.*

The differential $\mathrm{d}\phi(c; u)$ is thus the linear part of the increment $\phi(c+u)-\phi(c)$. This is expressed geometrically by replacing the curve at point c by its tangent (see Figure 1).

Conversely, if there exists a quantity α, depending on c but not on u, such that

$$\phi(c+u) = \phi(c) + u\alpha + r(u), \tag{5}$$

where $r(u)/u$ tends to 0 with u, that is if we can approximate $\phi(c+u)$ by an affine

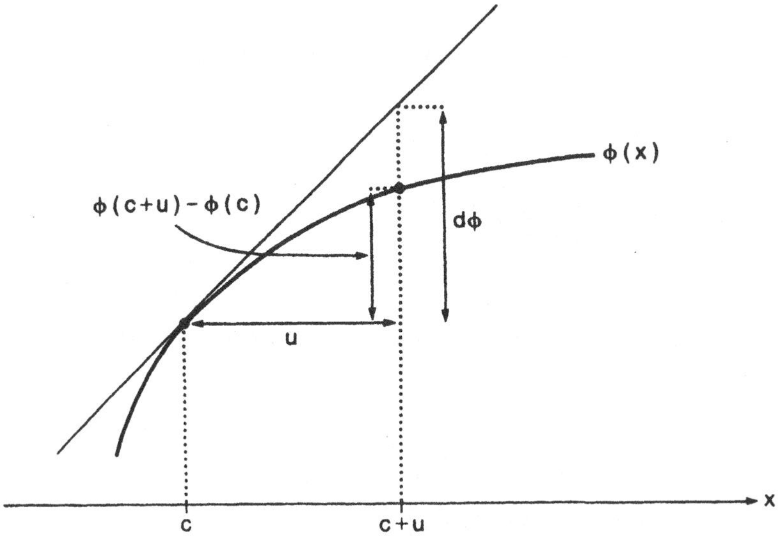

Figure 1 Geometric interpretation of the differential

function (in u) such that the difference between the function and the approximation function vanishes to a higher order than the increment u, then ϕ is differentiable at c. The quantity α must then be the derivative $\phi'(c)$. We see this immediately if we rewrite condition (5) in the form

$$\frac{\phi(c+u)-\phi(c)}{u}=\alpha+\frac{r(u)}{u} \tag{6}$$

and then let u tend to 0. Differentiability of a function and the possibility of approximating a function by means of an affine function are therefore equivalent properties.

4. THE DIFFERENTIAL OF A VECTOR FUNCTION

These ideas can be extended in a perfectly natural way to vector functions of two or more variables.

Definition

Let $f : S \rightarrow \mathbb{R}^m$ be a function defined on a set S in $\mathbb{R}^n$. Let c be an interior point of S, and let $B(c; r)$ be an n-ball lying in S. Let u be a point in $\mathbb{R}^n$ with $\|u\| < r$, so that $c+u \in B(c;r)$. If there exists a real $m \times n$ matrix A, depending on c but not on u, such that

$$f(c+u)=f(c)+A(c)u+r_c(u) \tag{1}$$

for all $u \in \mathbb{R}^n$ with $\|u\| < r$ and

$$\lim_{u \to 0} \frac{r_c(u)}{\|u\|}=0, \tag{2}$$

then the function f is said to be *differentiable at* c; the $m \times n$ matrix $A(c)$ is then called the *(first) derivative of f at* c, and the $m \times 1$ vector

$$\mathrm{d}f(c;u)=A(c)u, \tag{3}$$

which is a linear function of u, is called the *(first) differential of f at* c (with increment u). If f is differentiable at every point of an open subset E of S, we say f is *differentiable on* (or *in*) E.

In other words, f is differentiable at the point c if $f(c+u)$ can be approximated by an affine function of u. Note that a function f can only be differentiated at an interior point or on an open set.

Example

Let $\phi : \mathbb{R}^2 \rightarrow \mathbb{R}$ be a real-valued function defined by $\phi(x, y) = xy^2$. Then

$$\begin{aligned}\phi(x+u,\ y+v)&=(x+u)(y+v)^2\\&=xy^2+(y^2u+2xyv)+(xv^2+2yuv+uv^2)\\&=\phi(x,y)+\mathrm{d}\phi(x,y;\ u,v)+r(u,v)\end{aligned} \tag{4}$$

with

$$d\phi(x, y;\ u, v) = (y^2, 2xy)\begin{pmatrix} u \\ v \end{pmatrix} \tag{5}$$

and

$$r(u, v) = xv^2 + 2yuv + uv^2. \tag{6}$$

Since $r(u, v)/(u^2+v^2)^{1/2} \to 0$ as $(u, v) \to (0, 0)$, ϕ is differentiable at every point of $\mathbb{R}^2$.

We have seen before (section 2) that a function can be continuous at a point c, but fail to be continuous at points near c; indeed, the function may not even exist near c. If a function is differentiable at c, then *it must exist in a neighbourhood of* c, but the function need not be differentiable or continuous in that neighbourhood. For example, the real-valued function $\phi : \mathbb{R} \to \mathbb{R}$ defined by

$$\phi(x) = \begin{cases} x^2 & (x \text{ rational}), \\ 0 & (x \text{ irrational}) \end{cases} \tag{7}$$

is differentiable (and continuous) at $x = 0$, but neither differentiable nor continuous at any other point.

Let us return to equation (1). It consists of m equations,

$$f_i(c+u) = f_i(c) + \sum_{j=1}^{n} a_{ij}(c)u_j + r_c^i(u) \qquad (i = 1, \ldots, m) \tag{8}$$

with

$$\lim_{u \to 0} \frac{r_c^i(u)}{\|u\|} = 0 \qquad (i = 1, \ldots, m). \tag{9}$$

Hence we obtain our next theorem.

Theorem 2

Let S be a subset of $\mathbb{R}^n$. A function $f : S \to \mathbb{R}^m$ is differentiable at an interior point c of S if and only if each of its component functions f_i is differentiable at c. In that case, the i-th component of $df(c; u)$ is $df_i(c; u)$ $(i = 1, \ldots, m)$.

In view of Theorems 1 and 2, it is not surprising to find that many of the theorems on continuity and differentiation that are valid for real-valued functions remain valid for vector functions. It appears therefore that we need only study real-valued functions. This is, however, not so, because in practical applications real-valued functions are often expressed in terms of vector functions (and, indeed, matrix functions). Another reason for studying vector functions, rather than merely real-valued functions, is to obtain a meaningful chain rule (section 12).

If $f : S \to \mathbb{R}^m$, $S \subset \mathbb{R}^n$, is differentiable on an open subset E of S, there must exist real-valued *functions* $a_{ij} : E \to \mathbb{R}$ $(i = 1, \ldots, m;\ j = 1, \ldots, n)$ such that (1) holds for every point of E. We have, however, no guarantee that, for given f, *any* such function a_{ij} exists. We shall prove later (section 10) that, when f is suitably restricted, the functions a_{ij} exist. But first we prove that, if such functions exist, they are unique.

Exercise

1. Let $f: S \to \mathbb{R}^m$ and $g: S \to \mathbb{R}^m$ be differentiable at a point $c \in S \subset \mathbb{R}^n$. Then the function $h = f + g$ is differentiable at c with $\mathrm{d}h(c; u) = \mathrm{d}f(c; u) + \mathrm{d}g(c; u)$.

5. UNIQUENESS OF THE DIFFERENTIAL

Theorem 3

Let $f: S \to \mathbb{R}^m$, $S \subset \mathbb{R}^n$, be differentiable at a point $c \in S$ with differential $\mathrm{d}f(c; u) = A(c)u$. Suppose a second matrix $A^*(c)$ exists such that $\mathrm{d}f(c; u) = A^*(c)u$. Then $A(c) = A^*(c)$.

Proof. From the definition of differentiability we have

$$f(c+u) = f(c) + A(c)u + r_c(u) \tag{1}$$

and also

$$f(c+u) = f(c) + A^*(c)u + r_c^*(u), \tag{2}$$

where $r_c(u)/\|u\|$ and $r_c^*(u)/\|u\|$ both tend to 0 with u. Let $B(c) = A(c) - A^*(c)$. Subtracting (2) from (1) gives

$$B(c)u = r_c^*(u) - r_c(u). \tag{3}$$

Hence,

$$\frac{B(c)u}{\|u\|} \to 0 \qquad \text{as } u \to 0. \tag{4}$$

For fixed $u \neq 0$ it follows that

$$\frac{B(c)(tu)}{\|tu\|} \to 0 \qquad \text{as } t \to 0. \tag{5}$$

The left side of (5) is independent of t. Thus $B(c)u = 0$ for all $u \in \mathbb{R}^n$. The theorem follows. □

6. CONTINUITY OF DIFFERENTIABLE FUNCTIONS

Next we prove that the existence of the differential $\mathrm{d}f(c; u)$ implies continuity of f at c. In other words, that continuity is a necessary condition for differentiability.

Theorem 4

If f is differentiable at c, then f is continuous at c.

Proof. Since f is differentiable, we write

$$f(c+u)=f(c)+A(c)u+r_c(u). \tag{1}$$

Now, both $A(c)u$ and $r_c(u)$ tend to 0 with u. Hence

$$f(c+u)\to f(c) \qquad \text{as } u\to 0. \tag{2}$$

□

The converse of Theorem 4 is, of course, false. For example, the function $\phi : \mathsf{R} \to \mathsf{R}$ defined by the equation $\phi(x) = |x|$ is continuous but not differentiable at 0.

Exercise

1. Let $\phi : S \to \mathsf{R}$ be a real-valued function defined on a set S in R^n, and differentiable at an interior point c of S. Show that (a) there exists a non-negative number M, depending on c but not on u, such that $|\mathrm{d}\phi(c; u)| \leqslant M\|u\|$; (b) there exists a positive number η, again depending on c but not on u, such that $|r_c(u)| < \|u\|$ for all $u \neq 0$ with $\|u\| < \eta$. Conclude that (c) $|\phi(c + u) - \phi(c)| < (1 + M)\|u\|$ for all $u \neq 0$ with $\|u\| < \eta$. A function with this property is said to satisfy a *Lipschitz condition* at c. Of course, if ϕ satisfies a Lipschitz condition at c, then it must be continuous at c.

7. PARTIAL DERIVATIVES

Before we develop the theory of differentials any further, we introduce an important concept in multivariable calculus, the *partial derivative*.

Let $f : S \to \mathsf{R}^m$ be a function defined on a set S in R^n with values in R^m, and let $f_i : S \to \mathsf{R}\ (i = 1, \ldots, m)$ be the i-th component function of f. Let c be an interior point of S, and let e_j be the j-th unit vector in R^n, that is the vector whose j-th component is one and whose remaining components are zero. Consider another point $c + te_j$ in R^n, all of whose components except the j-th are the same as those of c. Since c is an interior point of S, $c + te_j$ is, for small enough t, also a point of S. Now consider the limit

$$\lim_{t\to 0} \frac{f_i(c+te_j)-f_i(c)}{t}. \tag{1}$$

When this limit exists, it is called the *partial derivative* of f_i with respect to the j-th coordinate (of the j-th partial derivative of f_i) at c and is denoted by $\mathsf{D}_j f_i(c)$. (Other notations include $[\partial f_i(x)/\partial x_j]_{x=c}$ or even $\partial f_i(c)/\partial x_j$.) Partial differentiation thus produces, from a given function f_i, n further functions $\mathsf{D}_1 f_i, \ldots, \mathsf{D}_n f_i$ defined at those points in S where the corresponding limits exist.

In fact, the concept of partial differentiation reduces the discussion of real-valued functions of several variables to the one-dimensional case. We are merely treating f_i as a function of one variable at a time. Thus $\mathsf{D}_j f_i$ is the derivative of f_i with respect to the j-th variable, holding the other variables fixed.

Theorem 5

If f is differentiable at c, then all partial derivatives $\mathsf{D}_j f_i(c)$ exist.

Proof. Since f is differentiable at c, there exists a real matrix $A(c)$ with elements $a_{ij}(c)$ such that, for all $\|u\| < r$,

$$f(c+u) = f(c) + A(c)u + r_c(u), \tag{2}$$

where

$$r_c(u)/\|u\| \to 0 \qquad \text{as } u \to 0. \tag{3}$$

Since (2) is true for all $\|u\| < r$, it is true in particular if we choose $u = te_j$ with $|t| < r$. This gives

$$f(c+te_j) = f(c) + tA(c)e_j + r_c(te_j) \tag{4}$$

where

$$r_c(te_j)/t \to 0 \qquad \text{as } t \to 0. \tag{5}$$

If we divide both sides of (4) by t and let t tend to 0, we find that

$$a_{ij}(c) = \lim_{t \to 0} \frac{f_i(c+te_j) - f_i(c)}{t}. \tag{6}$$

Since $a_{ij}(c)$ exists, so does the limit on the right-hand side of (6). But, by (1), this is precisely the partial derivative $\mathsf{D}_j f_i(c)$. □

The converse of Theorem 5 is false. Indeed, the existence of the partial derivatives with respect to each variable separately does not even imply continuity in all the variables simultaneously (although it does imply continuity in each variable separately, Theorem 4). Consider the following example of a function of two variables:

$$\phi(x, y) = \begin{cases} x + y, & \text{if } x = 0 \text{ or } y = 0 \text{ or both,} \\ 1, & \text{otherwise.} \end{cases} \tag{7}$$

This function is clearly not continuous at (0, 0), but the partial derivatives $\mathsf{D}_1\phi(0, 0)$ and $\mathsf{D}_2\phi(0, 0)$ both exist. In fact,

$$\mathsf{D}_1\phi(0, 0) = \lim_{t \to 0} \frac{\phi(t, 0) - \phi(0, 0)}{t - 0} = \lim_{t \to 0} \frac{t}{t} = 1 \tag{8}$$

and, similarly, $\mathsf{D}_2\phi(0, 0) = 1$.

A partial converse of Theorem 5 exists, however (Theorem 7).

Exercise

1. Show in the example given by (7) that $\mathsf{D}_1\phi$ and $\mathsf{D}_2\phi$, while existing at (0, 0), are not continuous there, and that every disc $B(0)$ contains points where the partials both exist and points where the partials both do not exist.

8. THE FIRST IDENTIFICATION THEOREM

If f is differentiable at c, then a matrix $A(c)$ exists such that for all $\|u\| < r$,

$$f(c+u) = f(c) + A(c)u + r_c(u), \tag{1}$$

where $r_c(u)/\|u\| \to 0$ as $u \to 0$. The proof of Theorem 5 reveals that the elements $a_{ij}(c)$ of the matrix $A(c)$ are, in fact, precisely the partial derivatives $\mathsf{D}_j f_i(c)$. This, in conjunction with the uniqueness theorem (Theorem 3), establishes the following central result.

Theorem 6 (first identification theorem)

Let $f : S \to \mathbb{R}^m$ be a vector function defined on a set S in $\mathbb{R}^n$, and differentiable at an interior point c of S. Let u be a real $n \times 1$ vector. Then

$$\mathrm{d}f(c; u) = (\mathsf{D}f(c))u, \tag{2}$$

where $\mathsf{D}f(c)$ is an $m \times n$ matrix whose elements $\mathsf{D}_j f_i(c)$ are the partial derivatives of f evaluated at c. Conversely, if $A(c)$ is a matrix such that

$$\mathrm{d}f(c; u) = A(c)u \tag{3}$$

for all real $n \times 1$ vectors u, then $A(c) = \mathsf{D}f(c)$.

The $m \times n$ matrix $\mathsf{D}f(c)$ in (2), whose ij-th element is $\mathsf{D}_j f_i(c)$, is called the *Jacobian matrix* of f at c. It is defined at each point where the partials $\mathsf{D}_j f_i$ $(i = 1, \ldots, m; j = 1, \ldots, n)$ exist. (Hence the Jacobian matrix $\mathsf{D}f(c)$ may exist even when the function f is not differentiable at c.) When $m = n$, the determinant of the Jacobian matrix of f is called the *Jacobian* of f. The transpose of the $m \times n$ Jacobian matrix $\mathsf{D}f(c)$ is an $n \times m$ matrix called the *gradient* of f at c; it is denoted by $\nabla f(c)$. (The symbol ∇ is pronounced 'del'.) Thus

$$\nabla f(c) = (\mathsf{D}f(c))'. \tag{4}$$

In particular, when $m = 1$, the vector function $f : S \to \mathbb{R}^m$ specializes to a real-valued function $\phi : S \to \mathbb{R}$, the Jacobian matrix specializes to a $1 \times n$ row vector $\mathsf{D}\phi(c)$ and the gradient specializes to an $n \times 1$ column vector $\nabla\phi(c)$.

The first identification theorem will be used throughout this book. Its great practical value lies in the fact that if f is differentiable at c and we have found a differential $\mathrm{d}f$ at c, then the value of the partials at c can be immediately determined.

Some caution is required when interpreting equation (2). The right side of (2) exists if (and only if) all the partial derivatives $\mathsf{D}_j f_i(c)$ exist. But this does not mean that the differential $\mathrm{d}f(c; u)$ exists if all partials exist. We know that $\mathrm{d}f(c; u)$ exists if and only if f is differentiable at c (section 4). We also know from Theorem 5 that the existence of all the partials is a necessary but not a sufficient condition for differentiability. Hence, equation (2) is only valid when f is differentiable at c.

9. EXISTENCE OF THE DIFFERENTIAL, I

So far we have derived some theorems concerning differentials on the assumption that the differential exists, or, what is the same, that the function is differentiable. We have seen (section 7) that the existence of all partial derivatives at a point is necessary but not sufficient for differentiability (in fact, it is not even sufficient for continuity).

What, then, is a sufficient condition for differentiability at a point? Before we answer this question, we pose four preliminary questions in order to gain further insight into the properties of differentiable functions.

(i) If f is differentiable at c, does it follow that each of the partials is continuous at c?
(ii) If each of the partials is continuous at c, does it follow that f is differentiable at c?
(iii) If f is differentiable at c, does it follow that each of the partials exists in some n-ball $B(c)$?
(iv) If each of the partials exists in some n-ball $B(c)$, does it follow that f is differentiable at c?

The answer to all four questions is, in general, 'No'. Let us see why.

Example 1

Let $\phi : \mathsf{R}^2 \to \mathsf{R}$ be a real-valued function defined by

$$\phi(x, y) = \begin{cases} x^2[y + \sin(1/x)], & \text{if } x \neq 0, \\ 0, & \text{if } x = 0. \end{cases} \tag{1}$$

Then ϕ is differentiable at every point in R^2 with partial derivatives

$$\mathsf{D}_1\phi(x, y) = \begin{cases} 2x[y + \sin(1/x)] - \cos(1/x), & \text{if } x \neq 0, \\ 0, & \text{if } x = 0, \end{cases} \tag{2}$$

and $\mathsf{D}_2\phi(x, y) = x^2$. We see that $\mathsf{D}_1\phi$ is not continuous at any point on the y-axis, since $\cos(1/x)$ in (2) does not tend to a limit as $x \to 0$.

Example 2

Let $A = \{(x, y) : x = y, x > 0\}$ be a subset of R^2, and let $\phi : \mathsf{R}^2 \to \mathsf{R}$ be defined by

$$\phi(x, y) = \begin{cases} x^{2/3}, & \text{if } (x, y) \in A, \\ 0, & \text{if } (x, y) \notin A. \end{cases} \tag{3}$$

Then $\mathsf{D}_1\phi$ and $\mathsf{D}_2\phi$ are both zero everywhere except on A, where they are not defined. Thus both partials are continuous at the origin. But ϕ is not differentiable at the origin.

Example 3

Let $\phi : \mathbb{R}^2 \to \mathbb{R}$ be defined by

$$\phi(x, y) = \begin{cases} x^2 + y^2, & \text{if } x \text{ and } y \text{ are rational,} \\ 0, & \text{otherwise.} \end{cases} \tag{4}$$

Here ϕ is differentiable at only one point, namely the origin. The partial derivative $D_1\phi$ is zero at the origin and at every point $(x, y) \in \mathbb{R}^2$ where y is irrational; it is undefined elsewhere. Similarly, $D_2\phi$ is zero at the origin and every point $(x, y) \in \mathbb{R}^2$ where x is irrational; it is undefined elsewhere. Hence every disc with centre 0 contains points where the partials do not exist.

Example 4

Let $\phi : \mathbb{R}^2 \to \mathbb{R}$ be defined by the equation

$$\phi(x, y) = \begin{cases} \dfrac{x^3}{x^2 + y^2}, & \text{if } (x, y) \neq (0, 0), \\ 0, & \text{if } (x, y) = (0, 0). \end{cases} \tag{5}$$

Here ϕ is continuous everywhere, both partials exist everywhere, but ϕ is not differentiable at the origin.

10. EXISTENCE OF THE DIFFERENTIAL, II

These examples show that neither the continuity of all partial derivatives at a point c nor the existence of all partial derivatives in some n-ball $B(c)$ is, in general, a sufficient condition for differentiability. With this knowledge the reader can now appreciate the following theorem.

Theorem 7

Let $f : S \to \mathbb{R}^m$ be a function defined on a set S in $\mathbb{R}^n$, and let c be an interior point of S. If each of the partial derivatives $D_j f_i$ exists in some n-ball $B(c)$ and is continuous at c, then f is differentiable at c.

Proof. In view of Theorem 2, it suffices to consider the case $m = 1$. The vector function $f : S \to \mathbb{R}^m$ then specializes to a real-valued function $\phi : S \to \mathbb{R}$.

Let $r > 0$ be the radius of the ball $B(c)$, and let u be a point in $\mathbb{R}^n$ with $\|u\| < r$, so that $c + u \in B(c)$. Expressing u in terms of its components we have

$$u = u_1 e_1 + \cdots + u_n e_n, \tag{1}$$

where e_j is the j-th unit vector in $\mathbb{R}^n$. Let $v_0 = 0$, and define the partial sums

$$v_k = u_1 e_1 + \cdots + u_k e_k \qquad (k = 1, \ldots, n). \tag{2}$$

Thus v_k is a point in $\mathbb{R}^n$ whose first k components are the same as those of u and whose last $n-k$ components are zero. Since $\|u\|<r$, we have $\|v_k\|<r$, so that $c+v_k \in B(c)$ for $k=1, \ldots, n$.

We now write the difference $\phi(c+u)-\phi(c)$ as a sum of n terms as follows:

$$\phi(c+u)-\phi(c)=\sum_{j=1}^{n}[\phi(c+v_j)-\phi(c+v_{j-1})]. \tag{3}$$

The k-th term in the sum is $\phi(c+v_k)-\phi(c+v_{k-1})$. Since $B(c)$ is convex, the line segment with endpoints $c+v_{k-1}$ and $c+v_k$ lies in $B(c)$. Further, since $v_k=v_{k-1}+u_k e_k$, the two points $c+v_{k-1}$ and $c+v_k$ differ only in their k-th component, and we can apply the one-dimensional mean-value theorem. This gives

$$\phi(c+v_k)-\phi(c+v_{k-1})=u_k \mathsf{D}_k\phi(c+v_{k-1}+\theta_k u_k e_k) \tag{4}$$

for some $\theta_k \in (0, 1)$. Now each partial derivative $\mathsf{D}_k\phi$ is continuous at c, so that

$$\mathsf{D}_k\phi(c+v_{k-1}+\theta_k u_k e_k)=\mathsf{D}_k\phi(c)+R_k(v_k, \theta_k), \tag{5}$$

where $R_k(v_k, \theta_k)\to 0$ as $v_k \to 0$. Substituting (5) in (4) and then (4) in (3) gives, after some rearrangement,

$$\phi(c+u)-\phi(c)-\sum_{j=1}^{n} u_j \mathsf{D}_j\phi(c)=\sum_{j=1}^{n} u_j R_j(v_j, \theta_j). \tag{6}$$

It follows that

$$\left|\phi(c+u)-\phi(c)-\sum_{j=1}^{n} u_j \mathsf{D}_j\phi(c)\right| \leqslant \|u\| \sum_{j=1}^{n} |R_j|, \tag{7}$$

where $R_k \to 0$ as $u \to 0$, $k=1, \ldots, n$. This completes the proof. □

Note. The examples 1 and 3 of the previous section show that neither the existence of all partials in an n-ball $B(c)$ nor the continuity of all partials at c is a *necessary* condition for differentiability of f at c.

Exercises

1. Prove equation (5).
2. Show that, in fact, only the existence of all the partials and continuity of *all but one* of them is sufficient for differentiability.
3. The condition that the n partials be continuous at c, although sufficient, is by no means a necessary condition for the existence of the differential at c. Consider, for example, the case where ϕ can be expressed as a sum of n functions,

$$\phi(x)=\phi_1(x_1)+\cdots+\phi_n(x_n),$$

where the j-th function ϕ_j is a function of the one-dimensional variable x_j alone. Prove that the mere existence of the partials $\mathsf{D}_1\phi, \ldots, \mathsf{D}_n\phi$ is sufficient for the existence of the differential at c.

11. CONTINUOUS DIFFERENTIABILITY

Definition

Let $f : S \to \mathbb{R}^m$ be a function defined on an open set S in $\mathbb{R}^n$. If all the first-order partial derivatives $\mathsf{D}_j f_i(x)$ exist and are continuous at every point x in S, then the function f is said to be *continuously differentiable on S.*

Notice that while we defined continuity and differentiability of a function at a *point*, continuous differentiability is only defined on an open *set*. In view of Theorem 7, continuous differentiability implies differentiability.

12. THE CHAIN RULE

A very important result is the so-called *chain rule.* In one dimension, the chain rule gives us a formula for differentiating a composite function $h = g \circ f$ defined by the equation

$$(g \circ f)(x) = g(f(x)). \tag{1}$$

The formula states that

$$h'(c) = g'(f(c)) \cdot f'(c) \tag{2}$$

and thus expresses the derivative of h in terms of the derivatives of g and f. Its extension to the multivariable case is as follows.

Theorem 8 (chain rule)

Let S be a subset of $\mathbb{R}^n$, and assume that $f : S \to \mathbb{R}^m$ is differentiable at an interior point c of S. Let T be a subset of $\mathbb{R}^m$ such that $f(x) \in T$ for all $x \in S$, and assume that $g : T \to \mathbb{R}^p$ is differentiable at an interior point $b = f(c)$ of T. Then the composite function $h : S \to \mathbb{R}^p$ defined by

$$h(x) = g(f(x)) \tag{3}$$

is differentiable at c, and

$$\mathsf{D}h(c) = (\mathsf{D}g(b))(\mathsf{D}f(c)). \tag{4}$$

Proof. We put $A = \mathsf{D}f(c)$, $B = \mathsf{D}g(b)$ and define the set $E_r^n = \{x : x \in \mathbb{R}^n, \|x\| < r\}$. Since $c \in S$ and $b \in T$ are interior points, there is an $r > 0$ such that $c + u \in S$ for all $u \in E_r^n$, and $b + v \in T$ for all $v \in E_r^m$. We may therefore define vector functions $r_1 : E_r^n \to \mathbb{R}^m$, $r_2 : E_r^m \to \mathbb{R}^p$ and $R : E_r^n \to \mathbb{R}^p$ by

$$f(c+u) = f(c) + Au + r_1(u), \tag{5}$$

$$g(b+v) = g(b) + Bv + r_2(v), \tag{6}$$

$$h(c+u) = h(c) + BAu + R(u). \tag{7}$$

Since f is differentiable at c, and g is differentiable at b, we have

$$\lim_{u\to 0} r_1(u)/\|u\| = 0 \qquad \text{and} \qquad \lim_{v\to 0} r_2(v)/\|v\| = 0. \tag{8}$$

We have to prove that

$$\lim_{u\to 0} R(u)/\|u\| = 0. \tag{9}$$

Defining a new vector function $z: E_r^n \to \mathbb{R}^m$ by

$$z(u) = f(c+u) - f(c), \tag{10}$$

and using the definitions of R and h, we obtain

$$R(u) = g(b+z(u)) - g(b) - Bz(u) + B[f(c+u) - f(c) - Au], \tag{11}$$

so that, in view of (5) and (6),

$$R(u) = r_2(z(u)) + Br_1(u). \tag{12}$$

Now, let μ_A and μ_B be constants such that

$$\|Ax\| \leqslant \mu_A \|x\| \qquad \text{and} \qquad \|By\| \leqslant \mu_B \|y\| \tag{13}$$

for every $x \in \mathbb{R}^n$ and $y \in \mathbb{R}^m$ (see exercise 2), and observe from (5) and (10) that $z(u) = Au + r_1(u)$. Repeated application of the triangle inequality then shows that

$$\begin{aligned} \|R(u)\| &\leqslant \|r_2(z(u))\| + \|Br_1(u)\| \\ &= \frac{\|r_2(z(u))\|}{\|z(u)\|} \cdot \|Au + r_1(u)\| + \|Br_1(u)\| \\ &\leqslant \frac{\|r_2(z)\|}{\|z\|} [\mu_A \|u\| + \|r_1(u)\|] + \mu_B \|r_1(u)\|. \end{aligned} \tag{14}$$

Dividing both sides of (14) by $\|u\|$ yields

$$\frac{\|R(u)\|}{\|u\|} \leqslant \mu_A \frac{\|r_2(z)\|}{\|z\|} + \mu_B \frac{\|r_1(u)\|}{\|u\|} + \frac{\|r_1(u)\|}{\|u\|} \cdot \frac{\|r_2(z)\|}{\|z\|}. \tag{15}$$

Now, $r_2(z)/\|z\| \to 0$ as $z \to 0$ by (8), and since $z(u)$ tends to 0 with u, it follows that $r_2(z)/\|z\| \to 0$ as $u \to 0$. Also by (8), $r_1(u)/\|u\| \to 0$ as $u \to 0$. This shows that (9) holds. □

Exercises

1. What is the order of the matrices A and B? Is the matrix product BA defined?
2. Show that the constants μ_A and μ_B in (13) exist. [*Hint*: Use exercise 1.14.2.]
3. Write out the chain rule as a system of np equations

$$\mathsf{D}_j h_i(c) = \sum_{k=1}^{m} \mathsf{D}_k g_i(b)\, \mathsf{D}_j f_k(c)$$

where $j = 1, \ldots, n$ and $i = 1, \ldots, p$.

13. CAUCHY INVARIANCE

The chain rule relates the *partial derivatives* of a composite function $h = g \circ f$ to the partial derivatives of g and f. We shall now discuss an immediate consequence of the chain rule, which relates the *differential* of h to the differentials of g and f. This result (known as *Cauchy's rule of invariance*) is particularly useful in performing computations with differentials.

Let $h = g \circ f$ be a composite function, as before, such that

$$h(x) = g(f(x)), \qquad x \in S. \tag{1}$$

If f is differentiable at c and g is differentiable at $b = f(c)$, then h is differentiable at c with

$$\mathrm{d}h(c; u) = (\mathrm{D}h(c))u. \tag{2}$$

Using the chain rule, (2) becomes

$$\begin{aligned} \mathrm{d}h(c; u) &= (\mathrm{D}g(b))(\mathrm{D}f(c))u \\ &= (\mathrm{D}g(b))\,\mathrm{d}f(c; u) = \mathrm{d}g(b; \mathrm{d}f(c; u)). \end{aligned} \tag{3}$$

We have thus proved the following.

Theorem 9 (Cauchy's rule of invariance)

If f is differentiable at c and g is differentiable at $b = f(c)$, then the differential of the composite function $h = g \circ f$ is

$$\mathrm{d}h(c; u) = \mathrm{d}g(b; \mathrm{d}f(c; u)) \tag{4}$$

for every u in $\mathbb{R}^n$.

Cauchy's rule of invariance justifies the use of a simpler notation for differentials in practical applications, which adds greatly to the ease and elegance of performing computations with differentials. We shall discuss notational matters in more detail in section 16.

14. THE MEAN-VALUE THEOREM FOR REAL-VALUED FUNCTIONS

The mean-value theorem for functions from $\mathbb{R}^1$ to $\mathbb{R}^1$ states that

$$\phi(c + u) = \phi(c) + (\mathrm{D}\phi(c + \theta u))u \tag{1}$$

for some $\theta \in (0, 1)$. This equation is false, in general, for vector functions. Consider, for example, the vector function $f : \mathbb{R}^1 \to \mathbb{R}^2$ defined by

$$f(t) = \begin{pmatrix} t^2 \\ t^3 \end{pmatrix}. \tag{2}$$

Then no value of $\theta \in (0, 1)$ exists such that

$$f(1) = f(0) + \mathrm{D}f(\theta), \tag{3}$$

as can be easily verified. Several modified versions of the mean-value theorem exist for vector functions, but here we only need the (straightforward) generalization of the one-dimensional mean-value theorem to real-valued functions of two or more variables.

Theorem 10 (mean-value theorem)

Let $\phi : S \to \mathbb{R}$ be a real-valued function, defined and differentiable on an open set S in $\mathbb{R}^n$. Let c be a point of S, and u a point in $\mathbb{R}^n$ such that $c + tu \in S$ for all $t \in [0, 1]$. Then

$$\phi(c+u) = \phi(c) + \mathrm{d}\phi(c+\theta u; u) \tag{4}$$

for some $\theta \in (0, 1)$.

Proof. Consider the real-valued function $\psi : [0, 1] \to \mathbb{R}$ defined by the equation

$$\psi(t) = \phi(c+tu). \tag{5}$$

Then ψ is differentiable at each point of $(0, 1)$ and its derivative is given by

$$\mathsf{D}\psi(t) = (\mathsf{D}\phi(c+tu))u = \mathrm{d}\phi(c+tu; u). \tag{6}$$

By the one-dimensional mean-value theorem we have

$$\frac{\psi(1)-\psi(0)}{1-0} = \mathsf{D}\psi(\theta) \tag{7}$$

for some $\theta \in (0, 1)$. Thus

$$\phi(c+u) - \phi(c) = \mathrm{d}\phi(c+\theta u; u). \qquad \square$$

Exercise

1. Let $\phi : S \to \mathbb{R}$ be a real-valued function, defined and differentiable on an open interval S in $\mathbb{R}^n$. If $\mathsf{D}\phi(c) = 0$ for each $c \in S$, then ϕ is constant on S.

15. MATRIX FUNCTIONS

Hitherto we have only considered vector functions. The following are examples of *matrix* functions:

$$F(\xi) = \begin{pmatrix} \xi & 0 \\ 0 & \xi^2 \end{pmatrix}, \qquad F(x) = xx', \qquad F(X) = X'. \tag{1}$$

The first example maps a scalar ξ into a matrix, the second example maps a vector x into a matrix xx', and the third example maps a matrix X into its transpose matrix X'.

To extend the calculus of vector functions to matrix functions is straightforward. Let us consider a matrix function $F : S \to \mathbb{R}^{m\times p}$ defined on a set S in $\mathbb{R}^{n\times q}$. That is, F maps an $n \times q$ matrix X in S into an $m \times p$ matrix $F(X)$.

Definition

Let $F : S \rightarrow \mathbb{R}^{m\times p}$ be a matrix function defined on a set S in $\mathbb{R}^{n\times q}$. Let C be an interior point of S, and let $B(C; r) \subset S$ be a ball with centre C and radius r. Let U be a point in $\mathbb{R}^{n\times q}$ with $\|U\| < r$, so that $C + U \in B(C; r)$. If there exists a real $mp \times nq$ matrix A, depending on C but not on U, such that

$$\operatorname{vec} F(C+U) = \operatorname{vec} F(C) + A(C) \operatorname{vec} U + \operatorname{vec} R_C(U) \tag{2}$$

for all $U \in \mathbb{R}^{n\times q}$ with $\|U\| < r$ and

$$\lim_{U\to 0} \frac{R_C(U)}{\|U\|} = 0, \tag{3}$$

then the function F is said to be *differentiable at* C; the $m \times p$ matrix $\mathrm{d}F(C; U)$ defined by

$$\operatorname{vec} \mathrm{d}F(C; U) = A(C) \operatorname{vec} U \tag{4}$$

is then called the *(first) differential of F at C with increment U* and the $mp \times nq$ matrix $A(C)$ is called the *(first) derivative of F at C*.

Note. Recall that the *norm* of a real matrix X is defined by

$$\|X\| = (\operatorname{tr} X'X)^{1/2} \tag{5}$$

and a *ball* in $\mathbb{R}^{n\times q}$ by

$$B(C; r) = \{X : X \in \mathbb{R}^{n\times q}, \|X - C\| < r\}. \tag{6}$$

In view of the above definition all calculus properties of matrix functions follow immediately from the corresponding properties of vector functions because, instead of the matrix function F, we can consider the vector function $f: \operatorname{vec} S \rightarrow \mathbb{R}^{mp}$ defined by

$$f(\operatorname{vec} X) = \operatorname{vec} F(X). \tag{7}$$

It is easy to see from (2) and (3) that the differentials of F and f are related by

$$\operatorname{vec} \mathrm{d}F(C; U) = \mathrm{d}f(\operatorname{vec} C; \operatorname{vec} U). \tag{8}$$

We then define the *Jacobian matrix of F at C* as

$$\mathsf{D}F(C) = \mathsf{D}f(\operatorname{vec} C). \tag{9}$$

This is an $mp \times nq$ matrix, whose ij-th element is the partial derivative of the i-th component of $\operatorname{vec} F(X)$ with respect to the j-th element of $\operatorname{vec} X$, evaluated at $X = C$.

The following three theorems are now straightforward generalizations of Theorems 6, 8 and 9.

Theorem 11 (first identification theorem for matrix functions)

Let $F : S \to \mathbb{R}^{m\times p}$ be a matrix function defined on a set S in $\mathbb{R}^{n\times q}$, and differentiable at an interior point C of S. Then

$$\operatorname{vec} \mathrm{d}F(C; U) = A(C) \operatorname{vec} U \tag{10}$$

for all $U \in \mathbb{R}^{n\times q}$ if and only if

$$\mathsf{D}F(C) = A(C). \tag{11}$$

Theorem 12 (chain rule)

Let S be a subset of $\mathbb{R}^{n\times q}$, and assume that $F : S \to \mathbb{R}^{m\times p}$ is differentiable at an interior point C of S. Let T be a subset of $\mathbb{R}^{m\times p}$ such that $F(X) \in T$ for all $X \in S$, and assume that $G : T \to \mathbb{R}^{r\times s}$ is differentiable at an interior point $B = F(C)$ of T. Then the composite function $H : S \to \mathbb{R}^{r\times s}$ defined by

$$H(X) = G(F(X)) \tag{12}$$

is differentiable at C, and

$$\mathsf{D}H(C) = (\mathsf{D}G(B))(\mathsf{D}F(C)). \tag{13}$$

Theorem 13 (Cauchy's rule of invariance)

If F is differentiable at C and G is differentiable at $B = F(C)$, then the differential of the composite function $H = G \circ F$ is

$$\mathrm{d}H(C; U) = \mathrm{d}G(B; \mathrm{d}F(C; U)) \tag{14}$$

for every U in $\mathbb{R}^{n\times q}$.

Exercise

1. Let S be a subset of $\mathbb{R}^n$ and assume that $F : S \to \mathbb{R}^{m\times p}$ is continuous at an interior point c of S. Assume also that $F(c)$ has full rank (that is, $F(c)$ has either full column rank p or full row rank m). Prove that $F(x)$ has full rank for all x in some neighbourhood of $x = c$.

16. SOME REMARKS ON NOTATION

We remarked in section 13 that Cauchy's rule of invariance justifies the use of a simpler notation for differentials in practical applications. (In the theoretical Chapters 4–7 we shall not use this simplified notation.) Let us now see what this simplification involves and how it is justified.

Let $g : \mathbb{R}^m \to \mathbb{R}^p$ be a given differentiable vector function and consider the equation

$$y = g(t). \tag{1}$$

We shall now use the symbol $\mathrm{d}y$ to denote the differential

$$\mathrm{d}y = \mathrm{d}g(t; \mathrm{d}t). \tag{2}$$

In this expression dt (previously u) denotes an arbitrary vector in $\mathbb{R}^m$, and dy denotes the corresponding vector in $\mathbb{R}^p$. Thus dt and dy are vectors of *variables*.

Suppose now that the variables $t_1, t_2, \ldots, t_m$ depend on certain other variables, say $x_1, x_2, \ldots, x_n$:

$$t = f(x). \tag{3}$$

Substituting $f(x)$ for t in (1), we obtain

$$y = g(f(x)) \equiv h(x), \tag{4}$$

and therefore

$$\mathrm{d}y = \mathrm{d}h(x; \mathrm{d}x). \tag{5}$$

The double use of the symbol dy in (2) and (5) is justified by Cauchy's rule of invariance. This is easy to see: from (3) we have

$$\mathrm{d}t = \mathrm{d}f(x; \mathrm{d}x), \tag{6}$$

where dx is an arbitrary vector in $\mathbb{R}^n$. Then (5) gives (by Theorem 9)

$$\mathrm{d}y = \mathrm{d}g(f(x); \mathrm{d}f(x; \mathrm{d}x)) = \mathrm{d}g(t; \mathrm{d}t) \tag{7}$$

using (3) and (6). We conclude that formula (2) is valid even when $t_1, \ldots, t_m$ *depend on other variables* $x_1, \ldots, x_n$, although (6) shows that dt is then no longer an *arbitrary* vector in $\mathbb{R}^m$.

We can economize still further with notation by replacing y in (1) with g itself, thus writing (2) as

$$\mathrm{d}g = \mathrm{d}g(t; \mathrm{d}t) \tag{8}$$

and calling dg the *differential of* g *at* t. This type of conceptually ambiguous usage (of g as both function symbol and variable) will assist practical work with differentials in Part 3.

Example 1

Let

$$y = \phi(x) = e^{x'x}. \tag{9}$$

Then

$$\begin{aligned}\mathrm{d}y = \mathrm{d}e^{x'x} = e^{x'x}(\mathrm{d}x'x) &= e^{x'x}((\mathrm{d}x)'x + x'\mathrm{d}x) \\ &= (2e^{x'x}x')\mathrm{d}x.\end{aligned} \tag{10}$$

Example 2

Let

$$z = \phi(\beta) = (y - X\beta)'(y - X\beta). \tag{11}$$

Then, letting $e = y - X\beta$, we have

$$\begin{aligned}\mathrm{d}z = \mathrm{d}e'e = 2e'\mathrm{d}e &= 2e'\mathrm{d}(y - X\beta) \\ &= -2e'X\mathrm{d}\beta = -2(y - X\beta)'X\mathrm{d}\beta.\end{aligned} \tag{12}$$

MISCELLANEOUS EXERCISES

1. Consider a vector-valued function $f(t) = (\cos t, \sin t)'$, $t \in \mathbb{R}$. Show that $f(2\pi) - f(0) = 0$, and that $\|\mathsf{D}f(t)\| = 1$ for all t. Conclude that the mean-value theorem does not hold for vector-valued functions.
2. Let S be an open subset of $\mathbb{R}^n$ and assume that $f : S \to \mathbb{R}^m$ is differentiable at each point of S. Let c be a point of S, and u a point in $\mathbb{R}^n$ such that $c + tu \in S$ for all $t \in [0, 1]$. Then for every vector a in $\mathbb{R}^m$ there exists a $\theta \in (0, 1)$ such that

$$a'[f(c+u) - f(c)] = a'(\mathsf{D}f(c + \theta u))u,$$

where $\mathsf{D}f$ is the $m \times n$ matrix of partial derivatives $\mathsf{D}_j f_i$ $(i = 1, \ldots, m; j = 1, \ldots, n)$. This is the *mean-value theorem for vector-valued functions*.
3. Now formulate the correct mean-value theorem for the example under exercise 1, and determine θ as a function of a.

BIBLIOGRAPHICAL NOTES

§1. See also Binmore (1982), Apostol (1974) and Dieudonné (1969). For a discussion of the origins of the differential calculus, see Baron (1969).
§6. There even exist functions which are continuous everywhere without being differentiable at any point. See Rudin (1964, p. 141) for an example of such a function.
§14. For modified versions of the mean-value theorem, see Dieudonné (1969, sec. 8.5). Dieudonné regards the mean-value theorem as the most useful theorem in analysis and argues (p. 148) that its real nature is exhibited by writing it as an inequality, and not as an equality.

CHAPTER 6

The second differential

1. INTRODUCTION

In this chapter we discuss second-order partial derivatives, twice differentiability and the second differential. Special attention is given to the relationship between twice differentiability and second-order approximation. We define the Hessian matrix (for vector functions) and find conditions for its (column) symmetry. We also obtain a chain rule for Hessian matrices, and its analogue for second differentials. Taylor's theorem for real-valued functions is proved. Finally, we discuss very briefly higher-order differentials, and show how the calculus of vector functions can be extended to matrix functions.

2. SECOND-ORDER PARTIAL DERIVATIVES

Consider a vector function $f : S \to \mathbb{R}^m$ defined on a set S in $\mathbb{R}^n$ with values in $\mathbb{R}^m$. Let $f_i : S \to \mathbb{R}$ ($i = 1, \ldots, m$) be the i-th component function of f, and assume that f_i has partial derivatives not only at an interior point c of S but also at each point of an open neighbourhood of c. Then we can also consider *their* partial derivatives, i.e. we can consider the limit

$$\lim_{t \to 0} \frac{(\mathsf{D}_j f_i)(c + te_k) - (\mathsf{D}_j f_i)(c)}{t} \tag{1}$$

where e_k is the k-th unit vector in $\mathbb{R}^n$. When this limit exists, it is called the (k, j)-th *second-order* partial derivative of f_i at c and is denoted $\mathsf{D}^2_{kj} f_i(c)$. (Other notations include $[\partial^2 f_i(x)/\partial x_k \partial x_j]_{x=c}$ or even $\partial^2 f_i(c)/\partial x_k \partial x_j$.) Thus $\mathsf{D}^2_{kj} f_i$ is obtained by first partially differentiating f_i with respect to the j-th variable, and then partially differentiating the resulting function $\mathsf{D}_j f_i$ with respect to the k-th variable.

Example 1

Let $\phi : \mathbb{R}^2 \to \mathbb{R}$ be a real-valued function defined by the equation

$$\phi(x, y) = xy^2(x^2 + y). \tag{2}$$

The two (first-order) partial derivatives are given by the derivative

$$\mathsf{D}\phi(x, y) = (3x^2y^2 + y^3,\ 2x^3y + 3xy^2) \tag{3}$$

and so the four second-order partial derivatives are

$$\begin{aligned} \mathsf{D}^2_{11}\phi(x, y) &= 6xy^2, & \mathsf{D}^2_{12}\phi(x, y) &= 6x^2y + 3y^2, \\ \mathsf{D}^2_{21}\phi(x, y) &= 6x^2y + 3y^2, & \mathsf{D}^2_{22}\phi(x, y) &= 2x^3 + 6xy. \end{aligned} \tag{4}$$

Notice that in this example $\mathsf{D}^2_{12}\phi = \mathsf{D}^2_{21}\phi$, but this is not always the case. The standard counter-example follows.

Example 2

Let $\phi : \mathbb{R}^2 \to \mathbb{R}$ be a real-valued function defined by

$$\phi(x, y) = \begin{cases} xy(x^2 - y^2)/(x^2 + y^2), & \text{if } (x, y) \neq (0, 0), \\ 0 & \text{if } (x, y) = (0, 0). \end{cases} \tag{5}$$

Here the function ϕ is differentiable on $\mathbb{R}^2$, the first-order partial derivatives are continuous on $\mathbb{R}^2$ (even differentiable, except at the origin), and the second-order partial derivatives exist at every point of $\mathbb{R}^2$ (and are continuous except at the origin). But

$$(\mathsf{D}^2_{12}\phi)(0, 0) = 1, \qquad (\mathsf{D}^2_{21}\phi)(0, 0) = -1. \tag{6}$$

3. THE HESSIAN MATRIX

Earlier we have defined a matrix which contains all the *first-order* partial derivatives. This is the *Jacobian matrix*. We now define a matrix (called the *Hessian matrix*) which contains all *second-order* partial derivatives. We define this matrix first for real-valued functions, then for vector functions.

Definition 1

Let $\phi : S \to \mathbb{R}$, $S \subset \mathbb{R}^n$, be a real-valued function, and let c be a point of S where the n^2 second-order partials $\mathsf{D}^2_{kj}\phi(c)$ exist. Then we define the $n \times n$ Hessian matrix $\mathsf{H}\phi(c)$ by

$$\mathsf{H}\phi(c) = \begin{bmatrix} \mathsf{D}^2_{11}\phi(c) & \mathsf{D}^2_{21}\phi(c) & \dots & \mathsf{D}^2_{n1}\phi(c) \\ \mathsf{D}^2_{12}\phi(c) & \mathsf{D}^2_{22}\phi(c) & \dots & \mathsf{D}^2_{n2}\phi(c) \\ \vdots & \vdots & & \vdots \\ \mathsf{D}^2_{1n}\phi(c) & \mathsf{D}^2_{2n}\phi(c) & \dots & \mathsf{D}^2_{nn}\phi(c) \end{bmatrix}. \tag{1}$$

Note that the ij-th element of $\mathsf{H}\phi(c)$ is $\mathsf{D}^2_{ji}\phi(c)$ and *not* $\mathsf{D}^2_{ij}\phi(c)$.

Definition 2

Let $f : S \to \mathbb{R}^m$, $S \subset \mathbb{R}^n$, be a vector function, and let c be a point of S where the mn^2 second-order partials $\mathsf{D}^2_{kj}f_i(c)$ exist. Then we define the $mn \times n$ Hessian matrix

$\mathsf{H}f(c)$ by

$$\mathsf{H}f(c)=\begin{bmatrix}\mathsf{H}f_1(c)\\ \mathsf{H}f_2(c)\\ \vdots\\ \mathsf{H}f_m(c)\end{bmatrix}. \tag{2}$$

Referring to the examples in the previous section, we have for the function in example 1:

$$\mathsf{H}\phi(x, y)=\begin{pmatrix} 6xy^2 & 6x^2y+3y^2\\ 6x^2y+3y^2 & 2x^3+6xy\end{pmatrix}, \tag{3}$$

and for the function in example 2:

$$\mathsf{H}\phi(0, 0) = \begin{pmatrix}0 & -1\\ 1 & 0\end{pmatrix}. \tag{4}$$

The first matrix is symmetric; the second is not. Sufficient conditions for the symmetry of the Hessian matrix of a real-valued function are derived in section 7. The Hessian matrix of a vector function f cannot, of course, be symmetric if $m \geqslant 2$. We shall say that $\mathsf{H}f(c)$ is *column symmetric* if the Hessian matrix of each of its component functions f_i $(i = 1, \ldots, m)$ is symmetric at c.

4. TWICE DIFFERENTIABILITY AND SECOND-ORDER APPROXIMATION, I

Consider a real-valued function $\phi : S \to \mathbb{R}$ which is differentiable at a point c in $S \subset \mathbb{R}^n$, i.e. there exists a vector a, depending on c but not on u, such that

$$\phi(c+u)=\phi(c)+a'u+r(u), \tag{1}$$

where

$$\lim_{u\to 0}\frac{r(u)}{\|u\|}=0. \tag{2}$$

The vector a', if it exists, is of course the derivative $\mathsf{D}\phi(c)$. Thus, differentiability is defined by means of a first-order Taylor formula.

Suppose now that there exists a symmetric matrix B, depending on c but not on u, such that

$$\phi(c+u)=\phi(c)+(\mathsf{D}\phi(c))u+\tfrac{1}{2}u'Bu+r(u), \tag{3}$$

where

$$\lim_{u\to 0}\frac{r(u)}{\|u\|^2}=0. \tag{4}$$

Equation (3) is called the *second-order Taylor formula*. The question naturally arises whether it is appropriate to define twice differentiability as the existence of a second-order Taylor formula. This question must be answered in the negative. To see why, we consider the function $\phi : \mathbb{R}^2 \to \mathbb{R}$ defined by the equation

$$\phi(x, y) = \begin{cases} x^3 + y^3 & (x \text{ and } y \text{ rational}), \\ 0 & (\text{otherwise}). \end{cases} \tag{5}$$

Then ϕ is differentiable at (0,0), but at no other point in $\mathbb{R}^2$. The partial derivative $\mathsf{D}_1\phi$ is zero at the origin and at every point in $\mathbb{R}^2$ where y is irrational; it is undefined elsewhere. Similarly, $\mathsf{D}_2\phi$ is zero at the origin and at every point in $\mathbb{R}^2$ where x is irrational; it is undefined elsewhere. Hence, neither of the partial derivatives is differentiable at any point in $\mathbb{R}^2$. In spite of this, a unique matrix B exists (the null matrix), such that the second-order Taylor formula (3) holds at $c=0$. Surely we do not want to say that ϕ is twice differentiable at a point, when its partial derivatives are not differentiable there!

5. DEFINITION OF TWICE DIFFERENTIABILITY

So, the existence of a second-order Taylor formula at a point c is not sufficient, in general, for all partial derivatives to be differentiable at that point. Neither is it necessary. That is, the fact that all partials are differentiable at c does not, in general, imply a second-order Taylor formula at that point. We shall return to this issue in section 9.

Motivated by these facts, we define twice differentiability in such a way that it implies both the existence of a second-order Taylor formula and differentiability of all the partials.

Definition

Let $f : S \to \mathbb{R}^m$ be a function defined on a set S in $\mathbb{R}^n$, and let c be an interior point of S. If f is differentiable in some n-ball $B(c)$ and each of the partial derivatives $\mathsf{D}_j f_i$ is differentiable at c, then we say that f is *twice differentiable at c*. If f is twice differentiable at every point of an open subset E of S, we say f is *twice differentiable on E*.

In the one-dimensional case ($n=1$), the requirement that the derivatives $\mathsf{D} f_i$ are differentiable at c necessitates the existence of $\mathsf{D} f_i(x)$ in a neighbourhood of c, and hence the differentiability of f itself in that neighbourhood. But for $n \geqslant 2$, the mere fact that each of the partials is differentiable at c, necessitating as it does the continuity of each of the partials at c, involves the differentiability of f at c, but not necessarily in the neighbourhood of that point. Hence the differentiability of each of the partials at c is necessary *but not sufficient*, in general, for f to be twice differentiable at c. However, if the partials are differentiable not only at c, but also at each point of an open neighbourhood of c, then f is twice differentiable in that neighbourhood. This follows from Theorems 5.4 and 5.7. In fact, we have the following theorem.

Theorem 1

Let S be an open subset of $\mathbb{R}^n$. Then $f : S \to \mathbb{R}^m$ is twice differentiable on S if and only if all partial derivatives are differentiable on S.

The non-trivial fact that twice differentiability implies (but is not implied by) the existence of a second-order Taylor formula will be proved in section 9.

Without difficulty we can prove the analogue of Theorems 5.1 and 5.2.

Theorem 2

Let S be a subset of $\mathbb{R}^n$. A function $f : S \to \mathbb{R}^m$ is twice differentiable at an interior point c of S if and only if each of its component functions is twice differentiable at c.

Let us summarize. If f is twice differentiable at c, then

(a) f is differentiable (and continuous) at c, *and* in a suitable neighbourhood $B(c)$;
(b) the first-order partials exist in $B(c)$ and are differentiable (and continuous) at c; and
(c) the second-order partials exist at c.

But

(d) the first-order partials need not be continuous at any point of $B(c)$, other than c itself;
(e) the second-order partials need not be continuous at c; and
(f) the second-order partials need not exist at any point of $B(c)$, other than c itself.

Exercise

1. Show that the real-valued function $\phi : \mathbb{R} \to \mathbb{R}$ defined by $\phi(x) = |x|x$ is differentiable everywhere, but not twice differentiable at the origin.

6. THE SECOND DIFFERENTIAL

The second differential is simply the differential of the (first) differential,

$$\mathrm{d}^2 f = \mathrm{d}(\mathrm{d}f). \tag{1}$$

Since $\mathrm{d}f$ is by definition a function of two sets of variables, say x and u, the expression $\mathrm{d}(\mathrm{d}f)$, with whose help the second differential $\mathrm{d}^2 f$ is determined, requires some explanation. While performing the operation $\mathrm{d}(\mathrm{d}f)$ we always consider df *as a function of* x *alone* by assuming u to be constant; furthermore, the same value of u is assumed for the first and second differential.

More formally, we propose the following definition.

Definition

Let $f : S \to \mathbb{R}^m$ be twice differentiable at an interior point c of $S \subset \mathbb{R}^n$. Let $B(c)$ be an n-ball lying in S such that f is differentiable at every point in $B(c)$, and let $g : B(c) \to \mathbb{R}^m$ be defined by the equation

$$g(x) = \mathrm{d}f(x; u). \tag{2}$$

Then the differential of g at c with increment u, i.e. d$g(c; u)$, is called the *second differential of* f *at* c *(with increment* u*)*, and is denoted by $\mathrm{d}^2 f(c; u)$.

We first settle the existence question.

Theorem 3

Let $f : S \to \mathbb{R}^m$ be a function defined on a set S in $\mathbb{R}^n$, and let c be an interior point of S. If each of the first-order partial derivatives is continuous in some n-ball $B(c)$, and if each of the second-order partial derivatives exists in $B(c)$ and is continuous at c, then f is twice differentiable at c and the second differential of f at c exists.

Proof. This is an immediate consequence of Theorem 5.7. □

Let us now evaluate the second differential of a real-valued function $\phi : S \to \mathbb{R}$, where S is a subset of $\mathbb{R}^n$. On the assumption that ϕ is twice differentiable at a point $c \in S$, we can define

$$\psi(x) = \mathrm{d}\phi(x; u) \tag{3}$$

for every x in a suitable n-ball $B(c)$. Hence

$$\psi(x) = \sum_{j=1}^{n} u_j \mathrm{D}_j \phi(x) \tag{4}$$

with partial derivatives

$$\mathrm{D}_i \psi(x) = \sum_{j=1}^{n} u_j \mathrm{D}_{ij}^2 \phi(x) \qquad (i = 1, \ldots, n), \tag{5}$$

and first differential (at u)

$$\mathrm{d}\psi(x; u) = \sum_{i=1}^{n} u_i \mathrm{D}_i \psi(x) = \sum_{i,j=1}^{n} u_i u_j \mathrm{D}_{ij}^2 \phi(x). \tag{6}$$

By definition, the second differential of ϕ equals the first differential of ψ, so that

$$\mathrm{d}^2\phi(x; u) = u'(\mathrm{H}\phi(x))u, \tag{7}$$

where $\mathrm{H}\phi(x)$ is the $n \times n$ Hessian matrix of ϕ at x.

Equation (7) shows that, while the first differential of a real-valued function ϕ is a *linear function* of u, the second differential is a *quadratic form* in u.

We now consider the uniqueness question. We are given a real-valued function ϕ, twice differentiable at c, and we evaluate its first and second differential at c. We find

$$\mathrm{d}\phi(c; u) = a'u, \qquad \mathrm{d}^2\phi(c; u) = u'Bu. \tag{8}$$

Suppose that another vector a^* and another matrix B^* exist such that also

$$\mathrm{d}\phi(c; u) = a^{*\prime}u, \qquad \mathrm{d}^2\phi(c; u) = u'B^*u. \tag{9}$$

Then the uniqueness theorem for first differentials (Theorem 5.3) tells us that $a = a^*$. But a similar uniqueness result does not hold, in general, for second differentials. We can only conclude that

$$B + B' = B^* + B^{*\prime}, \tag{10}$$

because, putting $A = B - B^*$, the fact that $u'Au = 0$ for every u does not imply that A is the null matrix, but only that A is skew symmetric ($A' = -A$); see Theorem 1.2 (c).

The symmetry of the Hessian matrix, which we will discuss in the next section, is therefore of fundamental importance, because without it we could not extract the Hessian matrix from the second differential.

Before we turn to proving this result, we note that the second differential of a *vector function* $f : S \to \mathbb{R}^m$, $S \subset \mathbb{R}^n$, is easily obtained from (7). In fact, we have

$$\mathrm{d}^2 f(c; u) = \begin{bmatrix} \mathrm{d}^2 f_1(c; u) \\ \vdots \\ \mathrm{d}^2 f_m(c; u) \end{bmatrix} = \begin{bmatrix} u'(\mathrm{H}f_1(c))u \\ \vdots \\ u'(\mathrm{H}f_m(c))u \end{bmatrix} = (I_m \otimes u') \begin{bmatrix} \mathrm{H}f_1(c) \\ \vdots \\ \mathrm{H}f_m(c) \end{bmatrix} u, \tag{11}$$

so that, in view of the definition of the Hessian matrix of a vector function (section 3),

$$\mathrm{d}^2 f(c; u) = (I_m \otimes u')(\mathrm{H}f(c))u. \tag{12}$$

7. (COLUMN) SYMMETRY OF THE HESSIAN MATRIX

We have already seen (section 3) that a Hessian matrix $\mathrm{H}\phi$ is not, in general, symmetric. The next theorem gives us a sufficient condition for symmetry of the Hessian matrix.

Theorem 4

Let $\phi : S \to \mathbb{R}$ be a real-valued function defined on a set S in $\mathbb{R}^n$. If ϕ is twice differentiable at an interior point c of S, then the $n \times n$ Hessian matrix $\mathrm{H}\phi$ is symmetric at c, i.e.

$$\mathsf{D}^2_{kj}\phi(c)=\mathsf{D}^2_{jk}\phi(c) \qquad (k, j=1, \ldots, n). \tag{1}$$

Proof. Let $B(c; r)$ be an n-ball such that for any point x in $B(c; r)$ all partial derivatives $\mathsf{D}_j\phi(x)$ exist. Let $A(r)$ be the open interval $(-\frac{1}{2}r\sqrt{2}, \frac{1}{2}r\sqrt{2})$, and t a point in $A(r)$. We consider real-valued functions $\tau_{ij} : A(r) \to \mathbb{R}$ defined by

$$\tau_{ij}(\xi)=\phi(c+te_i+\xi e_j)-\phi(c+\xi e_j), \tag{2}$$

where e_i and e_j are unit vectors in $\mathbb{R}^n$. The functions τ_{ij} are differentiable at each point of $A(r)$ with derivative

$$(\mathsf{D}\tau_{ij})(\xi)=\mathsf{D}_j\phi(c+te_i+\xi e_j)-\mathsf{D}_j\phi(c+\xi e_j). \tag{3}$$

Since $\mathsf{D}_j\phi$ is differentiable at c, we have the first-order Taylor formulae

$$\mathsf{D}_j\phi(c+te_i+\xi e_j)=\mathsf{D}_j\phi(c)+t\,\mathsf{D}^2_{ij}\phi(c)+\xi\mathsf{D}^2_{jj}\phi(c)+R_{ij}(t, \xi) \tag{4}$$

and

$$\mathsf{D}_j\phi(c+\xi e_j)=\mathsf{D}_j\phi(c)+\xi\mathsf{D}^2_{jj}\phi(c)+r_j(\xi), \tag{5}$$

where

$$\lim_{(t,\xi)\to(0,0)}\frac{R_{ij}(t, \xi)}{(t^2+\xi^2)^{1/2}}=0 \qquad \text{and} \qquad \lim_{\xi\to 0}\frac{r_j(\xi)}{\xi}=0. \tag{6}$$

Hence (3) becomes

$$(\mathsf{D}\tau_{ij})(\xi)=t\mathsf{D}^2_{ij}\phi(c)+R_{ij}(t, \xi)-r_j(\xi). \tag{7}$$

We now consider real-valued functions $\delta_{ij} : A(r) \to \mathbb{R}$ defined by

$$\delta_{ij}(\xi)=\tau_{ij}(\xi)-\tau_{ij}(0). \tag{8}$$

By the one-dimensional mean-value theorem we have

$$\delta_{ij}(\xi) = \xi(\mathsf{D}\tau_{ij})(\theta_{ij}\xi) \tag{9}$$

for some $\theta_{ij}\in(0, 1)$. (Of course, the point θ_{ij} depends on the value of ξ and on the function δ_{ij}.) Using (7) we thus obtain

$$\delta_{ij}(\xi) = \xi t\mathsf{D}^2_{ij}\phi(c) + \xi[R_{ij}(t, \theta_{ij}\xi) - r_j(\theta_{ij}\xi)]. \tag{10}$$

Now, since $\delta_{ij}(t)=\delta_{ji}(t)$ it follows that

$$\mathsf{D}^2_{ij}\phi(c) - \mathsf{D}^2_{ji}\phi(c) = \frac{R_{ji}(t, \theta_{ji}t) - R_{ij}(t, \theta_{ij}t) + r_j(\theta_{ij}t) - r_i(\theta_{ji}t)}{t} \tag{11}$$

for some θ_{ij} and θ_{ji} in the interval (0, 1). The left side of (11) is independent of t; the right side tends to 0 with t, by (6). Hence $\mathsf{D}^2_{ij}\phi(c) = \mathsf{D}^2_{ji}\phi(c)$. □

Note. The requirement in Theorem 4 that ϕ is twice differentiable at c is in fact stronger than necessary. The reader may verify that in the proof we have merely used the fact that each of the partial derivatives $\mathsf{D}_j\phi$ is differentiable at c.

The generalization of Theorem 4 to vector functions is simple.

Theorem 5

Let $f: S \rightarrow \mathbb{R}^m$ be a function defined on a set S in $\mathbb{R}^n$. If f is twice differentiable at an interior point c of S, then the $mn \times n$ Hessian matrix $\mathsf{H}f$ is column symmetric at c, i.e.

$$\mathsf{D}^2_{kj} f_i(c) = \mathsf{D}^2_{jk} f_i(c) \qquad (k, j = 1, \ldots, n;\ i = 1, \ldots, m). \tag{12}$$

The column symmetry of $\mathsf{H}f(c)$ is, as we recall from section 3, equivalent to the symmetry of each of the matrices $\mathsf{H}f_i(c)$, i.e. of the Hessian matrices of the component functions f_i.

8. THE SECOND IDENTIFICATION THEOREM

We now have all the ingredients for the following theorem which states that once we know the second differential, the Hessian matrix is uniquely determined (and vice versa).

Theorem 6 (second identification theorem for real-valued functions)

Let $\phi : S \rightarrow \mathbb{R}$ be a real-valued function defined on a set S in $\mathbb{R}^n$, and twice differentiable at an interior point c of S. Let u be a real $n \times 1$ vector. Then

$$\mathrm{d}^2\phi(c; u) = u'(\mathsf{H}\,\phi(c))u, \tag{1}$$

where $\mathsf{H}\phi(c)$ is the $n \times n$ symmetric Hessian matrix of ϕ with elements $\mathsf{D}^2_{ji}\phi(c)$. Furthermore, if $B(c)$ is a matrix such that

$$\mathrm{d}^2\phi(c; u) = u'B(c)u \tag{2}$$

for all real $n \times 1$ vectors u, then

$$\mathsf{H}\phi(c) = \tfrac{1}{2}[B(c) + B(c)']. \tag{3}$$

In order to state the second identification theorem for *vector functions*, of which Theorem 6 is a special case, we require some more notation.

Definition

Let $A_1, A_2, \ldots, A_m$ be square $n \times n$ matrices, and let

$$A = (A_1, A_2, \ldots, A_m). \tag{4}$$

Then we define the $mn \times n$ matrix A_v by

$$A_v = \begin{bmatrix} A_1 \\ A_2 \\ \vdots \\ A_m \end{bmatrix}. \tag{5}$$

As a result of this definition, if $B_1, B_2, \ldots, B_m$ are square matrices, then

$$B = \begin{bmatrix} B_1 \\ B_2 \\ \vdots \\ B_m \end{bmatrix} \leftrightarrow (B')_v = \begin{bmatrix} B'_1 \\ B'_2 \\ \vdots \\ B'_m \end{bmatrix}. \tag{6}$$

Theorem 7 (second identification theorem)

Let $f : S \to \mathbb{R}^m$ be a vector function defined on a set S in $\mathbb{R}^n$, and twice differentiable at an interior point c of S. Let u be a real $n \times 1$ vector. Then

$$\mathrm{d}^2 f(c; u) = (I_m \otimes u')(\mathsf{H}f(c))u, \tag{7}$$

where $\mathsf{H}f(c)$ is the $mn \times n$ column symmetric Hessian matrix of f with elements $\mathsf{D}^2_{kj} f_i(c)$. Furthermore, if $B(c)$ is a matrix such that

$$\mathrm{d}^2 f(c; u) = (I_m \otimes u')B(c)u \tag{8}$$

for all real $n \times 1$ vectors u, then

$$\mathsf{H}f(c) = \tfrac{1}{2}[B(c) + B(c)')_v]. \tag{9}$$

9. TWICE DIFFERENTIABILITY AND SECOND-ORDER APPROXIMATION, II

In section 5 the definition of twice differentiability was motivated, in part, by the claim that it implies the existence of a second-order Taylor formula. Let us now prove this assertion.

Theorem 8

Let $f : S \to \mathbb{R}^m$ be a function defined on a set S in $\mathbb{R}^n$. Let c be an interior point of S, and let $B(c; r)$ be an n-ball lying in S. Let u be a point in $\mathbb{R}^n$ with $\|u\| < r$, so that $c + u \in B(c; r)$. If f is twice differentiable at c, then

$$f(c+u) = f(c) + \mathrm{d}f(c; u) + \tfrac{1}{2}\mathrm{d}^2 f(c; u) + r_c(u), \tag{1}$$

where

$$\lim_{u \to 0} \frac{r_c(u)}{\|u\|^2} = 0. \tag{2}$$

Proof. It suffices to consider the case $m = 1$ (Why?), in which case the vector function f specializes to a real-valued function ϕ. Let $M = (m_{ij})$ be a symmetric $n \times n$ matrix, depending on c *and* u, such that

$$\phi(c+u) = \phi(c) + \mathrm{d}\phi(c; u) + \tfrac{1}{2}u'Mu. \tag{3}$$

Since ϕ is twice differentiable at c, there exists an n-ball $B(c; \rho) \subset B(c; r)$ such that ϕ

is differentiable at each point of $B(c;\rho)$. Let $A(\rho)=\{x : x\in\mathbb{R}^n, \|x\|<\rho\}$, and define a real-valued function $\psi : A(\rho)\to\mathbb{R}$ by the equation

$$\psi(x)=\phi(c+x)-\phi(c)-\mathrm{d}\phi(c;x)-\tfrac{1}{2}x'Mx. \tag{4}$$

Note that M depends on u (and c), but not on x. Then

$$\psi(0)=\psi(u)=0. \tag{5}$$

Also, since ϕ is differentiable in $B(c;\rho)$, ψ is differentiable in $A(\rho)$, so that, by the mean-value theorem (Theorem 5.10),

$$\mathrm{d}\psi(\theta u;u)=0 \tag{6}$$

for some $\theta\in(0,1)$. Now, since each $\mathsf{D}_j\phi$ is differentiable at c, we have the first-order Taylor formula

$$\mathsf{D}_j\phi(c+x)=\mathsf{D}_j\phi(c)+\sum_{i=1}^{n}x_i\mathsf{D}^2_{ij}\phi(c)+R_j(x), \tag{7}$$

where

$$R_j(x)/\|x\|\to 0 \qquad \text{as } x\to 0. \tag{8}$$

The partial derivatives of ψ are thus given by

$$\begin{aligned}\mathsf{D}_j\psi(x)&=\mathsf{D}_j\phi(c+x)-\mathsf{D}_j\phi(c)-\sum_{i=1}^{n}x_im_{ij}\\ &=\sum_{i=1}^{n}x_i[\mathsf{D}^2_{ij}\phi(c)-m_{ij}]+R_j(x),\end{aligned} \tag{9}$$

using (4) and (7). Hence, by (6),

$$\begin{aligned}0=\mathrm{d}\psi(\theta u;u)&=\sum_{j=1}^{n}u_j\mathsf{D}_j\psi(\theta u)\\ &=\theta\sum_{i=1}^{n}\sum_{j=1}^{n}u_iu_j[\mathsf{D}^2_{ij}\phi(c)-m_{ij}]+\sum_{j=1}^{n}u_jR_j(\theta u)\\ &=\theta[\mathrm{d}^2\phi(c;u)-u'Mu]+\sum_{j=1}^{n}u_jR_j(\theta u),\end{aligned} \tag{10}$$

so that

$$u'Mu=\mathrm{d}^2\phi(c;u)+(1/\theta)\sum_{j=1}^{n}u_jR_j(\theta u). \tag{11}$$

Substituting (11) in (3) and noting that

$$\sum_{j=1}^{n}\frac{u_jR_j(\theta u)}{\theta\|u\|^2}=\sum_{j=1}^{n}\frac{u_j}{\|u\|}\cdot\frac{R_j(\theta u)}{\|\theta u\|}\to 0 \qquad \text{as } u\to 0,$$

using (8), completes the proof. □

The example in section 4 shows that the existence of a second-order Taylor formula at a point does not imply, in general, twice differentiability there (in fact, not even differentiability of all partial derivatives).

It is worth remarking that if in Theorem 8 we replace the requirement that f is twice differentiable at c by the weaker condition that all first-order partials of f are differentiable at c, the theorem remains valid for $n=1$ (trivially) and $n=2$, but not, in general, for $n \geq 3$.

Exercise

1. Prove Theorem 8 for $n=2$, assuming that all first-order partials of f are differentiable at c, but without assuming that f is twice differentiable at c.

10. CHAIN RULE FOR HESSIAN MATRICES

In one dimension the first and second derivatives of the composite function $h = g \circ f$, defined by the equation

$$(g \circ f)(x) = g(f(x)), \tag{1}$$

can be expressed in terms of the first and second derivatives of g and f as follows:

$$h'(c) = g'(f(c)) \cdot f'(c) \tag{2}$$

and

$$h''(c) = g''(f(c)) \cdot (f'(c))^2 + g'(f(c)) \cdot f''(c). \tag{3}$$

The following theorem generalizes equation (3) to vector functions of several variables.

Theorem 9 (chain rule for Hessian matrices)

Let S be a subset of $\mathbb{R}^n$, and assume that $f : S \to \mathbb{R}^m$ is twice differentiable at an interior point c of S. Let T be a subset of $\mathbb{R}^m$ such that $f(x) \in T$ for all $x \in S$, and assume that $g : T \to \mathbb{R}^p$ is twice differentiable at an interior point $b = f(c)$ of T. Then the composite function $h : S \to \mathbb{R}^p$ defined by

$$h(x) = g(f(x)) \tag{4}$$

is twice differentiable at c, and

$$\mathsf{H}h(c) = (I_p \otimes \mathsf{D}f(c))'(\mathsf{H}g(b))\mathsf{D}f(c) + (\mathsf{D}g(b) \otimes I_n)\mathsf{H}f(c). \tag{5}$$

Proof. Since g is twice differentiable at b, it is differentiable in some m-ball $B_m(b)$. Also, since f is twice differentiable at c, we can choose an n-ball $B_n(c)$ such that (i) f is differentiable in $B_n(c)$, and (ii) $f(x) \in B_m(b)$ for all $x \in B_n(c)$. Hence, by Theorem 5.8, h is differentiable in $B_n(c)$. Further, since the partials $\mathsf{D}_j h_i$, given by

$$\mathsf{D}_j h_i(x) = \sum_{s=1}^{m} [(\mathsf{D}_s g_i)(f(x))][(\mathsf{D}_j f_s)(x)], \tag{6}$$

are differentiable at c (because the partials $\mathsf{D}_s g_i$ are differentiable at b and the partials $\mathsf{D}_j f_s$ are differentiable at c), the composite function h is twice differentiable at c.

The second-order partials of h_i evaluated at c are then given by

$$\mathsf{D}^2_{kj}h_i(c) = \sum_{s=1}^{m}\sum_{t=1}^{m}(\mathsf{D}^2_{ts}g_i(b))(\mathsf{D}_k f_t(c))(\mathsf{D}_j f_s(c)) + \sum_{s=1}^{m}(\mathsf{D}_s g_i(b))(\mathsf{D}^2_{kj}f_s(c)). \tag{7}$$

Thus, the Hessian matrix of the i-th component function h_i is

$$\begin{aligned}\mathsf{H}h_i(c) &= \sum_{s=1}^{m}\sum_{t=1}^{m}(\mathsf{D}^2_{ts}g_i(b))(\mathsf{D}f_t(c))'(\mathsf{D}f_s(c)) \\ &\quad + \sum_{s=1}^{m}(\mathsf{D}_s g_i(b))(\mathsf{H}f_s(c)) \\ &= (\mathsf{D}f(c))'(\mathsf{H}g_i(b))(\mathsf{D}f(c)) + ((\mathsf{D}g_i(b))\otimes I_n)(\mathsf{H}f(c)),\end{aligned} \tag{8}$$

and the result follows. □

11. THE ANALOGUE FOR SECOND DIFFERENTIALS

The chain rule for Hessian matrices expresses the second-order partial derivatives of the composite function $h = g \circ f$ in terms of the first-order and second-order partial derivatives of g and f. The next theorem expresses the second differential of h in terms of the first and second differentials of g and f.

Theorem 10

If f is twice differentiable at c and g is twice differentiable at $b = f(c)$, then the second differential of the composite function $h = g \circ f$ is

$$\mathrm{d}^2h(c;u) = \mathrm{d}^2g(b;\mathrm{d}f(c;u)) + \mathrm{d}g(b;\mathrm{d}^2f(c;u)) \tag{1}$$

for every u in $\mathbb{R}^n$.

Proof. By Theorems 7 and 9, we have

$$\begin{aligned}\mathrm{d}^2h(c;u) &= (I_p \otimes u')(\mathsf{H}h(c))u \\ &= (I_p \otimes u')(I_p \otimes \mathsf{D}f(c))'(\mathsf{H}g(b))(\mathsf{D}f(c))u \\ &\quad + (I_p \otimes u')(\mathsf{D}g(b)\otimes I_n)(\mathsf{H}f(c))u.\end{aligned} \tag{2}$$

The first term at the right side of (2) is

$$\begin{aligned}(I_p \otimes u')&(I_p \otimes \mathsf{D}f(c))'(\mathsf{H}g(b))(\mathsf{D}f(c))u \\ &= (I_p \otimes (\mathsf{D}f(c)u))'(\mathsf{H}g(b))(\mathsf{D}f(c))u \\ &= (I_p \otimes \mathrm{d}f(c;u))'(\mathsf{H}g(b))\mathrm{d}f(c;u) \\ &= \mathrm{d}^2g(b;\mathrm{d}f(c;u)).\end{aligned} \tag{3}$$

The second term is

$$
\begin{aligned}
(I_p \otimes u')(\mathsf{D}g(b) \otimes I_n)(\mathsf{H}f(c))u
&= (\mathsf{D}g(b) \otimes u')(\mathsf{H}f(c))u \\
&= (\mathsf{D}g(b))(I_m \otimes u')(\mathsf{H}f(c))u \\
&= (\mathsf{D}g(b))\, \mathrm{d}^2 f(c; u) \\
&= \mathrm{d}g(b; \mathrm{d}^2 f(c; u)). \qquad (4)
\end{aligned}
$$

The result follows. □

The most important lesson to be learned from Theorem 10 is that the second differential does *not*, in general, satisfy Cauchy's rule of invariance. By this we mean that, while the first differential of a composite function satisfies

$$\mathrm{d}h(c; u) = \mathrm{d}g(b; \mathrm{d}f(c; u)), \qquad (5)$$

by Theorem 5.9, it is not true, in general, that

$$\mathrm{d}^2 h(c; u) = \mathrm{d}^2 g(b; \mathrm{d}f(c; u)), \qquad (6)$$

unless f is an *affine* function. (A function f is called affine if $f(x) = Ax + b$ for some matrix A and vector b.) This case is of sufficient importance to state as a separate theorem.

Theorem 11

If f is an affine function and g is twice differentiable at $b = f(c)$, then the second differential of the composite function $h = g \circ f$ is

$$\mathrm{d}^2 h(c; u) = \mathrm{d}^2 g(b; \mathrm{d}f(c; u)) \qquad (7)$$

for every u in $\mathbb{R}^n$.

Proof. Since f is affine, $\mathrm{d}^2 f(c; u) = 0$. The result then follows from Theorem 10. □

12. TAYLOR'S THEOREM FOR REAL-VALUED FUNCTIONS

Let ϕ be a real-valued function defined on a subset S of $\mathbb{R}^n$, and let c be an interior point of S. If ϕ is continuous at c, then

$$\phi(c + u) = \phi(c) + R(u), \qquad (1)$$

and the error $R(u)$ made in this approximation will tend to zero as $u \to 0$.

If we make the stronger assumption that ϕ is differentiable in a neighbourhood of c, we obtain, by the mean-value theorem,

$$\phi(c + u) = \phi(c) + \mathrm{d}\phi(c + \theta u; u) \qquad (2)$$

for some $\theta \in (0, 1)$. This provides an explicit and very useful expression for the error $R(u)$ in (1).

If ϕ is differentiable at c, we also have the first-order Taylor formula

$$\phi(c+u) = \phi(c) + d\phi(c; u) + r(u), \tag{3}$$

where $r(u)/\|u\|$ tends to zero as $u \to 0$. Naturally the question arises whether it is possible to obtain an explicit form for the error $r(u)$. The following result (known as *Taylor's theorem*) answers this question.

Theorem 12 (Taylor)

Let $\phi : S \to \mathbb{R}$ be a real-valued function defined and twice differentiable on an open set S in $\mathbb{R}^n$. Let c be a point of S, and u a point in $\mathbb{R}^n$ such that $c + tu \in S$ for all $t \in [0, 1]$. Then

$$\phi(c+u) = \phi(c) + d\phi(c; u) + \tfrac{1}{2} d^2\phi(c + \theta u; u) \tag{4}$$

for some $\theta \in (0, 1)$.

Proof. As in the proof of the mean-value theorem 5.10, we consider a real-valued function $\psi : [0, 1] \to \mathbb{R}$ defined by

$$\psi(t) = \phi(c + tu). \tag{5}$$

The hypothesis of the theorem implies that ψ is twice differentiable at each point in (0, 1) with

$$D\psi(t) = d\phi(c + tu; u) \tag{6}$$

and

$$D^2\psi(t) = d^2\phi(c + tu; u). \tag{7}$$

By the one-dimensional Taylor theorem we have

$$\psi(1) = \psi(0) + D\psi(0) + \tfrac{1}{2} D^2\psi(\theta) \tag{8}$$

for some $\theta \in (0, 1)$. Thus

$$\phi(c+u) = \phi(c) + d\phi(c; u) + \tfrac{1}{2} d^2\phi(c + \theta u; u). \tag{9}$$

□

13. HIGHER-ORDER DIFFERENTIALS

Higher-order differentials are defined recursively. Let $f : S \to \mathbb{R}^m$ be a function defined on a set S in $\mathbb{R}^n$, and let c be an interior point of S. If f is $n - 1$ times differentiable in some n-ball $B(c)$ and each of the $(n - 1)$th-order partial derivatives is differentiable at c, then we say that f is *n times differentiable at c*.

Now consider the function $g : B(c) \to \mathbb{R}^m$ defined by the equation

$$g(x) = d^{n-1}f(x; u). \tag{1}$$

Then we define the *nth-order differential of f at c* as

$$\mathrm{d}^n f(c; u) = \mathrm{d}g(c; u). \tag{2}$$

We note from this definition that if f has an nth-order differential at c, then f itself has all the differentials up to the $(n-1)$th inclusive, not only at c, but also in a neighbourhood of c.

Third- and higher-order differentials will play no role of significance in this book.

14. MATRIX FUNCTIONS

As in the previous chapter, the extension to matrix functions is straightforward. Consider a matrix function $F : S \to \mathbb{R}^{m\times p}$ defined on a set S in $\mathbb{R}^{n\times q}$. Corresponding to the matrix function F we define a vector function $f : \operatorname{vec} S \to \mathbb{R}^{mp}$ by

$$f(\operatorname{vec} X) = \operatorname{vec} F(X). \tag{1}$$

In section 5.15 we defined the Jacobian matrix of F at C as the $mp \times nq$ matrix

$$\mathsf{D}F(C) = \mathsf{D}f(\operatorname{vec} C). \tag{2}$$

We now define the *Hessian matrix of F at C* as

$$\mathsf{H}F(C) = \mathsf{H}f(\operatorname{vec} C). \tag{3}$$

This is an $mnpq \times nq$ matrix stacking the Hessian matrices of the mp component functions F_{st} as follows:

$$\mathsf{H}F(C) = \begin{bmatrix} \mathsf{H}F_{11}(C) \\ \vdots \\ \mathsf{H}F_{m1}(C) \\ \vdots \\ \mathsf{H}F_{1p}(C) \\ \vdots \\ \mathsf{H}F_{mp}(C) \end{bmatrix}. \tag{4}$$

The matrices $\mathsf{H}F_{st}(C)$ are $nq \times nq$, and the ij-th element of $\mathsf{H}F_{st}(C)$ is the second-order partial derivative of $F_{st}(X)$ with respect to the elements of vec X, evaluated at $X = C$. That is, $(\mathsf{H}F_{st}(C))_{ij} = \mathsf{D}^2_{ji}F_{st}(C)$.

The second differential of F is the differential of the first differential:

$$\mathrm{d}^2F = \mathrm{d}(\mathrm{d}F). \tag{5}$$

More precisely, if we let

$$G(X) = \mathrm{d}F(X; U) \tag{6}$$

for all X in some ball $B(C)$, then

$$\mathrm{d}^2F(C; U) = \mathrm{d}G(C; U). \tag{7}$$

Since the differentials of F and f are related by

$$\operatorname{vec} \mathrm{d}F(C; U) = \mathrm{d}f(\operatorname{vec} C; \operatorname{vec} U), \tag{8}$$

the second differentials are related by

$$\operatorname{vec} \mathrm{d}^2F(C; U) = \mathrm{d}^2f(\operatorname{vec} C; \operatorname{vec} U). \tag{9}$$

The following two theorems are now straightforward generalizations of Theorems 7 and 10.

Theorem 13 (second identification theorem for matrix functions)

Let $F : S \to \mathbb{R}^{m \times p}$ be a matrix function defined on a set S in $\mathbb{R}^{n \times q}$, and twice differentiable at an interior point C of S. Then

$$\operatorname{vec} \mathrm{d}^2F(C; U) = (I_{mp} \otimes \operatorname{vec} U)' B(C) \operatorname{vec} U \tag{10}$$

for all $U \in \mathbb{R}^{n \times q}$ if and only if

$$\mathsf{H}F(C) = \tfrac{1}{2}[B(C) + (B(C)')_v]. \tag{11}$$

Note. Recall the definition of $(B(C)')_v$ from section 8.

Theorem 14

If F is twice differentiable at C and G is twice differentiable at $B = F(C)$, then the second differential of the composite function $H = G \circ F$ is

$$\mathrm{d}^2H(C; U) = \mathrm{d}^2G(B; \mathrm{d}F(C; U)) + \mathrm{d}G(B; \mathrm{d}^2F(C; U)) \tag{12}$$

for every U in $\mathbb{R}^{n \times q}$.

BIBLIOGRAPHICAL NOTES

§9. The fact that, for $n = 2$, the requirement that f is twice differentiable at c can be replaced by the weaker condition that all first-order partial derivatives are differentiable at c, is proved in Young (1910, sec. 23).

CHAPTER 7

Static optimization

1. INTRODUCTION

Static optimization theory is concerned with finding those points (if any) at which a real-valued function ϕ, defined on a subset S of $\mathbb{R}^n$, has a minimum or a maximum.

Two types of problems will be investigated in this chapter:

(i) *Unconstrained optimization* (sections 2–10) is concerned with the problem

$$\min_{x \in S}(\max)\ \phi(x), \tag{1}$$

where the point at which the extremum occurs is an *interior* point of S.

(ii) *Optimization subject to constraints* (sections 11–16) is concerned with the problem of optimizing ϕ subject to m nonlinear equality constraints, say $g_1(x)=0, \ldots, g_m(x)=0$. Letting $g=(g_1, g_2, \ldots, g_m)'$ and

$$\Gamma=\{x: x \in S,\ g(x)=0\}, \tag{2}$$

the problem can be written as

$$\min_{x \in \Gamma}(\max)\ \phi(x), \tag{3}$$

or, equivalently, as

$$\min_{x \in S}(\max)\quad \phi(x) \tag{4}$$

$$\text{subject to}\quad g(x)=0. \tag{5}$$

We shall *not* deal with inequality constraints.

2. UNCONSTRAINED OPTIMIZATION

In sections 2–10 we wish to show how the one-dimensional theory of maxima and minima of differentiable functions generalizes to functions of more than one variable. We start with some definitions.

Let $\phi : S \to \mathbb{R}$ be a real-valued function defined on a set S in $\mathbb{R}^n$, and let c be a point of S. We say that ϕ has a *local minimum* at c if there exists an n-ball $B(c)$ such that

$$\phi(x) \geqslant \phi(c) \qquad \text{for all } x \in S \cap B(c). \tag{1}$$

ϕ has a *strict local minimum* at c if we can choose $B(c)$ such that

$$\phi(x) > \phi(c) \qquad \text{for all } x \in S \cap B(c),\ x \neq c. \tag{2}$$

ϕ has an *absolute minimum* at c if

$$\phi(x) \geqslant \phi(c) \qquad \text{for all } x \in S. \tag{3}$$

ϕ has a *strict absolute minimum* at c if

$$\phi(x) > \phi(c) \qquad \text{for all } x \in S,\ x \neq c. \tag{4}$$

The point c at which the minimum is attained is called a (strict) local minimum point for ϕ, or a (strict) absolute minimum point for ϕ on S, depending on the nature of the minimum.

If ϕ has a minimum at c, then the function $\psi \equiv -\phi$ has a maximum at c. Each maximization problem can thus be converted to a minimization problem (and vice versa). For this reason we lose no generality by treating minimization problems only.

If c is an interior point of S, and ϕ is differentiable at c, then we say that c is a *critical point* (stationary point) of ϕ if

$$\mathrm{d}\phi(c; u) = 0 \qquad \text{for all } u \text{ in } \mathbb{R}^n. \tag{5}$$

The function value $\phi(c)$ is then called the *critical value* of ϕ at c.

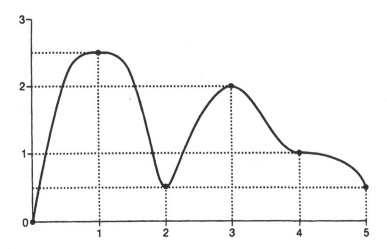

***Figure* 1** Unconstrained optimization in one variable

A critical point is called a *saddle point* if every n-ball $B(c)$ contains points x such that $\phi(x) > \phi(c)$ and other points such that $\phi(x) < \phi(c)$. In other words, a saddle point is a critical point which is neither a local minimum point nor a local maximum point.

Figure 1 illustrates some of these concepts. The function ϕ is defined and continuous on [0, 5]. It has a strict absolute minimum at $x=0$, and a (not strict) absolute maximum at $x=1$. There are strict local minima at $x=2$ and $x=5$, and a strict local maximum at $x=3$. At $x=4$ the derivative ϕ' is zero, but this is not an extremum point of ϕ; it is a saddle point.

3. THE EXISTENCE OF ABSOLUTE EXTREMA

In the example of Figure 1 the function ϕ is continuous on the compact interval [0, 5], and has an absolute minimum (at $x=0$) and an absolute maximum (at $x=1$). That this is typical for continuous functions on compact sets is shown by the following fundamental result.

Theorem 1 (Weierstrass)

Let $\phi : S \to \mathbb{R}$ be a real-valued function defined on a compact set S in $\mathbb{R}^n$. If ϕ is continuous on S, then ϕ attains its maximum and minimum values on S. Thus, there exist points c_1 and c_2 in S such that

$$\phi(c_1) \leq \phi(x) \leq \phi(c_2) \qquad \text{for all } x \in S. \tag{1}$$

Note. The Weierstrass theorem is an *existence theorem.* It tells us that certain conditions are sufficient to ensure the existence of extrema. The theorem does not tell us how to find these extrema.

Proof. By Theorem 4.9, ϕ is bounded on S. Hence the set

$$M = \{m \in \mathbb{R}; \phi(x) > m \text{ for all } x \in S\} \tag{2}$$

is not empty; indeed, M contains infinitely many points. Let

$$m_0 = \sup M. \tag{3}$$

Then,

$$\phi(x) \geq m_0 \qquad \text{for all } x \in S. \tag{4}$$

Now suppose that ϕ does not attain its infimum on S. Then

$$\phi(x) > m_0 \qquad \text{for all } x \in S \tag{5}$$

and the real-valued function $\psi : S \to \mathbb{R}$ defined by

$$\psi(x) = [\phi(x) - m_0]^{-1} \tag{6}$$

is continuous (and positive) on S. Again by Theorem 4.9, ψ is bounded on S, say by μ. Thus,

$$\psi(x) \leq \mu \qquad \text{for all } x \in S, \tag{7}$$

that is,

$$\phi(x) \geq m_0 + 1/\mu \qquad \text{for all } x \in S. \tag{8}$$

It follows that $m_0 + 1/(2\mu)$ is an element of M. But this is impossible, because no element of M can exceed m_0, the supremum of M. Hence, ϕ attains its minimum (and similarly its maximum) and the theorem is proved. □

Exercises

1. The Weierstrass theorem is not, in general, correct if we drop any of the conditions, as the following three counter-examples demonstrate.
   (a) $\phi(x) = x$, $x \in (-1, 1)$, $\phi(-1) = \phi(1) = 0$;
   (b) $\phi(x) = x$, $x \in (-\infty, \infty)$;
   (c) $\phi(x) = x/(1 - |x|)$, $x \in (-1, 1)$.
2. Consider the real-valued function $\phi : (0, \infty) \to \mathbb{R}$ defined by

$$\phi(x) = \begin{cases} x, & x \in (0, 2] \\ 1, & x \in (2, \infty). \end{cases}$$

   The set $(0, \infty)$ is neither bounded nor closed, and the function ϕ is not continuous on $(0, \infty)$. Nevertheless, ϕ attains its maximum on $(0, \infty)$. This shows that none of the conditions of the Weierstrass theorem is necessary.

4. NECESSARY CONDITIONS FOR A LOCAL MINIMUM

In the one-dimensional case, if a real-valued function ϕ, defined on an interval (a, b), has a local minimum at an interior point c of (a, b), and if ϕ has a derivative at c, then $\phi'(c)$ must be zero. This result, which relates zero derivatives and local extrema at interior points, can be generalized to the multivariable case as follows.

Theorem 2

Let $\phi : S \to \mathbb{R}$ be a real-valued function defined on a set S in $\mathbb{R}^n$, and assume that ϕ has a local minimum at an interior point c of S. If ϕ is differentiable at c, then

$$\mathrm{d}\phi(c; u) = 0 \tag{1}$$

for every u in $\mathbb{R}^n$. If ϕ is twice differentiable at c, then also

$$\mathrm{d}^2\phi(c; u) \geq 0 \tag{2}$$

for every u in $\mathbb{R}^n$.

Note 1. If ϕ has a local maximum (rather than a minimum) at c, then condition (2) is replaced by $\mathrm{d}^2\phi(c; u) \leq 0$ for every u in $\mathbb{R}^n$.

Note 2. The necessary conditions (1) and (2) are of course equivalent to the conditions

$$\frac{\partial \phi(c)}{\partial x_1} = \frac{\partial \phi(c)}{\partial x_2} = \ldots = \frac{\partial \phi(c)}{\partial x_n} = 0 \tag{3}$$

and

$$\mathsf{H}\phi(c) = \left(\frac{\partial^2 \phi(c)}{\partial x_i \partial x_j}\right) \quad \text{is positive semidefinite.} \tag{4}$$

Note 3. The example $\phi(x) = x^3$ shows (at $x = 0$) that the converse of Theorem 2 is not true. The example $\phi(x) = |x|$ shows (again at $x = 0$) that ϕ can have a local extremum without the derivative being zero.

Proof. Since ϕ has a local minimum at an interior point c, there exists an n-ball $B(c; \delta) \subset S$ such that

$$\phi(x) \geqslant \phi(c) \qquad \text{for all } x \in B(c; \delta). \tag{5}$$

Let u be a point in $\mathbb{R}^n$, $u \neq 0$ and choose $\varepsilon > 0$ such that $c + \varepsilon u \in B(c, \delta)$. From the definition of differentiability we have for every $|t| < \varepsilon$,

$$\phi(c + tu) = \phi(c) + t\mathrm{d}\phi(c; u) + r(t), \tag{6}$$

where $r(t)/t \to 0$ as $t \to 0$. Therefore

$$t\mathrm{d}\phi(c; u) + r(t) \geqslant 0. \tag{7}$$

Replacing t by $-t$ in (7), we obtain

$$-r(t) \leqslant t\mathrm{d}\phi(c; u) \leqslant r(-t). \tag{8}$$

Dividing by $t \neq 0$, and letting $t \to 0$, we find $\mathrm{d}\phi(c; u) = 0$ for all u in $\mathbb{R}^n$. This establishes the first part of the theorem.

To prove the second part, assume that ϕ is twice differentiable at c. Then, by the second-order Taylor formula (Theorem 6.8),

$$\phi(c + tu) = \phi(c) + t\mathrm{d}\phi(c; u) + \tfrac{1}{2}t^2\mathrm{d}^2\phi(c; u) + R(t), \tag{9}$$

where $R(t)/t^2 \to 0$ as $t \to 0$. Therefore

$$\tfrac{1}{2}t^2\mathrm{d}^2\phi(c; u) + R(t) \geqslant 0. \tag{10}$$

Dividing by $t^2 \neq 0$, and letting $t \to 0$, we find $\mathrm{d}^2\phi(c; u) \geqslant 0$ for all u in $\mathbb{R}^n$. □

Exercises

1. Find the extreme value(s) of the following real-valued functions defined on $\mathbb{R}^2$, and determine whether they are minima or maxima:
   (i) $\phi(x, y) = x^2 + xy + 2y^2 + 3$
   (ii) $\phi(x, y) = -x^2 + xy - y^2 + 2x + y$
   (iii) $\phi(x, y) = (x - y + 1)^2$.

2. Answer the same questions as above for the following real-valued functions defined for $0 \leqslant x \leqslant 2, 0 \leqslant y \leqslant 1$:
   (i) $\phi(x, y) = x^3 + 8y^3 - 9xy + 1$
   (ii) $\phi(x, y) = (x-2)(y-1) \exp(x^2 + \frac{1}{4}y^2 - x - y + 1)$.

5. SUFFICIENT CONDITIONS FOR A LOCAL MINIMUM: FIRST-DERIVATIVE TEST

In the one-dimensional case, a sufficient condition for a differentiable function ϕ to have a minimum at an interior point c is that $\phi'(c) = 0$ and that there exists an interval (a, b) containing c such that $\phi'(x) < 0$ in (a, c) and $\phi'(x) > 0$ in (c, b). (These conditions are *not* necessary, see exercise 1.)

The multivariable generalization is as follows.

Theorem 3 (the first-derivative test)

Let $\phi : S \to \mathbb{R}$ be a real-valued function defined on a set S in $\mathbb{R}^n$, and let c be an interior point of S. If ϕ is differentiable in some n-ball $B(c)$, and

$$\mathrm{d}\phi(x; x-c) \geqslant 0 \tag{1}$$

for every x in $B(c)$, then ϕ has a local minimum at c. Moreover, if the inequality in (1) is strict for every x in $B(c)$, $x \neq c$, then ϕ has a *strict* local minimum at c.

Proof. Let $u \neq 0$ be a point in $\mathbb{R}^n$ such that $c + u \in B(c)$. Then, by the mean-value theorem for real-valued functions,

$$\phi(c+u) = \phi(c) + \mathrm{d}\phi(c + \theta u; u) \tag{2}$$

for some $\theta \in (0, 1)$. Hence

$$\begin{aligned} \theta(\phi(c+u) - \phi(c)) &= \theta \mathrm{d}\phi(c + \theta u; u) \\ &= \mathrm{d}\phi(c + \theta u; \theta u) \\ &= \mathrm{d}\phi(c + \theta u; c + \theta u - c) \geqslant 0. \end{aligned} \tag{3}$$

Since $\theta > 0$, it follows that $\phi(c+u) \geqslant \phi(c)$. This proves the first part of the theorem; the second part is proved in the same way. □

Example

Let A be a positive definite (hence symmetric) $n \times n$ matrix, and let $\phi : \mathbb{R}^n \to \mathbb{R}$ be defined by $\phi(x) = x'Ax$. We find

$$\mathrm{d}\phi(x; u) = 2x'Au, \tag{4}$$

and since A is non-singular, the only critical point is the origin $x = 0$. To prove that this is a local minimum point, we compute

$$\mathrm{d}\phi(x; x-0) = 2x'Ax > 0 \tag{5}$$

for all $x \neq 0$. Hence ϕ has a strict local minimum at $x=0$. (In fact, ϕ has a strict absolute minimum at $x=0$.) In this example the function ϕ is strictly convex on $\mathbb{R}^n$, so that the condition of Theorem 3 is automatically fulfilled. We shall explore this in more detail in section 7.

Exercises

1. Consider the function $\phi(x)=x^2[2+\sin(1/x)]$ when $x \neq 0$, and $\phi(0)=0$. The function ϕ clearly has an absolute minimum at $x=0$. Show that the derivative is $\phi'(x)=4x+2x\sin(1/x)-\cos(1/x)$ when $x \neq 0$, and $\phi'(0)=0$. Show further that we can find values of x arbitrarily close to the origin such that $x\phi'(x)<0$. Conclude that the converse of Theorem 3 is, in general, not true.
2. Consider the function $\phi : \mathbb{R}^2 \to \mathbb{R}$ given by $\phi(x, y) = x^2 + (1+x)^3y^2$. Prove that it has one local minimum (at the origin), no other critical points and no absolute minimum.

6. SUFFICIENT CONDITIONS FOR A LOCAL MINIMUM: SECOND-DERIVATIVE TEST

Another test for local extrema is based on the Hessian matrix.

Theorem 4 (the second-derivative test)

Let $\phi : S \to \mathbb{R}$ be a real-valued function defined on a set S in $\mathbb{R}^n$. Assume that ϕ is twice differentiable at an interior point c of S. If

$$\mathrm{d}\phi(c; u) = 0 \qquad \text{for all } u \text{ in } \mathbb{R}^n \tag{1}$$

and

$$\mathrm{d}^2\phi(c; u) > 0 \qquad \text{for all } u \neq 0 \text{ in } \mathbb{R}^n, \tag{2}$$

then ϕ has a strict local minimum at c.

Proof. Since ϕ is twice differentiable at c, we have the second-order Taylor formula (Theorem 6.8)

$$\phi(c+u)=\phi(c)+\mathrm{d}\phi(c; u)+\tfrac{1}{2}\mathrm{d}^2\phi(c; u)+r(u), \tag{3}$$

where $r(u)/\|u\|^2 \to 0$ as $u \to 0$. Now, $\mathrm{d}\phi(c; u)=0$. Further, since the Hessian matrix $\mathsf{H}\phi(c)$ is positive definite by assumption, all its eigenvalues are positive (Theorem 1.8). In particular, if λ denotes the smallest eigenvalue of $\mathsf{H}\phi(c)$, then $\lambda > 0$ and (by exercise 1.14.1)

$$\mathrm{d}^2\phi(c; u)=u'(\mathsf{H}\phi(c))u \geqslant \lambda \|u\|^2. \tag{4}$$

It follows that, for $u \neq 0$,

$$\frac{\phi(c+u)-\phi(c)}{\|u\|^2} \geqslant \frac{\lambda}{2}+\frac{r(u)}{\|u\|^2}. \tag{5}$$

Choose $\delta>0$ such that $|r(u)|/\|u\|^2\leqslant\lambda/4$ for every $u\neq 0$ with $\|u\|<\delta$. Then

$$\phi(c+u)-\phi(c)\geqslant(\lambda/4)\|u\|^2>0 \tag{6}$$

for every $u\neq 0$ with $\|u\|<\delta$. Hence ϕ has a strict local minimum at c. □

In other words, Theorem 4 tells us that the conditions

$$\frac{\partial\phi(c)}{\partial x_1}=\frac{\partial\phi(c)}{\partial x_2}=\ldots=\frac{\partial\phi(c)}{\partial x_n}=0 \tag{7}$$

and

$$\mathsf{H}\phi(c)=\left(\frac{\partial^2\phi(c)}{\partial x_i\partial x_j}\right)\quad\text{is positive definite} \tag{8}$$

together are *sufficient* for ϕ to have a strict local minimum at c. If we replace (8) by the condition that $\mathsf{H}\phi(c)$ is negative definite, then we obtain sufficient conditions for a strict local *maximum*.

If the Hessian matrix $\mathsf{H}\phi(c)$ is neither positive definite nor negative definite, but is *non-singular*, then c cannot be a local extremum point (see Theorem 2); thus c is a *saddle point*.

In the case where $\mathsf{H}\phi(c)$ is singular, we cannot tell whether c is a maximum point, a minimum point, or a saddle point (see exercise 3). This shows that the converse of Theorem 4 is not true.

Example

Let $\phi:\mathbb{R}^2\to\mathbb{R}$ be twice differentiable at a critical point c in $\mathbb{R}^2$ of ϕ. Denote the second-order partial derivatives by $\mathsf{D}_{11}\phi(c)$, $\mathsf{D}_{12}\phi(c)$ and $\mathsf{D}_{22}\phi(c)$, and let Δ be the determinant of the Hessian matrix, i.e. $\Delta=\mathsf{D}_{11}\phi(c)\cdot\mathsf{D}_{22}\phi(c)-(\mathsf{D}_{12}\phi(c))^2$. Then Theorem 4 implies that

(i) if $\Delta>0$ and $\mathsf{D}_{11}\phi(c)>0$, ϕ has a strict local minimum at c;
(ii) if $\Delta>0$ and $\mathsf{D}_{11}\phi(c)<0$, ϕ has a strict local maximum at c;
(iii) if $\Delta<0$, ϕ has a saddle point at c;
(iv) if $\Delta=0$, ϕ may have a local minimum, maximum, or saddle point at c.

Exercises

1. Show that the function $\phi:\mathbb{R}^2\to\mathbb{R}$ defined by $\phi(x,y)=x^4+y^4-2(x-y)^2$ has strict local minima at $(\sqrt{2},-\sqrt{2})$ and $(-\sqrt{2},\sqrt{2})$, and a saddle point at (0,0).
2. Show that the function $\phi:\mathbb{R}^2\to\mathbb{R}$ defined by $\phi(x,y)=(y-x^2)(y-2x^2)$ has a local minimum along each straight line through the origin, but that ϕ has no local minimum at the origin. In fact, the origin is a saddle point.
3. Consider the functions (i) $\phi(x,y)=x^4+y^4$, (ii) $\phi(x,y)=-x^4-y^4$ and (iii) $\phi(x,y)=x^3+y^3$. For each of these functions show that the origin is a critical point and that the Hessian matrix is singular at the origin. Then prove that the origin is a minimum point, a maximum point and a saddle point, respectively.

4. Show that the function $\phi : \mathbb{R}^3 \to \mathbb{R}$ defined by $\phi(x, y, z) = xy + yz + zx$ has a saddle point at the origin, and no other critical points.
5. Consider the function $\phi : \mathbb{R}^2 \to \mathbb{R}$ defined by $\phi(x, y) = x^3 - 3xy^2 + y^4$. Find the critical points of ϕ and show that ϕ has two strict local minima and one saddle point.

7. CHARACTERIZATION OF DIFFERENTIABLE CONVEX FUNCTIONS

So far we have dealt only with local extrema. However, in the optimization problems that arise in economics (among other disciplines) we are usually interested in finding absolute extrema. The importance of convex (and concave) functions in optimization comes from the fact that every local minimum (maximum) of such a function is an absolute minimum (maximum). Before we prove this statement (Theorem 8), let us study convex (concave) functions in some more detail.

Recall that a *set* S in $\mathbb{R}^n$ is convex if, for all x, y in S and all $\lambda \in (0, 1)$,

$$\lambda x + (1 - \lambda) y \in S, \tag{1}$$

and a real-valued *function* ϕ, defined on a convex set S in $\mathbb{R}^n$, is convex if for all $x, y \in S$ and all $\lambda \in (0, 1)$,

$$\phi(\lambda x + (1 - \lambda) y) \leqslant \lambda \phi(x) + (1 - \lambda) \phi(y). \tag{2}$$

If (2) is satisfied with strict inequality for $x \neq y$, then we call ϕ strictly convex. If ϕ is (strictly) convex, then $\psi \equiv -\phi$ is (strictly) concave.

In this section we consider (strictly) convex functions that are differentiable, but not necessarily twice differentiable. In the next section we consider twice differentiable convex functions.

We first show that ϕ is convex if, and only if, at any point the tangent hyperplane is below the graph of ϕ (or coincides with it).

Theorem 5

Let $\phi : S \to \mathbb{R}$ be a real-valued function, defined and differentiable on an open convex set S in $\mathbb{R}^n$. Then ϕ is convex on S if and only if

$$\phi(x) \geqslant \phi(y) + \mathrm{d}\phi(y; x - y) \qquad \text{for every } x, y \in S. \tag{3}$$

Furthermore, ϕ is strictly convex on S if and only if the inequality in (3) is strict for every $x \neq y \in S$.

Proof. Assume that ϕ is convex on S. Let x be a point of S, and let u be a point in $\mathbb{R}^n$ such that $x + u \in S$. Then the point $x + tu$, $t \in (0,1)$, lies on the line segment joining

x and $x+u$. Since ϕ is differentiable at x, we have

$$\phi(x+tu)=\phi(x)+\mathrm{d}\phi(x;tu)+r(t), \tag{4}$$

where $r(t)/t \to 0$ as $t \to 0$. Also, since ϕ is convex on S, we have

$$\begin{aligned}\phi(x+tu)&=\phi((1-t)x+t(x+u))\leq(1-t)\,\phi(x)+t\phi(x+u)\\ &=\phi(x)+t[\phi(x+u)-\phi(x)].\end{aligned} \tag{5}$$

Combining (4) and (5) and dividing by t, we obtain

$$\phi(x+u)\geq\phi(x)+\mathrm{d}\phi(x;u)+r(t)/t. \tag{6}$$

Let $t \to 0$ and (3) follows.

To prove the converse, assume that (3) holds. Let x and y be two points in S, and let z be a point on the line segment joining x and y, that is, $z=tx+(1-t)y$ for some $t\in[0, 1]$. Using our assumption (3), we have

$$\phi(x)-\phi(z)\geq\mathrm{d}\phi(z;x-z), \qquad \phi(y)-\phi(z)\geq\mathrm{d}\phi(z;y-z). \tag{7}$$

Multiply the first inequality in (7) with t and the second with $(1-t)$, and add the resulting inequalities. This gives

$$t[\phi(x)-\phi(z)]+(1-t)[\phi(y)-\phi(z)]\geq\mathrm{d}\phi(z;t(x-z)+(1-t)(y-z))=0, \tag{8}$$

because

$$t(x-z)+(1-t)(y-z)=tx+(1-t)y-z=0.$$

By rearranging, (8) simplifies to

$$\phi(z)\leq t\phi(x)+(1-t)\phi(y), \tag{9}$$

which shows that ϕ is convex.

Next assume that ϕ is strictly convex. Let x be a point of S, and let u be a point in $\mathbb{R}^n$ such that $x+u\in S$. Since ϕ is strictly convex on S, ϕ is convex on S. Thus,

$$\phi(x+tu)\geq\phi(x)+t\mathrm{d}\phi(x;u) \tag{10}$$

for every $t\in(0, 1)$. Also, using the definition of strict convexity,

$$\phi(x+tu)<\phi(x)+t[\phi(x+u)-\phi(x)]. \tag{11}$$

(This is (5) with strict inequality.) Combining (10) and (11) and dividing by t, we obtain

$$\mathrm{d}\phi(x;u)\leq\frac{\phi(x+tu)-\phi(x)}{t}<\phi(x+u)-\phi(x), \tag{12}$$

and the strict version of inequality (3) follows.

Finally, the proof that the strict inequality (3) implies that ϕ is strictly convex is the same as the proof that (3) implies that ϕ is convex, all inequalities now being strict. □

Another characterization of differentiable functions exploits the fact that, in the one-dimensional case, the first derivative of a convex function is monotonically non-decreasing. The generalization of this property to the multivariable case is contained in Theorem 6.

Theorem 6

Let $\phi : S \rightarrow \mathbb{R}$ be a real-valued function, defined and differentiable on an open convex set S in $\mathbb{R}^n$. Then ϕ is convex on S if and only if

$$\mathrm{d}\phi(x; x-y) - \mathrm{d}\phi(y; x-y) \geqslant 0 \qquad \text{for every } x, y \in S. \tag{13}$$

Furthermore, ϕ is strictly convex on S if and only if the inequality in (13) is strict for every $x \neq y \in S$.

Proof. Assume that ϕ is convex on S. Let x and y be two points in S. Then, using Theorem 5,

$$\begin{aligned} \mathrm{d}\phi(x; x-y) = -\mathrm{d}\phi(x; y-x) &\geqslant \phi(x) - \phi(y) \\ &\geqslant \mathrm{d}\phi(y; x-y). \end{aligned} \tag{14}$$

To prove the converse, assume that (13) holds. Let x and y be two distinct points in S. Let $L(x, y)$ denote the line segment joining x and y, that is,

$$L(x, y) = \{tx + (1-t)y : 0 \leqslant t \leqslant 1\}, \tag{15}$$

and let z be a point in $L(x, y)$. By the mean-value theorem there exists a point $\xi = \alpha x + (1-\alpha)z$, $0 < \alpha < 1$, on the line segment joining x and z (hence in $L(x, y)$), such that

$$\phi(x) - \phi(z) = \mathrm{d}\phi(\xi; x-z). \tag{16}$$

Noting that $\xi - z = \alpha(x-z)$ and assuming (13), we have

$$\begin{aligned} \mathrm{d}\phi(\xi; x-z) &= (1/\alpha)\mathrm{d}\phi(\xi; \xi-z) \\ &\geqslant (1/\alpha)\mathrm{d}\phi(z; \xi-z) = \mathrm{d}\phi(z; x-z). \end{aligned} \tag{17}$$

Further, if $z = tx + (1-t)y$, then $x - z = (1-t)(x-y)$. It follows that

$$\phi(x) - \phi(z) \geqslant (1-t)\mathrm{d}\phi(z; x-y). \tag{18}$$

In precisely the same way we can show that

$$\phi(z) - \phi(y) \leqslant t\mathrm{d}\phi(z; x-y). \tag{19}$$

From (18) and (19) we obtain

$$t[\phi(x) - \phi(z)] - (1-t)[\phi(z) - \phi(y)] \geqslant 0. \tag{20}$$

By rearranging, (20) simplifies to

$$\phi(z) \leqslant t\phi(x) + (1-t)\phi(y), \tag{21}$$

which shows that ϕ is convex.

The corresponding result for ϕ strictly convex is obtained in precisely the same way, all inequalities now being strict. □

Exercises

1. Show that the function $\phi(x, y) = x + y(y-1)$ is convex. Is ϕ strictly convex?
2. Prove that $\phi(x) = x^4$ is strictly convex.

8. CHARACTERIZATION OF TWICE DIFFERENTIABLE CONVEX FUNCTIONS

Both characterizations of differentiable convex functions (Theorems 5 and 6) involved conditions on *two* points. For twice differentiable functions there is a characterization that involves only *one* point.

Theorem 7

Let $\phi : S \to \mathbb{R}$ be a real-valued function, defined and twice differentiable on an open convex set S in $\mathbb{R}^n$. Then ϕ is convex on S if and only if

$$d^2\phi(x; u) \geq 0 \qquad \text{for all } x \in S \text{ and } u \in \mathbb{R}^n. \tag{1}$$

Furthermore, if the inequality in (1) is strict for all $x \in S$ and $u \neq 0$ in $\mathbb{R}^n$, then ϕ is strictly convex on S.

Note 1. The 'strict' part of Theorem 7 is a one-way implication, and not an equivalence, i.e. if ϕ is twice differentiable and strictly convex, then by (1) the Hessian matrix $H\phi(x)$ is positive semidefinite, but not necessarily positive definite for every x. For example, the function $\phi(x) = x^4$ is strictly convex but its second derivative $\phi''(x) = 12x^2$ vanishes at $x = 0$.

Note 2. Theorem 7 tells us that ϕ is convex (strictly convex, concave, strictly concave) on S if the Hessian matrix $H\phi(x)$ is positive semidefinite (positive definite, negative semidefinite, negative definite) for all x in S.

Proof. Let c be a point of S, and let $u \neq 0$ be a point in $\mathbb{R}^n$ such that $c + u \in S$. By Taylor's theorem, we have

$$\phi(c+u) = \phi(c) + d\phi(c; u) + \tfrac{1}{2}d^2\phi(c + \theta u; u) \tag{2}$$

for some $\theta \in (0, 1)$. If $d^2\phi(x; u) \geq 0$ for every $x \in S$, then in particular $d^2\phi(c + \theta u; u) \geq 0$, so that

$$\phi(c+u) \geq \phi(c) + d\phi(c; u). \tag{3}$$

Then, by Theorem 5, ϕ is convex on S.

If $d^2\phi(x; u)>0$ for every $x\in S$, then

$$\phi(c+u)>\phi(c)+d\phi(c; u), \tag{4}$$

which shows, by Theorem 5, that ϕ is strictly convex on S.

To prove the 'only if' part of (1), assume that ϕ is convex on S. Let $t\in(0, 1)$. Then, by Theorem 5,

$$\phi(c+tu)\geq\phi(c)+td\phi(c; u). \tag{5}$$

Also, by the second-order Taylor formula (Theorem 6.8),

$$\phi(c+tu)=\phi(c)+td\phi(c; u)+\tfrac{1}{2}t^2d^2\phi(c; u)+r(t), \tag{6}$$

where $r(t)/t^2\to 0$ as $t\to 0$. Combining (5) and (6) and dividing by t^2 we obtain

$$\tfrac{1}{2}d^2\phi(c; u)\geq -r(t)/t^2. \tag{7}$$

The left side of (7) is independent of t; the right side tends to zero as $t\to 0$. Hence $d^2\phi(c; u)\geq 0$. □

Exercises

1. Repeat exercise 4.9.1 using Theorem 7.
2. Show that the function $\phi(x)=x^p$, $p>1$, is strictly convex on $[0, \infty)$.
3. Show that the function $\phi(x) = x'x$, defined on $\mathbb{R}^n$, is strictly convex.
4. Consider the CES (constant elasticity of substitution) production function

$$\phi(x, y)=A[\delta x^{-\rho}+(1-\delta)y^{-\rho}]^{-1/\rho} \qquad (A>0, 0\leq\delta\leq 1, \rho\neq 0)$$

defined for $x>0$ and $y>0$. Show that ϕ is convex if $\rho\leq -1$, and concave if $\rho\geq -1$ (and $\rho\neq 0$). What happens if $\rho=-1$?

9. SUFFICIENT CONDITIONS FOR AN ABSOLUTE MINIMUM

The convexity (concavity) of a function enables us to find the absolute minimum (maximum) of the function, since every local minimum (maximum) of such a function is an absolute minimum (maximum).

Theorem 8

Let $\phi : S \to \mathbb{R}$ be a real-valued function defined and differentiable on an open convex set S in $\mathbb{R}^n$, and let c be a point of S where

$$d\phi(c; u)=0 \tag{1}$$

for every $u \in \mathbb{R}^n$. If ϕ is (strictly) convex on S, then ϕ has a (strict) absolute minimum at c.

Proof. If ϕ is convex on S, then, by Theorem 5,

$$\phi(x) \geqslant \phi(c) + d\phi(c; x-c) = \phi(c) \tag{2}$$

for all x in S. If ϕ is strictly convex on S, then the inequality (2) is strict for all $x \neq c$ in S. □

To check whether a given differentiable function is (strictly) convex, we have four criteria at our disposal: the definition in section 4.9, Theorems 5 and 6, and, if the function is twice differentiable, Theorem 7.

Exercises

1. Let a be an $n \times 1$ vector and A a positive definite $n \times n$ matrix. Prove that
$$a'x + x'Ax \geqslant -\tfrac{1}{4}a'A^{-1}a$$
for every x in $\mathbb{R}^n$. For which value of x does the function $\phi(x) = a'x + x'Ax$ attain its minimum value?
2. (More difficult.) If A is positive semidefinite, under what condition is it true that
$$a'x + x'Ax \geqslant -\tfrac{1}{4}a'A^{+}a$$
for every x in $\mathbb{R}^n$?

10. MONOTONIC TRANSFORMATIONS

To complete our discussion of unconstrained optimization we shall prove the useful, if simple, fact that minimizing a function is equivalent to minimizing a monotonically increasing transformation of that function.

Theorem 9

Let S be a subset of $\mathbb{R}^n$, and let $\phi : S \to \mathbb{R}$ be a real-valued function defined on S. Let $T \subset \mathbb{R}$ be the range of ϕ (the set of all elements $\phi(x)$, for $x \in S$), and let $\eta : T \to \mathbb{R}$ be a real-valued function defined on T. Define the composite function $\psi : S \to \mathbb{R}$ by

$$\psi(x) = \eta(\phi(x)). \tag{1}$$

If η is increasing on T, and if ϕ has an absolute minimum (maximum) at a point c of S, then ψ has an absolute minimum (maximum) at c.

If η in addition is *strictly* increasing on T, then ϕ has an absolute minimum (maximum) at c if and only if ψ has an absolute minimum (maximum) at c.

Proof. Let η be an increasing function on T, and suppose that $\phi(x) \geqslant \phi(c)$ for all x in S. Then

$$\psi(x) = \eta(\phi(x)) \geqslant \eta(\phi(c)) = \psi(c) \tag{2}$$

for all x in S. Next, let η be strictly increasing on T, and suppose that $\phi(x_0) < \phi(c)$ for some x_0 in S. Then

$$\psi(x_0) = \eta(\phi(x_0)) < \eta(\phi(c)) = \psi(c). \tag{3}$$

Hence, if $\psi(x) \geqslant \psi(c)$ for all x in S, then $\phi(x) \geqslant \phi(c)$ for all x in S.

The case where ϕ has an absolute maximum is proved in the same way.

□

Note. Theorem 9 is clearly not affected by the presence of constraints. Thus, minimizing a function subject to certain constraints is equivalent to minimizing a monotonically increasing transformation of that function subject to the same constraints.

Exercise

1. Consider the likelihood function

$$L(\mu, \sigma^2) = (2\pi\sigma^2)^{-n/2} \exp\left(-\tfrac{1}{2}\sum_{i=1}^{n} (x_i - \mu)^2/\sigma^2\right).$$

Use Theorem 9 to maximize L with respect to μ and σ^2.

11. OPTIMIZATION SUBJECT TO CONSTRAINTS

Let $\phi : S \to \mathbb{R}$ be a real-valued function defined on a set S in $\mathbb{R}^n$. Hitherto we have considered optimization problems of the type

$$\underset{x \in S}{\text{minimize}}\ \phi(x). \tag{1}$$

It may happen, however, that the variables $x_1, \ldots, x_n$ are subject to certain constraints, say $g_1(x) = 0, \ldots, g_m(x) = 0$. Our problem is now

$$\text{minimize} \quad \phi(x) \tag{2}$$

$$\text{subject to} \quad g(x) = 0, \tag{3}$$

where $g : S \to \mathbb{R}^m$ is the vector function $g = (g_1, g_2, \ldots, g_m)'$. This is known as a *constrained minimization problem* (or a minimization problem subject to equality constraints), and the most convenient way of solving it is, in general, to use the *Lagrange multiplier theory*. In the remainder of this chapter we shall study that important theory in some detail.

We start our discussion with some definitions. The subset of S on which g vanishes, that is,

$$\Gamma = \{x : x \in S, g(x) = 0\}, \tag{4}$$

is known as the *opportunity set* (constraint set). Let c be a point of Γ. We say that ϕ

has a *local minimum* at c *under the constraint* $g(x)=0$ if there exists an n-ball $B(c)$ such that

$$\phi(x) \geqslant \phi(c) \qquad \text{for all } x \in \Gamma \cap B(c). \tag{5}$$

ϕ has a *strict local minimum* at c *under the constraint* $g(x)=0$ if we can choose $B(c)$ such that

$$\phi(x) > \phi(c) \qquad \text{for all } x \in \Gamma \cap B(c), x \neq c. \tag{6}$$

ϕ has an *absolute minimum* at c *under the constraint* $g(x)=0$ if

$$\phi(x) \geqslant \phi(c) \qquad \text{for all } x \in \Gamma. \tag{7}$$

ϕ has a *strict absolute minimum* at c *under the constraint* $g(x)=0$ if

$$\phi(x) > \phi(c) \qquad \text{for all } x \in \Gamma,\ x \neq c. \tag{8}$$

12. NECESSARY CONDITIONS FOR A LOCAL MINIMUM UNDER CONSTRAINTS

The next theorem gives a necessary condition for a constrained minimum to occur at a given point.

Theorem 10 (Lagrange)

Let $g : S \to \mathbb{R}^m$ be a vector function defined on a set S in $\mathbb{R}^n$ $(n > m)$, and let c be an interior point of S. Assume that

(i) $g(c)=0$,
(ii) g is differentiable in some n-ball $B(c)$,
(iii) the $m \times n$ Jacobian matrix $\mathsf{D}g$ is continuous at c, and
(iv) $\mathsf{D}g(c)$ has full row rank m.

Further, let $\phi : S \to \mathbb{R}$ be a real-valued function defined on S, and assume that

(v) ϕ is differentiable at c, and
(vi) $\phi(x) \geqslant \phi(c)$ for every $x \in B(c)$ satisfying $g(x)=0$.

Then there exists a unique vector l in $\mathbb{R}^m$ satisfying the n equations

$$\mathsf{D}\phi(c) - l'\mathsf{D}g(c) = 0. \tag{1}$$

Note. If condition (vi) is replaced by

(vi)′ $\phi(x) \leqslant \phi(c)$ for every $x \in B(c)$ satisfying $g(x)=0$,

then the conclusion of the theorem remains valid.

Lagrange's theorem establishes the validity of the following formal method ('Lagrange's multiplier method') for obtaining *necessary* conditions for an extremum subject to equality constraints. We first define the *Lagrangian function*

ψ by

$$\psi(x)=\phi(x)-l'g(x), \tag{2}$$

where l is an $m \times 1$ vector of constants $\lambda_1, \ldots, \lambda_m$, called the *Lagrange multipliers.* (One multiplier is introduced for each constraint. Notice that $\psi(x)$ equals $\phi(x)$ for every x that satisfies the constraint.) Next we differentiate ψ with respect to x and set the result equal to 0. Together with the m constraints we get the following system of $n+m$ equations (the *first-order conditions*):

$$d\psi(x; u) = 0 \qquad \text{for every } u \in \mathbb{R}^n, \tag{3}$$

$$g(x)=0. \tag{4}$$

We then try to solve this system of $n+m$ equations in $n+m$ unknowns, $\lambda_1, \ldots, \lambda_m$ and $x_1, \ldots, x_n$. The points $x=(x_1, \ldots, x_n)'$ obtained in this way are called *critical points*, and among them are any points of S at which constrained minima or maxima occur. (A critical point of the constrained problem is thus defined as 'a critical point of the function $\phi(x)$ defined on the surface $g(x)=0$', and *not* as 'a critical point of $\phi(x)$ whose coordinates satisfy $g(x)=0$'. Any critical point in the latter sense is also a critical point in the former, but not conversely.)

Of course, the question remains whether a given critical point actually yields a minimum, maximum, or neither.

Proof. Let us partition the $m \times n$ matrix $\mathsf{D}g(c)$ as

$$\mathsf{D}g(c)=(\mathsf{D}_1 g(c),\ \mathsf{D}_2 g(c)), \tag{5}$$

where $\mathsf{D}_1 g(c)$ is an $m \times m$ matrix, and $\mathsf{D}_2 g(c)$ is an $m \times (n-m)$ matrix. By renumbering the variables (if necessary), we may assume that

$$|\mathsf{D}_1 g(c)| \neq 0. \tag{6}$$

We shall denote points x in S by $(z; t)$, where $z \in \mathbb{R}^m$ and $t \in \mathbb{R}^{n-m}$, so that $z=(x_1, \ldots, x_m)'$ and $t=(x_{m+1}, \ldots, x_n)'$. Also, we write $c=(z_0; t_0)$.

By the implicit function theorem (Theorem A.1 in the appendix to this chapter) there exists an open set T in $\mathbb{R}^{n-m}$ containing t_0, and a unique function $h : T \to \mathbb{R}^m$ such that

(i) $h(t_0)=z_0$,
(ii) $g(h(t); t)=0$ for all $t \in T$, and
(iii) h is differentiable at t_0.

Since h is continuous at t_0, we can choose an $(n-m)$-ball $T_0 \subset T$ with centre t_0 such that

$$(h(t); t) \in B(c) \qquad \text{for all } t \in T_0. \tag{7}$$

Then the real-valued function $\psi : T_0 \to \mathbb{R}$ defined by

$$\psi(t)=\phi(h(t); t) \tag{8}$$

has the property

$$\psi(t) \geqslant \psi(t_0) \qquad \text{for all } t \in T_0, \tag{9}$$

that is, ψ has a local (unconstrained) minimum at t_0. Since h is differentiable at t_0 and ϕ is differentiable at $(z_0;\, t_0)$, it follows that ψ is differentiable at t_0. Hence, by Theorem 2, its derivative vanishes at t_0, and, using the chain rule, we find

$$0 = \mathsf{D}\psi(t_0) = \mathsf{D}\phi(c)\begin{pmatrix} \mathsf{D}h(t_0) \\ I_{n-m} \end{pmatrix}. \tag{10}$$

Next, consider the vector function $\kappa : T \to \mathbb{R}^m$ defined by

$$\kappa(t) = g(h(t);\, t). \tag{11}$$

The function κ is identically zero on the set T. Therefore, all its partial derivatives are zero on T. In particular, $\mathsf{D}\kappa(t_0) = 0$. Further, since h is differentiable at t_0 and g is differentiable at $(z_0;\, t_0)$, the chain rule yields

$$0 = \mathsf{D}\kappa(t_0) = \mathsf{D}g(c)\begin{pmatrix} \mathsf{D}h(t_0) \\ I_{n-m} \end{pmatrix}. \tag{12}$$

Combining (10) and (12), we obtain

$$E\begin{pmatrix} \mathsf{D}h(t_0) \\ I_{n-m} \end{pmatrix} = 0, \tag{13}$$

where E is the $(m+1) \times n$ matrix

$$E = \begin{pmatrix} \mathsf{D}\phi(c) \\ \mathsf{D}g(c) \end{pmatrix} = \begin{pmatrix} \mathsf{D}_1\phi(c) & \mathsf{D}_2\phi(c) \\ \mathsf{D}_1 g(c) & \mathsf{D}_2 g(c) \end{pmatrix}. \tag{14}$$

Equation (13) shows that the last $n-m$ columns of E are linear combinations of the first m columns. Hence $r(E) \leqslant m$. But since $\mathsf{D}_1 g(c)$ is a submatrix of E with rank m, the rank of E cannot be smaller than m. It follows that

$$r(E) = m. \tag{15}$$

The $m+1$ rows of E are therefore linearly dependent. By assumption, the m rows of $\mathsf{D}g(c)$ are linearly independent. Hence $\mathsf{D}\phi(c)$ is a linear combination of the m rows of $\mathsf{D}g(c)$, that is,

$$\mathsf{D}\phi(c) - l'\mathsf{D}g(c) = 0 \tag{16}$$

for some $l \in \mathbb{R}^m$. This proves the existence of l; its uniqueness follows immediately from the fact that $\mathsf{D}g(c)$ has full row rank. □

Example

To solve the problem

$$\text{minimize} \qquad x'x \tag{17}$$

$$\text{subject to} \qquad x'Ax = 1 \qquad (A \text{ positive definite}) \tag{18}$$

by Lagrange's method, we introduce one multiplier λ and define the Lagrangian function

$$\psi(x) = x'x - \lambda(x'Ax - 1). \tag{19}$$

Differentiating ψ with respect to x and setting the result equal to zero yields

$$x = \lambda Ax. \tag{20}$$

To this we add the constraint

$$x'Ax = 1. \tag{21}$$

Equations (20) and (21) are the first-order conditions, from which we shall solve for x and λ. Pre-multiplying both sides of (20) by x' gives

$$x'x = \lambda x'Ax = \lambda, \tag{22}$$

using (21), and since $x \neq 0$ we obtain from (20)

$$Ax = (1/x'x)x. \tag{23}$$

This shows that $(1/x'x)$ is an eigenvalue of A. Let $\mu(A)$ be the largest eigenvalue of A. Then the minimum value of $x'x$ under the constraint $x'Ax = 1$, is $1/\mu(A)$. The value of x for which the minimum is attained is the eigenvector of A associated with the eigenvalue $\mu(A)$.

Exercises

1. Consider the problem

$$\begin{array}{ll} \text{minimize} & (x-1)(y+1) \\ \text{subject to} & x - y = 0. \end{array}$$

   By using Lagrange's method, show that the minimum point is $(0, 0)$ with $\lambda = 1$. Next consider the Lagrangian function

$$\psi(x, y) = (x-1)(y+1) - 1(x-y),$$

   and show that ψ has a saddle point at $(0, 0)$. That is, the point $(0, 0)$ does not minimize ψ. (This shows that it is not correct to say that minimizing a function subject to constraints is equivalent to minimizing the Lagrangian function.)
2. Solve the following problems by using the Lagrange multiplier method:
   (i) max(min) xy subject to $x^2 + xy + y^2 = 1$.
   (ii) max(min) $(y-z)(z-x)(x-y)$ subject to $x^2 + y^2 + z^2 = 2$.
   (iii) max(min) $x^2 + y^2 + z^2 - yz - zx - xy$ subject to $x^2 + y^2 + z^2 - 2x + 2y + 6z + 9 = 0$.
3. Prove the inequality

$$(x_1 x_2 \ldots x_n)^{1/n} \leqslant \frac{x_1 + x_2 + \ldots + x_n}{n}$$

   for all positive real numbers $x_1, \ldots, x_n$.

4. Solve the problem

$$\begin{aligned} &\text{minimize} && x^2+y^2+z^2 \\ &\text{subject to} && 4x+3y+z=25. \end{aligned}$$

5. Solve the following utility maximization problem:

$$\begin{aligned} &\text{maximize} && x_1^{\alpha}x_2^{1-\alpha} \qquad (0<\alpha<1) \\ &\text{subject to} && p_1x_1+p_2x_2=y \qquad (p_1>0, p_2>0, y>0) \end{aligned}$$

with respect to x_1 and x_2 ($x_1>0, x_2>0$).

13. SUFFICIENT CONDITIONS FOR A LOCAL MINIMUM UNDER CONSTRAINTS

In the previous section we obtained conditions that are necessary for a function to achieve a local minimum or maximum subject to equality constraints. To investigate whether a given critical point actually yields a minimum, maximum, or neither, it is often practicable to proceed on an *ad hoc* basis. If this fails, the following theorem provides sufficient conditions to ensure the existence of a constrained minimum or maximum at a critical point.

Theorem 11

Let $\phi: S \to \mathbb{R}$ be a real-valued function defined on a set S in $\mathbb{R}^n$, and $g: S \to \mathbb{R}^m$ $(m<n)$ a vector function defined on S. Let c be an interior point of S and let l be a point in $\mathbb{R}^m$. Define the Lagrangian function $\psi: S \to \mathbb{R}$ by the equation

$$\psi(x)=\phi(x)-l'g(x), \tag{1}$$

and assume that

(i) ϕ is twice differentiable at c,
(ii) g is twice differentiable at c,
(iii) the $m\times n$ Jacobian matrix $\mathrm{D}g(c)$ has full row rank m,
(iv) (first-order conditions)

$$\mathrm{d}\psi(c;u)=0 \qquad \text{for all } u \text{ in } \mathbb{R}^n, \tag{2}$$

$$g(c)=0, \tag{3}$$

(v) (second-order condition)

$$\mathrm{d}^2\psi(c;u)>0 \qquad \text{for all } u\neq 0 \text{ satisfying } \mathrm{d}g(c;u)=0. \tag{4}$$

Then ϕ has a strict local minimum at c under the constraint $g(x)=0$.

The difficulty in applying Theorem 11 lies, of course, in the verification of the second-order condition. This condition requires that

$$u'Au>0 \qquad \text{for every } u\neq 0 \text{ such that } Bu=0, \tag{5}$$

where

$$A = \mathsf{H}\phi(c) - \sum_{i=1}^{m} \lambda_i \mathsf{H}g_i(c), \qquad B = \mathsf{D}g(c). \tag{6}$$

Several sets of necessary and sufficient conditions exist for a quadratic form to be positive definite under linear constraints, and one of these (the 'bordered determinantal criterion') is discussed in section 3.11. The following theorem is therefore easily proved.

Theorem 12 (bordered determinantal criterion)

Assume that conditions (i)–(iv) of Theorem 11 are satisfied, and let Δ_r be the symmetric $(m+r)\times(m+r)$ matrix

$$\Delta_r = \begin{pmatrix} 0 & B_r \\ B_r' & A_{rr} \end{pmatrix} \qquad (r=1, \ldots, n), \tag{7}$$

where A_{rr} is the $r\times r$ matrix in the top left corner of A, and B_r is the $m\times r$ matrix whose columns are the first r columns of B. Assume that $|B_m| \neq 0$. (This can always be achieved by renumbering the variables, if necessary.) If

$$(-1)^m|\Delta_r| > 0 \qquad (r=m+1, \ldots, n), \tag{8}$$

then ϕ has a strict local minimum at c under the constraint $g(x)=0$. If

$$(-1)^r|\Delta_r| > 0 \qquad (r=m+1, \ldots, n), \tag{9}$$

then ϕ has a strict local maximum at c under the constraint $g(x)=0$.

Proof of Theorem 11. Let us define the sets

$$U(\delta) = \{u \in \mathbb{R}^n : \|u\| < \delta\}, \qquad \delta > 0 \tag{10}$$

and

$$T = \{u \in \mathbb{R}^n : u \neq 0, c+u \in S, g(c+u) = 0\}. \tag{11}$$

We need to show that a $\delta > 0$ exists such that

$$\phi(c+u) - \phi(c) > 0 \qquad \text{for all } u \in T \cap U(\delta). \tag{12}$$

By assumption, ϕ and g are twice differentiable at c, and therefore differentiable at each point of an n-ball $B(c) \subset S$. Let δ_0 be the radius of $B(c)$. Since ψ is twice differentiable at c, we have for every $u \in U(\delta_0)$ the second-order Taylor formula (Theorem 6.8)

$$\psi(c+u) = \psi(c) + \mathrm{d}\psi(c; u) + \tfrac{1}{2}\mathrm{d}^2\psi(c; u) + r(u), \tag{13}$$

where $r(u)/\|u\|^2 \to 0$ as $u \to 0$. Now, $g(c)=0$ and $\mathrm{d}\psi(c; u)=0$ (first-order conditions). Further, $g(c+u)=0$ for $u \in T$. Hence (13) reduces to

$$\phi(c+u) - \phi(c) = \tfrac{1}{2}\mathrm{d}^2\psi(c; u) + r(u) \qquad \text{for all } u \in T \cap U(\delta_0). \tag{14}$$

Next, since g is differentiable at each point of $B(c)$, we may apply the mean-value theorem to each of its components $g_1, \ldots, g_m$. This yields, for every $u \in U(\delta_0)$,

$$g_i(c+u)=g_i(c)+\mathrm{d}g_i(c+\theta_i u;\, u), \tag{15}$$

where $\theta_i \in (0, 1)$, $i=1, \ldots, m$. Again, $g_i(c)=0$ and, for $u \in T$, $g_i(c+u)=0$. Hence

$$\mathrm{d}g_i(c+\theta_i u;\, u)=0 \qquad (i=1, \ldots, m) \qquad \text{for all } u \in T \cap U(\delta_0). \tag{16}$$

Let us denote by $\Delta(u)$, $u \in U(\delta_0)$, the $m \times n$ matrix whose ij-th element is the j-th first-order partial derivative of g_i evaluated at $c+\theta_i u$, that is,

$$\Delta_{ij}(u)=\mathrm{D}_j g_i(c+\theta_i u) \qquad (i=1, \ldots, m;\, j=1, \ldots, n). \tag{17}$$

(Notice that the rows of Δ are evaluated at possibly different points.) Then the m equations (16) can be written as one vector equation

$$\Delta(u)u=0 \qquad \text{for all } u \in T \cap U(\delta_0). \tag{18}$$

Since the functions $\mathrm{D}_j g_i$ are continuous at $u=0$, the Jacobian matrix Δ is continuous at $u=0$. By assumption $\Delta(0)$ has maximum rank m, and therefore there exists a $\delta_1 \in (0, \delta_0]$ such that

$$\operatorname{rank}(\Delta(u))=m \qquad \text{for all } u \in U(\delta_1) \tag{19}$$

(see exercise 5.15.1). Now, $\Delta(u)$ has n columns of which only m are linearly independent. Hence by exercise 1.14.3 there exists an $n \times (n-m)$ matrix $\Gamma(u)$ such that

$$\Delta(u)\Gamma(u)=0, \qquad \Gamma'(u)\Gamma(u)=I_{n-m}, \qquad \text{for all } u \in U(\delta_1). \tag{20}$$

(The columns of Γ are of course n-m normalized eigenvectors associated with the $n-m$ zero eigenvalues of $\Delta'\Delta$.) Further, since Δ is continuous at $u=0$, so is Γ.

From (18)–(20) it follows that u must be a linear combination of the columns of $\Gamma(u)$, that is, there exists, for every u in $T \cap U(\delta_1)$, a vector $q \in \mathbb{R}^{n-m}$ such that

$$u=\Gamma(u)q. \tag{21}$$

If we denote by $K(u)$ the symmetric $(n-m) \times (n-m)$ matrix

$$K(u)=\Gamma'(u)(\mathrm{H}\psi(c))\Gamma(u), \qquad u \in U(\delta_1), \tag{22}$$

and by $\lambda(u)$ its smallest eigenvalue, then

$$\begin{aligned}\mathrm{d}^2\psi(c;\, u)&=u'(\mathrm{H}\psi(c))u=q'\Gamma'(u)(\mathrm{H}\psi(c))\Gamma(u)q\\ &=q'K(u)q \geqslant \lambda(u)q'q \text{ (exercise 1.14.1)}\\ &=\lambda(u)q'\Gamma'(u)\Gamma(u)q=\lambda(u)\|u\|^2\end{aligned} \tag{23}$$

for every u in $T \cap U(\delta_1)$. Now, since Γ is continuous at $u=0$, so is K and so is λ. Hence we may write, for u in $U(\delta_1)$,

$$\lambda(u)=\lambda(0)+R(u), \tag{24}$$

where $R(u)\to 0$ as $u\to 0$. Combining (14), (23) and (24), we obtain

$$\phi(c+u)-\phi(c)\geq[\tfrac{1}{2}\lambda(0)+\tfrac{1}{2}R(u)+r(u)/\|u\|^2]\,\|u\|^2 \tag{25}$$

for every u in $T\cap U(\delta_1)$.

Let us now prove that $\lambda(0)>0$. By assumption,

$$u'(\mathrm{H}\psi(c))u>0 \qquad \text{for all } u\neq 0 \text{ satisfying } \Delta(0)u=0. \tag{26}$$

For $u\in U(\delta_1)$, the condition $\Delta(0)u=0$ is equivalent to $u=\Gamma(0)q$ for some $q\in\mathbb{R}^{n-m}$. Hence (26) is equivalent to

$$q'\Gamma'(0)(\mathrm{H}\psi(c))\Gamma(0)q>0 \qquad \text{for all } q\neq 0. \tag{27}$$

This shows that $K(0)$ is positive definite, and hence that its smallest eigenvalue $\lambda(0)$ is positive.

Finally, choose $\delta_2\in(0,\delta_1]$ such that

$$|\tfrac{1}{2}R(u)+r(u)/\|u\|^2|\leq\lambda(0)/4 \tag{28}$$

for every $u\neq 0$ with $\|u\|<\delta_2$. Then (25) and (28) imply

$$\phi(c+u)-\phi(c)\geq[\lambda(0)/4]\,\|u\|^2>0 \tag{29}$$

for every u in $T\cap U(\delta_2)$. Hence ϕ has a strict local minimum at c under the constraint $g(x)=0$. □

Example ($n=2, m=1$)

Solve the problem

$$\begin{array}{ll} \text{max(min)} & x^2+y^2 \\ \text{subject to} & x^2+xy+y^2=3. \end{array}$$

Let λ be a constant, and define the Lagrangian function

$$\psi(x,y)=x^2+y^2-\lambda(x^2+xy+y^2-3).$$

The first-order conditions are

$$\begin{aligned} 2x-\lambda(2x+y)&=0\\ 2y-\lambda(x+2y)&=0\\ x^2+xy+y^2&=3, \end{aligned}$$

from which we find the following four solutions: $(1, 1)$ and $(-1, -1)$ with $\lambda=\frac{2}{3}$; and $(\sqrt{3}, -\sqrt{3})$ and $(-\sqrt{3}, \sqrt{3})$ with $\lambda=2$. We now compute the bordered Hessian matrix

$$\Delta(x,y)=\begin{bmatrix} 0 & 2x+y & x+2y \\ 2x+y & 2-2\lambda & -\lambda \\ x+2y & -\lambda & 2-2\lambda \end{bmatrix}$$

whose determinant equals

$$|\Delta(x, y)| = \tfrac{1}{2}(3\lambda - 2)(x - y)^2 - \tfrac{9}{2}(2 - \lambda)(x + y)^2.$$

For $\lambda = \frac{2}{3}$ we find $|\Delta(1, 1)| = |\Delta(-1, -1)| = -24$, and for $\lambda = 2$ we find $|\Delta(\sqrt{3}, -\sqrt{3})| = |\Delta(-\sqrt{3}, \sqrt{3})| = 24$. We thus conclude, using Theorem 12, that $(1, 1)$ and $(-1, -1)$ are strict local minimum points, and that $(\sqrt{3}, -\sqrt{3})$ and $(-\sqrt{3}, \sqrt{3})$ are strict local maximum points. (These points are, in fact, absolute extreme points, as is evident geometrically.)

Exercises

1. Discuss the second-order conditions for the constrained optimization problems in exercise 12.2.
2. Answer the same question as above for exercises 12.4 and 12.5.
3. Compare the example and solution method of section 13 with that of section 12.

14. SUFFICIENT CONDITIONS FOR AN ABSOLUTE MINIMUM UNDER CONSTRAINTS

The Lagrange theorem (Theorem 10) gives *necessary* conditions for a local (and hence also for an absolute) constrained extremum to occur at a given point. In Theorem 11 we obtained *sufficient* conditions for a local constrained extremum. To find sufficient conditions for an absolute constrained extremum, we proceed as in the unconstrained case (section 9), and impose appropriate convexity (concavity) conditions.

Theorem 13

Let $\phi : S \to \mathbb{R}$ be a real-valued function defined and differentiable on an open convex set S in $\mathbb{R}^n$, and let $g : S \to \mathbb{R}^m$ $(m < n)$ be a vector function defined and differentiable on S. Let c be a point of S and let l be a point in $\mathbb{R}^m$. Define the Lagrangian function $\psi : S \to \mathbb{R}$ by the equation

$$\psi(x) = \phi(x) - l'g(x), \tag{1}$$

and assume that the first-order conditions are satisfied, that is,

$$\mathrm{d}\psi(c; u) = 0 \qquad \text{for all } u \text{ in } \mathbb{R}^n, \tag{2}$$

and

$$g(c) = 0. \tag{3}$$

If ψ is (strictly) convex on S, then ϕ has a (strict) absolute minimum at c under the constraint $g(x) = 0$.

Note. Under the same conditions, if ψ is (strictly) concave on S, then ϕ has a (strict) absolute maximum at c under the constraint $g(x) = 0$.

Proof. If ψ is convex on S and $d\psi(c; u) = 0$ for every $u \in \mathbb{R}^n$, then ψ has an absolute minimum at c (Theorem 8), that is,

$$\psi(x) \geqslant \psi(c) \qquad \text{for all } x \text{ in } S. \tag{4}$$

Since $\psi(x) = \phi(x) - l'g(x)$, it follows that

$$\phi(x) \geqslant \phi(c) + l'[g(x) - g(c)] \qquad \text{for all } x \text{ in } S. \tag{5}$$

But $g(c) = 0$ by assumption. Hence,

$$\phi(x) \geqslant \phi(c) \qquad \text{for all } x \text{ in } S \text{ satisfying } g(x) = 0, \tag{6}$$

that is, ϕ has an absolute minimum at c under the constraint $g(x) = 0$. The case in which ψ is strictly convex is treated similarly. □

Note. To prove that the Lagrangian function ψ is (strictly) convex or (strictly) concave, we can use the definition in section 4.9, Theorem 5 or Theorem 6, or (if ψ is twice differentiable) Theorem 7. In addition we observe that

(a) if the constraints $g_1(x), \ldots, g_m(x)$ are all *linear*, and $\phi(x)$ is (strictly) convex, then $\psi(x)$ is (strictly) convex.

In fact, (a) is a special case of

(b) if the functions, $\lambda_1 g_1(x), \ldots, \lambda_m g_m(x)$ are all *concave* (i.e. for $i = 1, 2, \ldots, m$, either $g_i(x)$ is concave and $\lambda_i \geqslant 0$, or $g_i(x)$ is convex and $\lambda_i \leqslant 0$) and if $\phi(x)$ is convex, then $\psi(x)$ is convex; furthermore, if at least one of these $m+1$ conditions is *strict*, then $\psi(x)$ is strictly convex.

15. A NOTE ON CONSTRAINTS IN MATRIX FORM

Let $\phi : S \to \mathbb{R}$ be a real-valued function defined on a set S in $\mathbb{R}^{n \times q}$, and let $G : S \to \mathbb{R}^{m \times p}$ be a matrix function defined on S. We shall frequently encounter the problem

$$\text{minimize} \qquad \phi(X) \tag{1}$$

$$\text{subject to} \qquad G(X) = 0. \tag{2}$$

This problem is, of course, mathematically equivalent to the case where X and G are vectors rather than matrices, so all theorems remain valid. We now introduce mp multipliers λ_{ij} (one for each constraint $g_{ij}(X) = 0$, $i = 1, \ldots, m$; $j = 1, \ldots, p$), and define the $m \times p$ *matrix* of Lagrange multipliers $L = (\lambda_{ij})$. The Lagrangian function then takes the convenient form

$$\psi(X) = \phi(X) - \operatorname{tr} L'G(X). \tag{3}$$

16. ECONOMIC INTERPRETATION OF LAGRANGE MULTIPLIERS

Consider the constrained minimization problem

$$\text{minimize} \quad \phi(x) \tag{1}$$
$$\text{subject to} \quad g(x)=b, \tag{2}$$

where ϕ is a real-valued function defined on an open set S in $\mathbb{R}^n$, g is a vector function defined on S with values in $\mathbb{R}^m$ $(m<n)$ and $b=(b_1, \ldots, b_m)'$ is a given $m \times 1$ vector of constants (parameters). In this section we shall examine how the optimal solution of this constrained minimization problem changes when the parameters change.

We shall assume that

(i) ϕ and g are twice *continuously* differentiable on S,
(ii) (first-order conditions) there exist points $x_0=(x_{01}, \ldots, x_{0n})'$ in S and $l_0=(\lambda_{01}, \ldots, \lambda_{0m})'$ in $\mathbb{R}^m$ such that

$$\mathrm{D}\phi(x_0)=l_0'\mathrm{D}g(x_0) \tag{3}$$

$$g(x_0)=b. \tag{4}$$

Now let

$$B_n=\mathrm{D}g(x_0), \qquad A_{nn}=\mathrm{H}\phi(x_0)-\sum_{i=1}^{m}\lambda_{0i}\mathrm{H}g_i(x_0), \tag{5}$$

and define, for $r=1,2, \ldots, n$, B_r as the $m \times r$ matrix whose columns are the first r columns of B_n, and A_{rr} as the $r \times r$ matrix in the top left corner of A_{nn}. In addition to (i) and (ii) we assume that

(iii) $|B_m| \neq 0$, (6)
(iv) (second-order conditions)

$$(-1)^m \begin{vmatrix} 0 & B_r \\ B_r' & A_{rr} \end{vmatrix} > 0 \qquad (r=m+1, \ldots, n). \tag{7}$$

These assumptions are sufficient (in fact, more than sufficient) for the function ϕ to have a strict local minimum at x_0 under the constraint $g(x)=b$ (see Theorem 12).

The vectors x_0 and l_0 for which the first-order conditions (3) and (4) are satisfied will, in general, depend on the parameter vector b. The question is whether x_0 and l_0 are differentiable functions of b. Given assumptions (i)–(iv), this question can be answered in the affirmative. By using the implicit function theorem (Theorem A.2 in the appendix to this chapter) we can show that there exists an m-ball $B(0)$ with the origin as its centre, and unique functions x^* and l^* defined on $B(0)$ with values in $\mathbb{R}^n$ and $\mathbb{R}^m$, respectively, such that

(a) $x^*(0)=x_0$, $l^*(0)=l_0$,
(b) $\mathrm{D}\phi(x^*(y))=(l^*(y))'\mathrm{D}g(x^*(y))$ for all y in $B(0)$,

(c) $g(x^*(y)) = b$ for all y in $B(0)$,
(d) the functions x^* and l^* are continuously differentiable on $B(0)$.

Now consider the real-valued function ϕ^* defined on $B(0)$ by the equation

$$\phi^*(y) = \phi(x^*(y)). \tag{8}$$

We first differentiate both sides of (c). This gives

$$(\mathrm{D}g(x^*(y)))(\mathrm{D}x^*(y)) = I_m, \tag{9}$$

using the chain rule. Next we differentiate ϕ^*. Using (again) the chain rule, (b) and (9), we obtain

$$\begin{aligned}\mathrm{D}\phi^*(y) &= (\mathrm{D}\phi(x^*(y)))(\mathrm{D}x^*(y)) \\ &= (l^*(y))'(\mathrm{D}g(x^*(y)))(\mathrm{D}x^*(y)) \\ &= (l^*(y))'I_m = (l^*(y))'. \end{aligned} \tag{10}$$

In particular, at $y = 0$,

$$\frac{\partial\phi^*(0)}{\partial b_j} = \lambda_{0j} \qquad (j = 1, \ldots, m). \tag{11}$$

Thus the Lagrange multiplier λ_{0j} measures the rate at which the optimal value of the objective function changes with respect to a small change in the right-hand side of the j-th constraint. For example, suppose we are maximizing a firm's profit subject to one resource limitation, then the Lagrange multiplier λ_0 is the extra profit that could be earned if the firm had one more unit of the resource, and therefore represents the maximum price the firm is willing to pay for this additional unit. For this reason λ_0 is often referred to as a *shadow price.*

Exercise

1. In exercise 12.2, find whether a small relaxation of the constraint will increase or decrease the optimal function value. At what rate?

APPENDIX: THE IMPLICIT FUNCTION THEOREM

Let $f : \mathbb{R}^{m+k} \to \mathbb{R}^m$ be a *linear* function defined by

$$f(x; t) = Ax + Bt, \tag{1}$$

where, as the notation indicates, points in $\mathbb{R}^{m+k}$ are denoted by $(x; t)$ with $x \in \mathbb{R}^m$ and $t \in \mathbb{R}^k$. If the $m \times m$ matrix A is non-singular, then there exists a unique function $g : \mathbb{R}^k \to \mathbb{R}^m$ such that

(a) $g(0) = 0$,
(b) $f(g(t); t) = 0$ for all $t \in \mathbb{R}^k$,
(c) g is infinitely times differentiable on $\mathbb{R}^k$.

This unique function is, of course,

$$g(t) = -A^{-1}Bt. \tag{2}$$

The implicit function theorem asserts that a similar conclusion holds for certain differentiable transformations which are not necessarily linear. In this appendix we present, without proof, three versions of the implicit function theorem, each one being useful in slightly different circumstances.

Theorem A.1

Let $f : S \to \mathbb{R}^m$ be a vector function defined on a set S in $\mathbb{R}^{m+k}$. Denote points in S by $(x; t)$ where $x \in \mathbb{R}^m$ and $t \in \mathbb{R}^k$, and let $(x_0; t_0)$ be an interior point of S. Assume that

(i) $f(x_0; t_0) = 0$,
(ii) f is differentiable at $(x_0; t_0)$,
(iii) f is differentiable with respect to x in some $(m+k)$-ball $B(x_0; t_0)$,
(iv) the $m \times m$ matrix $J(x; t) = \partial f(x; t)/\partial x'$ is continuous at $(x_0; t_0)$,
(v) $|J(x_0; t_0)| \neq 0$.

Then there exists an open set T in $\mathbb{R}^k$ containing t_0, and a unique function $g : T \to \mathbb{R}^m$ such that

(a) $g(t_0) = x_0$,
(b) $f(g(t); t) = 0$ for all $t \in T$,
(c) g is differentiable at t_0.

Theorem A.2

Let $f : S \to \mathbb{R}^m$ be a vector function defined on an open set S in $\mathbb{R}^{m+k}$, and let $(x_0; t_0)$ be a point of S. Assume that

(i) $f(x_0; t_0) = 0$,
(ii) f is continuously differentiable on S,
(iii) the $m \times m$ matrix $J(x; t) = \partial f(x; t)/\partial x'$ is non-singular at $(x_0; t_0)$.

Then there exists an open set T in $\mathbb{R}^k$ containing t_0, and a unique function $g : T \to \mathbb{R}^m$ such that

(a) $g(t_0) = x_0$,
(b) $f(g(t); t) = 0$ for all $t \in T$,
(c) g is continuously differentiable on T.

Theorem A.3

Let $f : S \to \mathbb{R}^m$ be a vector function defined on a set S in $\mathbb{R}^{m+k}$, and let $(x_0; t_0)$ be an interior point of S. Assume that

(i) $f(x_0; t_0) = 0$,
(ii) f is $p \geq 2$ times differentiable at $(x_0; t_0)$,
(iii) the $m \times m$ matrix $J(x; t) = \partial f(x; t)/\partial x'$ is non-singular at $(x_0; t_0)$.

Then there exists an open set T in $\mathbb{R}^k$ containing t_0, and a unique function $g : T \to \mathbb{R}^m$ such that

(a) $g(t_0) = x_0$,
(b) $f(g(t); t) = 0$ for all $t \in T$,
(c) g is $p-1$ times differentiable on T and p times differentiable at t_0.

BIBLIOGRAPHICAL NOTES

§1. Apostol (1974, chap. 13) has a good discussion of implicit functions and extremum problems. See also Luenberger (1969) and Sydsaeter (1981, chap. 5).
§9 and §14. For an interesting approach to absolute minima with applications in statistics, see Rolle (1996).
Appendix. There are many versions of the implicit function theorem, but Theorem A.2 is what most authors would call 'the' implicit function theorem. See Dieudonné (1969, th. 10.2.1) or Apostol (1974, th. 13.7). Theorems A.1 and A.3 are less often presented. See, however, Young (1910, sec. 38).

Part Three—

Differentials: the practice

CHAPTER 8

Some important differentials

1. INTRODUCTION

Now that we know what differentials are, and have adopted a convenient and simple notation for them, our next step is to determine the differentials of some important functions.

In this chapter, X always denotes a matrix (usually square) of real variables, and Z a matrix of complex variables. We shall discuss the differentials of some *scalar* functions of X (eigenvalue, determinant), a *vector* function of X (eigenvector) and some *matrix* functions of X (inverse, Moore–Penrose inverse, adjoint matrix).

But first we must list the basic rules.

2. FUNDAMENTAL RULES OF DIFFERENTIAL CALCULUS

The following rules are easily verified. If u and v are real-valued differentiable functions and α is a real constant, then we have

$$\mathrm{d}\alpha = 0 \tag{1}$$

$$\mathrm{d}(\alpha u) = \alpha\, \mathrm{d}u \tag{2}$$

$$\mathrm{d}(u+v) = \mathrm{d}u + \mathrm{d}v \tag{3}$$

$$\mathrm{d}(u-v) = \mathrm{d}u - \mathrm{d}v \tag{4}$$

$$\mathrm{d}(uv) = (\mathrm{d}u)v + u\, \mathrm{d}v \tag{5}$$

$$\mathrm{d}\left(\frac{u}{v}\right) = \frac{v\, \mathrm{d}u - u\, \mathrm{d}v}{v^2} \qquad (v \neq 0). \tag{6}$$

The differentials of the power function*, logarithmic function and exponential function are

$$\mathrm{d}u^{\alpha}=\alpha u^{\alpha-1}\,\mathrm{d}u \tag{7}$$

$$\mathrm{d}\log u=u^{-1}\,\mathrm{d}u \qquad (u>0) \tag{8}$$

$$\mathrm{d}e^{u}=e^{u}\,\mathrm{d}u \tag{9}$$

$$\mathrm{d}\alpha^{u}=\alpha^{u}\log\alpha\,\mathrm{d}u \qquad (\alpha>0). \tag{10}$$

Similar results hold if U and V are *matrix* functions, and A is a matrix of real constants:

$$\mathrm{d}A=0 \tag{11}$$

$$\mathrm{d}(\alpha U)=\alpha\,\mathrm{d}U \tag{12}$$

$$\mathrm{d}(U+V)=\mathrm{d}U+\mathrm{d}V \tag{13}$$

$$\mathrm{d}(U-V)=\mathrm{d}U-\mathrm{d}V \tag{14}$$

$$\mathrm{d}(UV)=(\mathrm{d}U)V+U\,\mathrm{d}V. \tag{15}$$

For the Kronecker product and Hadamard product the analogue of (15) holds:

$$\mathrm{d}(U\otimes V)=(\mathrm{d}U)\otimes V+U\otimes\mathrm{d}V \tag{16}$$

$$\mathrm{d}(U\odot V)=(\mathrm{d}U)\odot V+U\odot\mathrm{d}V. \tag{17}$$

Finally we have

$$\mathrm{d}U'=(\mathrm{d}U)' \tag{18}$$

$$\mathrm{d}\,\mathrm{vec}\,U=\mathrm{vec}\,\mathrm{d}U \tag{19}$$

$$\mathrm{d}\,\mathrm{tr}\,U=\mathrm{tr}\,\mathrm{d}U. \tag{20}$$

For example, to prove (3), let $\phi(x)=u(x)+v(x)$. Then,

$$\begin{aligned}\mathrm{d}\phi(x;h)&=\sum_j h_j\mathrm{D}_j\phi(x)=\sum_j h_j[\mathrm{D}_ju(x)+\mathrm{D}_jv(x)]\\&=\sum_j h_j\mathrm{D}_ju(x)+\sum_j h_j\mathrm{D}_jv(x)=\mathrm{d}u(x;h)+\mathrm{d}v(x;h).\end{aligned} \tag{21}$$

* The domain of definition of the power function u^{α} depends on the arithmetical nature of α. If α is a positive integer then u^{α} is defined for all real u; but if α is a negative integer or zero, the point $u=0$ must be excluded. If α is a rational fraction, e.g. $\alpha=p/q$ (where p and q are integers and we can always assume that $q>0$), then

$$u^{\alpha}=\sqrt[q]{(u^{p})},$$

so that the function is determined for all values of u when q is odd, and only for $u\geqslant 0$ when q is even. In cases where α is irrational, the function is defined for $u>0$.

As a second example, let us prove (15). Using only (3) and (5), we have

$$
\begin{aligned}
(\mathrm{d}(UV))_{ij} &= \mathrm{d}(UV)_{ij} \\
&= \mathrm{d}\sum_k u_{ik}v_{kj} = \sum_k \mathrm{d}(u_{ik}v_{kj}) \\
&= \sum_k [(\mathrm{d}u_{ik})v_{kj} + u_{ik}\,\mathrm{d}v_{kj}] \\
&= \sum_k (\mathrm{d}u_{ik})v_{kj} + \sum_k u_{ik}\,\mathrm{d}v_{kj} \\
&= ((\mathrm{d}U)V)_{ij} + (U\,\mathrm{d}V)_{ij}.
\end{aligned}
\tag{22}
$$

Hence (15) follows.

Exercises

1. Prove (16).
2. Show that $\mathrm{d}(UVW) = (\mathrm{d}U)VW + U(\mathrm{d}V)W + UV(\mathrm{d}W)$.
3. Show that $\mathrm{d}(AXB) = A(\mathrm{d}X)B$, A and B constant.
4. Show that $\mathrm{d}\,\mathrm{tr}\,X'X = 2\,\mathrm{tr}\,X'\,\mathrm{d}X$.
5. Let $u : S \to \mathbb{R}$ be a real-valued function defined on an open subset S of $\mathbb{R}^n$. If $u'u = 1$ on S, then $u'\mathrm{d}u = 0$ on S.

3. THE DIFFERENTIAL OF A DETERMINANT

Let us now apply these rules to obtain a number of useful results. The first of these follows.

Theorem 1

Let S be an open subset of $\mathbb{R}^{n\times q}$. If the matrix function $F : S \to \mathbb{R}^{m\times m}$ $(m \geq 2))$ is k times (continuously) differentiable on S, then so is the real-valued function $|F| : S \to \mathbb{R}$ given by $|F|(X) = |F(X)|$. Moreover,

$$\mathrm{d}|F| = \mathrm{tr}\,F^{\#}\,\mathrm{d}F, \tag{1}$$

where $F^{\#}(X) = (F(X))^{\#}$ denotes the adjoint matrix of $F(X)$. In particular,

$$\mathrm{d}|F| = |F|\,\mathrm{tr}\,F^{-1}\,\mathrm{d}F \tag{2}$$

at points X with $r(F(X)) = m$;

$$\mathrm{d}|F| = (-1)^{\kappa+1}\mu(F)\,\frac{v'(\mathrm{d}F)u}{v'(F^{\kappa-1})^{+}u} \tag{3}$$

at points X with $r(F(X)) = m-1$. Here κ denotes the multiplicity of the zero eigenvalue of $F(X)$, $1 \leq \kappa \leq m$, $\mu(F(X))$ is the product of the $m-\kappa$ non-zero eigenvalues of $F(X)$ if $\kappa<m$ and $\mu(F(X))=1$ if $\kappa=m$, and the $m\times 1$ vectors u and v satisfy $F(X)u=F'(X)v=0$; and

$$\mathrm{d}|F|=0 \tag{4}$$

at points X with $r(F(X))\leqslant m-2$.

Proof. Consider the real-valued function $\phi:\mathbb{R}^{m\times m}\rightarrow\mathbb{R}$ defined by $\phi(Y)=|Y|$. Clearly, ϕ is ∞ times differentiable at every point of $\mathbb{R}^{m\times m}$. If $Y=(y_{ij})$ and c_{ij} is the cofactor of y_{ij}, then, by (1.9.7),

$$\phi(Y)=|Y|=\sum_{i=1}^{m}c_{ij}y_{ij}, \tag{5}$$

and, since $c_{1j},\ldots,c_{mj}$ do not depend on y_{ij}, we have

$$\frac{\partial\phi(Y)}{\partial y_{ij}}=c_{ij}. \tag{6}$$

From these partial derivatives we obtain the differential

$$\mathrm{d}\phi(Y)=\sum_{i=1}^{m}\sum_{j=1}^{m}c_{ij}\mathrm{d}y_{ij}=\operatorname{tr}Y^{\#}\,\mathrm{d}Y. \tag{7}$$

Now, since the function $|F|$ is the composite of ϕ and F, Cauchy's rule of invariance (Theorem 5.9) applies, and

$$\mathrm{d}|F|=\operatorname{tr}F^{\#}\,\mathrm{d}F. \tag{8}$$

The remainder of the theorem follows from Theorem 3.1. □

It is worth stressing that at points where $r(F(X))=m-1$, $F(X)$ must have *at least* one zero eigenvalue. At points where $F(X)$ has a *simple* zero eigenvalue (and where, consequently, $r(F(X))=m-1$), (3) simplifies to

$$\mathrm{d}|F|=\mu(F)\,\frac{v'(\mathrm{d}F)u}{v'u}, \tag{9}$$

where $\mu(F(X))$ is the product of the $m-1$ non-zero eigenvalues of $F(X)$.

We do not, at this point, derive the second- and higher-order differentials of the determinant function. In section 4 (exercises 1 and 2) we obtain the differentials of $\log|F|$ assuming that $F(X)$ is non-singular. To obtain the general result we need the differential of the adjoint matrix. A formula for the first differential of the adjoint matrix will be obtained in section 6.

Result (2), the case where $F(X)$ is non-singular, is of great practical interest. At points where $|F(X)|$ is positive, its logarithm exists and we arrive at the following theorem.

Theorem 2

Let T_+ denote the set

$$T_+=\{Y:Y\in\mathbb{R}^{m\times m},|Y|>0\}. \tag{10}$$

Let S be an open subset of $\mathbb{R}^{n\times q}$. If the matrix function $F : S \to T_+$ is k times (continuously) differentiable on S, then so is the real-valued function $\log|F| : S \to \mathbb{R}$ given by $(\log|F|)(X) = \log|F(X)|$. Moreover

$$\mathrm{d}\log|F| = \operatorname{tr} F^{-1}\mathrm{d}F. \tag{11}$$

Proof. Immediate from (2) in Theorem 1. □

Exercises

1. Give an intuitive explanation of the fact that $\mathrm{d}|X| = 0$ at points $X \in \mathbb{R}^{n\times n}$ where $r(X) \leqslant n-2$.
2. Show that, if $F(X) \in \mathbb{R}^{m\times m}$ and $r(F(X)) = m-1$ for every X in some neighbourhood of X_0, then $\mathrm{d}|F(X)| = 0$ at X_0.
3. Show that $\mathrm{d}\log|X'X| = 2\operatorname{tr}(X'X)^{-1}X'\mathrm{d}X$ at every point where X has full column rank.

4. THE DIFFERENTIAL OF AN INVERSE

The next theorem deals with the differential of the inverse function.

Theorem 3

Let T be the set of non-singular real $m\times m$ matrices, i.e. $T = \{Y : Y \in \mathbb{R}^{m\times m}, |Y| \neq 0\}$. Let S be an open subset of $\mathbb{R}^{n\times q}$. If the matrix function $F : S \to T$ is k times (continuously) differentiable on S, then so is the matrix function $F^{-1}: S \to T$ defined by $F^{-1}(X) = (F(X))^{-1}$, and

$$\mathrm{d}F^{-1} = -F^{-1}(\mathrm{d}F)F^{-1}. \tag{1}$$

Proof. Let $A_{ij}(X)$ be the $(m-1)\times(m-1)$ submatrix of $F(X)$ obtained by deleting row i and column j of $F(X)$. The typical element of $F^{-1}(X)$ can then be expressed as

$$[F^{-1}(X)]_{ij} = (-1)^{i+j}|A_{ji}(X)|/|F(X)|. \tag{2}$$

Since both determinants $|A_{ji}|$ and $|F|$ are k times (continuously) differentiable on S, the same is true for their ratio and hence for the matrix function F^{-1}. To prove (1) we then write

$$0 = \mathrm{d}I = \mathrm{d}F^{-1}F = (\mathrm{d}F^{-1})F + F^{-1}\mathrm{d}F, \tag{3}$$

and post-multiply with F^{-1}. □

Let us consider the set T of non-singular real $m\times m$ matrices. T is an open subset of $\mathbb{R}^{m\times m}$, so that for every $Y_0 \in T$ there exists an open neighbourhood $N(Y_0)$, all of whose points are non-singular. This follows from the continuity of the determinant function $|Y|$. Put differently, if Y_0 is non-singular and $\{E_j\}$ is a sequence of real $m\times m$ matrices such that $E_j\to 0$ as $j\to\infty$, then

$$r(Y_0+E_j)=r(Y_0) \tag{4}$$

for every j greater than some fixed j_0, and

$$\lim_{j\to\infty}(Y_0+E_j)^{-1}=Y_0^{-1}. \tag{5}$$

Exercises

1. Let $T_+=\{Y:Y\in\mathbb{R}^{m\times m}, |Y|>0\}$. If $F:S\to T_+$, $S\subset\mathbb{R}^{n\times q}$, is twice differentiable on S, then show that
$$\mathrm{d}^2\log|F|=-\operatorname{tr}(F^{-1}\,\mathrm{d}F)^2+\operatorname{tr}F^{-1}\,\mathrm{d}^2F.$$
2. Show that, for $X\in T_+$, $\log|X|$ is ∞ times differentiable on T_+, and
$$\mathrm{d}^r\log|X|=(-1)^{r-1}(r-1)!\operatorname{tr}(X^{-1}\,\mathrm{d}X)^r \qquad (r=1,2,\ldots).$$
3. Let $T=\{Y:Y\in\mathbb{R}^{m\times m}, |Y|\neq 0\}$. If $F:S\to T$, $S\subset\mathbb{R}^{n\times q}$, is twice differentiable on S, then show
$$\mathrm{d}^2F^{-1}=2[F^{-1}(\mathrm{d}F)]^2F^{-1}-F^{-1}(\mathrm{d}^2F)F^{-1}.$$
4. Show that, for $X\in T$, X^{-1} is ∞ times differentiable on T, and
$$\mathrm{d}^rX^{-1}=(-1)^r r!\,(X^{-1}\,\mathrm{d}X)^rX^{-1} \qquad (r=1,2,\ldots).$$

5. THE DIFFERENTIAL OF THE MOORE–PENROSE INVERSE

Equation (4.4) above tells us that non-singular matrices have *locally constant rank*. Singular matrices (more precisely matrices of less than full row or column rank) do not share this property. Consider, for example, the matrices

$$Y_0=\begin{pmatrix}1&0\\0&0\end{pmatrix} \quad\text{and}\quad E_j=\begin{pmatrix}0&0\\0&1/j\end{pmatrix}, \tag{1}$$

and let $Y=Y(j)=Y_0+E_j$. Then $r(Y_0)=1$, but $r(Y)=2$ for all j. Moreover, $Y\to Y_0$ as $j\to\infty$, but

$$Y^+=\begin{pmatrix}1&0\\0&j\end{pmatrix} \tag{2}$$

does certainly not converge to Y_0^+, because it does not converge to anything. It follows that (i) $r(Y)$ is not constant in any neighbourhood of Y_0, and (ii) Y^+ is not continuous at Y_0. The following lemma shows that the conjoint occurrence of (i) and (ii) is typical.

Lemma 1

Let $Y_0 \in \mathbb{R}^{m\times p}$ and $\{E_j\}$ be a sequence of real $m \times p$ matrices such that $E_j \to 0$ as $j \to \infty$. Then

$$r(Y_0 + E_j) = r(Y_0) \qquad \text{for every } j \geqslant j_0 \tag{3}$$

if and only if

$$\lim_{j\to\infty}(Y_0 + E_j)^+ = Y_0^+. \tag{4}$$

Lemma 1 tells us that if $F : S \to \mathbb{R}^{m\times p}$, $S \subset \mathbb{R}^{n\times q}$, is a matrix function defined and continuous on S, then $F^+ : S \to \mathbb{R}^{p\times m}$ is continuous on S if and only if $r(F(X))$ is constant on S. If F^+ is to be differentiable at $X_0 \in S$ it must be continuous at X_0, hence of constant rank in some neighbourhood $N(X_0)$ of X_0. Provided that $r(F(X))$ is constant in $N(X_0)$, the differentiability of F at X_0 implies the differentiability of F^+ at X_0. In fact, we have the next lemma.

Lemma 2

Let X_0 be an interior point of a subset S of $\mathbb{R}^{n\times q}$. Let $F : S \to \mathbb{R}^{m\times p}$ be a matrix function defined on S and $k \geqslant 1$ times (continuously) differentiable at each point of some neighbourhood $N(X_0) \subset S$ of X_0. Then the following three statements are equivalent:

(i) the rank of $F(X)$ is constant on $N(X_0)$;
(ii) F^+ is continuous on $N(X_0)$;
(iii) F^+ is k times (continuously) differentiable on $N(X_0)$.

Having established the existence of differentiable MP inverses, we now want to find the relationship between $\mathrm{d}F^+$ and $\mathrm{d}F$. First, we find $\mathrm{d}F^+F$ and $\mathrm{d}FF^+$; then we use these results to obtain $\mathrm{d}F^+$.

Theorem 4

Let S be an open subset of $\mathbb{R}^{n\times q}$, and let $F : S \to \mathbb{R}^{m\times p}$ be a matrix function defined and $k \geqslant 1$ times (continuously) differentiable on S. If $r(F(X))$ is constant on S, then $F^+F : S \to \mathbb{R}^{p\times p}$ and $FF^+ : S \to \mathbb{R}^{m\times m}$ are k times (continuously) differentiable on S, and

$$\mathrm{d}F^+F = F^+(\mathrm{d}F)(I - F^+F) + (F^+(\mathrm{d}F)(I - F^+F))' \tag{5}$$

and

$$\mathrm{d}FF^+ = (I - FF^+)(\mathrm{d}F)F^+ + ((I - FF^+)(\mathrm{d}F)F^+)'. \tag{6}$$

Proof. Let us demonstrate the first result, leaving the second as an exercise for the reader.

Since the matrix F^+F is idempotent and symmetric, we have

$$\begin{aligned}\mathrm{d}F^+F &= \mathrm{d}(F^+FF^+F) = (\mathrm{d}F^+F)F^+F + F^+F(\mathrm{d}F^+F)\\ &= F^+F(\mathrm{d}F^+F) + (F^+F(\mathrm{d}F^+F))'.\end{aligned} \tag{7}$$

To find $\mathrm{d}F^+F$ it suffices therefore to find $F(\mathrm{d}F^+F)$. But this is easy, since the equality

$$\mathrm{d}F=\mathrm{d}(FF^+F)=(\mathrm{d}F)(F^+F)+F(\mathrm{d}F^+F) \tag{8}$$

can be rearranged as

$$F(\mathrm{d}F^+F)=(\mathrm{d}F)(I-F^+F). \tag{9}$$

The result follows by inserting (9) into (7). □

We now have all the ingredients for the main result.

Theorem 5

Let S be an open subset of $\mathbb{R}^{n\times q}$, and let $F: S \to \mathbb{R}^{m\times p}$ be a matrix function defined and $k \geqslant 1$ times (continuously) differentiable on S. If $r(F(X))$ is constant on S, then $F^+: S \to \mathbb{R}^{p\times m}$ is k times (continuously) differentiable on S, and

$$\mathrm{d}F^+ = -F^+(\mathrm{d}F)F^+ + F^+F^{+\prime}(\mathrm{d}F')(I-FF^+)+(I-F^+F)(\mathrm{d}F')F^{+\prime}F^+. \tag{10}$$

Proof. The strategy of the proof is to express $\mathrm{d}F^+$ in $\mathrm{d}FF^+$ and $\mathrm{d}F^+F$, and apply Theorem 4. We have

$$\mathrm{d}F^+ = \mathrm{d}(F^+FF^+)=(\mathrm{d}F^+F)F^+ + F^+F\,\mathrm{d}F^+ \tag{11}$$

and also

$$\mathrm{d}FF^+=(\mathrm{d}F)F^+ + F\mathrm{d}F^+. \tag{12}$$

Inserting the expression for $F\,\mathrm{d}F^+$ from (12) into the last term of (11), we obtain

$$\mathrm{d}F^+ = (\mathrm{d}F^+F)F^+ + F^+(\mathrm{d}FF^+) - F^+(\mathrm{d}F)F^+. \tag{13}$$

Application of Theorem 4 gives the desired result. □

Exercises

1. Prove (6).
2. If $F(X)$ is idempotent for every X in some neighbourhood of a point X_0, then F is said to be *locally idempotent* at X_0. Show that $F(\mathrm{d}F)F=0$ at points where F is differentiable and locally idempotent.
3. If F is locally idempotent at X_0 and continuous in a neighbourhood of X_0, then $\operatorname{tr} F$ is differentiable at X_0 with $\mathrm{d}(\operatorname{tr} F)(X_0)=0$.
4. We say that F has *locally constant rank* at a point X_0 if $r(F(X))$ is constant for every X in some neighbourhood of X_0. If F has locally constant rank at X_0 and is continuous in a neighbourhood of X_0, then $\operatorname{tr} F^+F$ and $\operatorname{tr} FF^+$ are differentiable at X_0 with $\mathrm{d}(\operatorname{tr} F^+F)(X_0)=\mathrm{d}(\operatorname{tr} FF^+)(X_0)=0$.
5. If F has locally constant rank at X_0 and is differentiable in a neighbourhood of X_0, then $\operatorname{tr} F\mathrm{d}F^+ = -\operatorname{tr} F^+\mathrm{d}F$.

6. THE DIFFERENTIAL OF THE ADJOINT MATRIX

If Y is a real $m \times m$ matrix, then by $Y^{\#}$ we denote the $m \times m$ adjoint matrix of Y. Given an $m \times m$ matrix function F we now define an $m \times m$ matrix function $F^{\#}$ by $F^{\#}(X) = (F(X))^{\#}$. The purpose of this section is to find the differential of $F^{\#}$. We first prove Theorem 6.

Theorem 6

Let S be a subset of $\mathbb{R}^{n\times q}$, and let $F : S \rightarrow \mathbb{R}^{m\times m} (m \geqslant 2)$ be a matrix function defined on S. If F is k times (continuously) differentiable at a point X_0 of S, then so is the matrix function $F^{\#} : S \rightarrow \mathbb{R}^{m\times m}$, and, at X_0,

$$(\mathrm{d}F^{\#})_{ij} = (-1)^{i+j} \operatorname{tr} E_i(E_j' F E_i)^{\#} E_j' \mathrm{d}F \qquad (i, j = 1, \ldots, m), \tag{1}$$

where E_i denotes the $m \times (m-1)$ matrix obtained from I_m by deleting column i.

Note. The matrix $E_j' F(X) E_i$ is obtained from $F(X)$ by deleting row j and column i; the matrix $E_i(E_j' F(X) E_i)^{\#} E_j'$ is obtained from $(E_j' F(X) E_i)^{\#}$ by inserting a row of zeros between row $i-1$ and i, and a column of zeros between column $j-1$ and j.

Proof. Since, by definition (see section 1.9),

$$(F^{\#}(X))_{ij} = (-1)^{i+j} |E_j' F(X) E_i|, \tag{2}$$

we have, from Theorem 1,

$$\begin{aligned}(\mathrm{d}F^{\#}(X))_{ij} &= (-1)^{i+j} \operatorname{tr} (E_j' F(X) E_i)^{\#} \,\mathrm{d}(E_j' F(X) E_i) \\ &= (-1)^{i+j} \operatorname{tr} (E_j' F(X) E_i)^{\#} E_j' (\mathrm{d}F(X)) E_i \\ &= (-1)^{i+j} \operatorname{tr} E_i (E_j' F(X) E_i)^{\#} E_j' \mathrm{d}F(X).\end{aligned} \tag{3}$$

□

Recall from Theorem 3.2 that if $Y = F(X)$ is an $m \times m$ matrix and $m \geqslant 2$, then the rank of $Y^{\#} = F^{\#}(X)$ is given by

$$r(Y^{\#}) = \begin{cases} m, & \text{if } r(Y) = m, \\ 1, & \text{if } r(Y) = m-1, \\ 0, & \text{if } r(Y) \leqslant m-2. \end{cases} \tag{4}$$

As a result, two special cases of Theorem 6 can be proved. The first relates to the situation where $F(X_0)$ is non-singular.

Corollary 1

If $F : S \rightarrow \mathbb{R}^{m\times m} (m \geqslant 2)$, $S \subset \mathbb{R}^{n\times q}$, is k times (continuously) differentiable at a point $X_0 \in S$ where $F(X_0)$ is non-singular, then $F^{\#} : S \rightarrow \mathbb{R}^{m\times m}$ is also k times (continuously) differentiable at X_0, and the differential at that point is given by

$$\mathrm{d}F^{\#}=|F|[(\operatorname{tr}F^{-1}\mathrm{d}F)F^{-1}-F^{-1}(\mathrm{d}F)F^{-1}] \tag{5}$$

or, equivalently,

$$\mathrm{d}\operatorname{vec}F^{\#}=|F|[(\operatorname{vec}F^{-1})(\operatorname{vec}(F')^{-1})'-(F')^{-1}\otimes F^{-1}]\mathrm{d}\operatorname{vec}F. \tag{6}$$

Proof. To demonstrate this result as a special case of Theorem 6 is somewhat involved, and is left to the reader. Much simpler is to write $F^{\#}=|F|F^{-1}$ and use the facts, established in Theorems 1 and 3, that $\mathrm{d}|F|=|F|\operatorname{tr}F^{-1}\mathrm{d}F$ and $\mathrm{d}F^{-1}=-F^{-1}(\mathrm{d}F)F^{-1}$. Details of the proof are left to the reader. □

The second special case of Theorem 6 concerns points where the rank of $F(X_0)$ does not exceed $m-3$.

Corollary 2

Let $F:S\to\mathbb{R}^{m\times m}(m\geq 3)$, $S\subset\mathbb{R}^{n\times q}$, be differentiable at a point $X_0\in S$. If $r(F(X_0))\leq m-3$, then

$$(\mathrm{d}F^{\#})(X_0)=0. \tag{7}$$

Proof. Since the rank of the $(m-1)\times(m-1)$ matrix $E_j'F(X_0)E_i$ in Theorem 6 cannot exceed $m-3$, it follows by (4) that its adjoint matrix is the null matrix. Inserting $(E_j'F(X_0)E_i)^{\#}=0$ in (1) gives $(\mathrm{d}F^{\#})(X_0)=0$. □

There is another, more illuminating, proof of Corollary 2—one which does not depend on Theorem 6. Let $Y_0\in\mathbb{R}^{m\times m}$ and assume Y_0 is singular. Then $r(Y)$ is not locally constant at Y_0. In fact, if $r(Y_0)=r$ $(1\leq r\leq m-1)$ and we perturb one element of Y_0, then the rank of $\tilde{Y}_0$ (the perturbed matrix) will be $r-1$, r, or $r+1$. An immediate consequence of this simple observation is that if $r(Y_0)$ does not exceed $m-3$, then $r(\tilde{Y}_0)$ will not exceed $m-2$. But this means that at points Y_0 with $r(Y_0)\leq m-3$,

$$(\tilde{Y}_0)^{\#}=Y_0^{\#}=0, \tag{8}$$

implying that the differential of $Y^{\#}$ at Y_0 must be the null matrix.

The two corollaries provide expressions for $\mathrm{d}F^{\#}$ at every point X where $r(F(X))=m$ or $r(F(X))\leq m-3$. The remaining points to consider are those where $r(F(X))$ is either $m-1$ or $m-2$. At such points we must, unfortunately, use Theorem 6, which holds irrespective of rank considerations.

Only if we know that the rank of $F(X)$ is *locally constant* can we say more. If $r(F(X))=m-2$ for every X in some neighbourhood $N(X_0)$ of X_0, then $F^{\#}(X)$ vanishes in that neighbourhood, and hence $(\mathrm{d}F^{\#})(X)=0$ for every $X\in N(X_0)$. More complicated is the situation where $r(F(X))=m-1$ in some neighbourhood of X_0. A discussion of this case is postponed to the miscellaneous exercises at the end of this chapter.

Exercise

1. The matrix function $F : \mathbb{R}^{n\times n} \to \mathbb{R}^{n\times n}$ defined by $F(X) = X^{\#}$ is ∞ times differentiable on $\mathbb{R}^{n\times n}$, and $(\mathrm{d}^j F)(X) = 0$ for every $j \leqslant n - 2 - r(X)$.

7. ON DIFFERENTIATING EIGENVALUES AND EIGENVECTORS

There are two problems involved in differentiating eigenvalues and eigenvectors. The first problem is that the eigenvalues of a real matrix A need not, in general, be real numbers—they may be complex. The second problem is the possible occurrence of multiple eigenvalues.

To appreciate the first point, consider the real 2×2 matrix function

$$A(\varepsilon) = \begin{pmatrix} 1 & \varepsilon \\ -\varepsilon & 1 \end{pmatrix}, \qquad \varepsilon \neq 0. \tag{1}$$

The matrix A is not symmetric, and its eigenvalues are $1 \pm \mathrm{i}\varepsilon$. Since both eigenvalues are complex, the corresponding eigenvectors must be complex as well; in fact, they can be chosen as

$$\begin{pmatrix} 1 \\ \mathrm{i} \end{pmatrix} \quad \text{and} \quad \begin{pmatrix} 1 \\ -\mathrm{i} \end{pmatrix}. \tag{2}$$

We know, however (Theorem 1.4), that if A is a *real symmetric* matrix, then its eigenvalues are real and its eigenvectors can always be taken to be real. Since the derivations in the real symmetric case are somewhat simpler, we consider this case first.

Thus, let X_0 be a real symmetric $n\times n$ matrix, and let u_0 be a (normalized) eigenvector associated with an eigenvalue λ_0 of X_0, so that the triple (X_0, u_0, λ_0) satisfies the equations

$$Xu = \lambda u, \qquad u'u = 1. \tag{3}$$

Since the $n+1$ equations in (3) are implicit relations rather than explicit functions, we must first show that there exist explicit unique functions $\lambda = \lambda(X)$ and $u = u(X)$ satisfying (3) in a neighbourhood of X_0 and such that $\lambda(X_0) = \lambda_0$ and $u(X_0) = u_0$. Here the second (and more serious) problem arises—the possible occurrence of multiple eigenvalues.

We shall see that the implicit function theorem (given in the appendix to Chapter 7) implies the existence of a neighbourhood $N(X_0) \subset \mathbb{R}^{n\times n}$ of X_0 where the functions λ and u both exist and are ∞ times (continuously) differentiable, *provided λ_0 is a simple eigenvalue of X_0*. If, however, λ_0 is a multiple eigenvalue of X_0, then the conditions of the implicit function theorem are not satisfied. The difficulty is illustrated by the following example. Consider the real 2×2 matrix function

$$A(\varepsilon, \delta) = \begin{pmatrix} 1+\varepsilon & \delta \\ \delta & 1-\varepsilon \end{pmatrix}. \tag{4}$$

The matrix A is symmetric for every value of ε and δ; its eigenvalues are $\lambda_1 = 1 + \sqrt{(\varepsilon^2+\delta^2)}$ and $\lambda_2 = 1 - \sqrt{(\varepsilon^2+\delta^2)}$. Both eigenvalue functions are continuous in ε and δ, but clearly not differentiable at (0, 0). (Strictly speaking we should also prove that λ_1 and λ_2 are the *only* two continuous eigenvalue functions.) The conical surface formed by the eigenvalues of $A(\varepsilon, \delta)$ has a singularity at $\varepsilon = \delta = 0$ (Figure 1). For a fixed ratio ε/δ, however, we can pass from one side of the surface to the other going through (0, 0) without noticing the singularity. This phenomenon is quite general and it indicates the need to restrict our study of differentiability of multiple eigenvalues to one-dimensional perturbations only. We shall delay a further discussion of multiple eigenvalues to section 12.

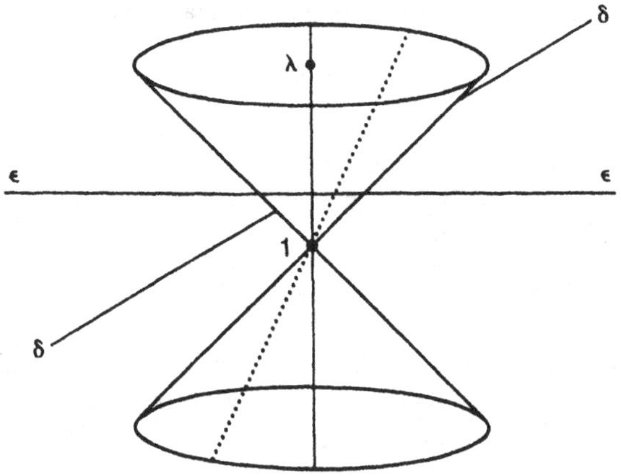

***Figure* 1** The eigenvalue functions $\lambda_{1,2} = 1 \pm \sqrt{(\varepsilon^2+\delta^2)}$

8. THE DIFFERENTIAL OF EIGENVALUES AND EIGENVECTORS: THE REAL SYMMETRIC CASE

Let us now demonstrate the following theorem.

Theorem 7

Let X_0 be a real symmetric $n \times n$ matrix. Let u_0 be a normalized eigenvector associated with a simple eigenvalue λ_0 of X_0. Then a real-valued function λ and a vector function u are defined for all X in some neighbourhood $N(X_0) \subset \mathbb{R}^{n\times n}$ of X_0, such that

$$\lambda(X_0) = \lambda_0, \qquad u(X_0) = u_0, \tag{1}$$

and

$$Xu = \lambda u, \qquad u'u = 1 \qquad (X \in N(X_0)). \tag{2}$$

Moreover, the functions λ and u are ∞ times differentiable on $N(X_0)$, and the differentials at X_0 are

$$\mathrm{d}\lambda = u_0'(\mathrm{d}X)u_0 \tag{3}$$

and

$$\mathrm{d}u = (\lambda_0 I_n - X_0)^+(\mathrm{d}X)u_0. \tag{4}$$

Note. In order for λ (and u) to be differentiable at X_0 we require that λ_0 is simple, but this does not, of course, exclude the possibility of multiplicities among the remaining $n-1$ eigenvalues of X_0.

Proof. Consider the vector function $f : \mathbb{R}^{n+1} \times \mathbb{R}^{n\times n} \to \mathbb{R}^{n+1}$ defined by the equation

$$f(u, \lambda; X) = \begin{pmatrix} (\lambda I_n - X)u \\ u'u - 1 \end{pmatrix}, \tag{5}$$

and observe that f is ∞ times differentiable on $\mathbb{R}^{n+1} \times \mathbb{R}^{n\times n}$. The point $(u_0, \lambda_0; X_0)$ in $\mathbb{R}^{n+1} \times \mathbb{R}^{n\times n}$ satisfies

$$f(u_0, \lambda_0; X_0) = 0 \tag{6}$$

and

$$\begin{vmatrix} \lambda_0 I_n - X_0 & u_0 \\ 2u_0' & 0 \end{vmatrix} \neq 0. \tag{7}$$

We note that the determinant in (7) is non-zero if and only if the eigenvalue λ_0 is *simple*, in which case it takes the value of -2 times the product of the $n-1$ non-zero eigenvalues of $\lambda_0 I_n - X_0$ (see Theorem 3.5).

The conditions of the implicit function theorem (Theorem A.3 in the appendix to Chapter 7) thus being satisfied, there exist a neighbourhood $N(X_0) \subset \mathbb{R}^{n\times n}$ of X_0, a unique real-valued function $\lambda : N(X_0) \to \mathbb{R}$, and a unique (apart from its sign) vector function $u : N(X_0) \to \mathbb{R}^n$, such that

(a) λ and u are ∞ times differentiable on $N(X_0)$,
(b) $\lambda(X_0) = \lambda_0$, $u(X_0) = u_0$,
(c) $Xu = \lambda u$, $u'u = 1$ for every $X \in N(X_0)$.

This completes the first part of our proof.

Let us now derive an explicit expression for $\mathrm{d}\lambda$. From $Xu = \lambda u$ we obtain

$$(\mathrm{d}X)u_0 + X_0\,\mathrm{d}u = (\mathrm{d}\lambda)u_0 + \lambda_0\,\mathrm{d}u, \tag{8}$$

where the differentials $\mathrm{d}\lambda$ and $\mathrm{d}u$ are defined at X_0. Pre-multiplying by u_0' gives

$$u_0'(\mathrm{d}X)u_0 + u_0'X_0\mathrm{d}u = (\mathrm{d}\lambda)u_0'u_0 + \lambda_0 u_0'\mathrm{d}u. \tag{9}$$

Since X_0 is symmetric we have $u_0'X_0 = \lambda_0 u_0'$. Hence

$$\mathrm{d}\lambda = u_0'(\mathrm{d}X)u_0, \tag{10}$$

because the eigenvector u_0 is normalized by $u_0'u_0 = 1$. The normalization of u

is not important here; it *is* important, however, in order to obtain an expression for du. To this we now turn. Let $Y_0 = \lambda_0 I - X_0$ and rewrite (8) as

$$Y_0 du = (dX)u_0 - (d\lambda)u_0. \tag{11}$$

Pre-multiplying by Y_0^+ we obtain

$$Y_0^+ Y_0 du = Y_0^+(dX)u_0, \tag{12}$$

because $Y_0^+ u_0 = 0$ (exercise 1). To complete the proof we need only show that

$$Y_0^+ Y_0 du = du. \tag{13}$$

To prove (13), let

$$C_0 = Y_0^+ Y_0 + u_0 u_0'. \tag{14}$$

The matrix C_0 is symmetric idempotent (because $Y_0 u_0 = Y_0^+ u_0 = 0$), so that $r(C_0) = r(Y_0) + 1 = n$. Hence, $C_0 = I_n$ and

$$du = C_0 du = (Y_0^+ Y_0 + u_0 u_0')du = Y_0^+ Y_0 du, \tag{15}$$

since $u_0' du = 0$ because of the normalization $u'u = 1$. This shows that (13) holds and concludes the proof. □

Note 1. We have chosen to normalize the eigenvector u by $u'u = 1$, which means that u is a point on the unit ball. This is, however, not the only possibility. Another normalization,

$$u_0' u = 1, \tag{16}$$

though less common, is in many ways more appropriate. The reason for this will become clear when we discuss the complex case (section 9). If the eigenvectors are normalized according to (16), then u is a point in the hyperplane tangent (at u_0) to the unit ball. In either case we obtain $u' du = 0$ at $X = X_0$, which is all that is needed in the proof.

Note 2. It is important to note that, while X_0 is symmetric, the perturbations are not assumed to be symmetric. For symmetric perturbations, application of Theorem 2.2 and the chain rule immediately yields

$$d\lambda = (u_0' \otimes u_0')D\, dv(X), \qquad du = (u_0' \otimes (\lambda_0 I - X_0)^+)D\, dv(X), \tag{17}$$

where D is the duplication matrix (see Chapter 3).

Exercises

1. If $A = A'$, then $Ab = 0$ if and only if $A^+ b = 0$.
2. Consider the symmetric 2×2 matrix

$$X_0 = \begin{pmatrix} 1 & 0 \\ 0 & -1 \end{pmatrix}.$$

When $\lambda_0 = 1$ show that, at X_0,

$$\mathrm{d}\lambda = \mathrm{d}x_{11} \qquad \text{and} \qquad \mathrm{d}u = \tfrac{1}{2}(\mathrm{d}x_{21})\begin{pmatrix} 0 \\ 1 \end{pmatrix},$$

and derive the corresponding result when $\lambda_0 = -1$. Interpret these results.

3. Now consider the matrix function

$$A(\varepsilon) = \begin{pmatrix} 1 & \varepsilon \\ \varepsilon & -1 \end{pmatrix}.$$

Plot a graph of the two eigenvalue functions $\lambda_1(\varepsilon)$ and $\lambda_2(\varepsilon)$, and show that the derivative at $\varepsilon = 0$ vanishes. Also obtain this result directly from the previous exercise.

4. Consider the symmetric matrix

$$X_0 = \begin{bmatrix} 3 & 0 & 0 \\ 0 & 4 & \sqrt{3} \\ 0 & \sqrt{3} & 6 \end{bmatrix}.$$

Show that the eigenvalues of X_0 are 3 (twice) and 7, and prove that at X_0 the differentials of the eigenvalue—and eigenvector—function associated with the eigenvalue 7 are

$$\mathrm{d}\lambda = \tfrac{1}{4}[\mathrm{d}x_{22} + (\mathrm{d}x_{23} + \mathrm{d}x_{32})\sqrt{3} + 3\mathrm{d}x_{33}]$$

and

$$\mathrm{d}u = \frac{1}{32}\begin{bmatrix} 4 & 0 & 0 & 4\sqrt{3} & 0 & 0 \\ 0 & 3 & -\sqrt{3} & 0 & 3\sqrt{3} & -3 \\ 0 & -\sqrt{3} & 1 & 0 & -3 & \sqrt{3} \end{bmatrix} \mathrm{d}p(X)$$

where

$$p(X) = (x_{12}, x_{22}, x_{32}, x_{13}, x_{23}, x_{33})'.$$

9. THE DIFFERENTIAL OF EIGENVALUES AND EIGENVECTORS: THE GENERAL COMPLEX CASE

Precisely the same techniques as used in establishing Theorem 7 enable us to establish Theorem 8.

Theorem 8

Let λ_0 be a simple eigenvalue (possible complex) of a matrix $Z_0 \in \mathbb{C}^{n \times n}$, and let u_0 be an associated eigenvector, so that $Z_0 u_0 = \lambda_0 u_0$. Then a complex-valued function λ and a (complex) vector function u are defined for all Z in some neighbourhood $N(Z_0) \in \mathbb{C}^{n \times n}$ of Z_0, such that

$$\lambda(Z_0) = \lambda_0, \qquad u(Z_0) = u_0, \tag{1}$$

and

$$Zu = \lambda u, \qquad u_0^* u = 1 \qquad (Z \in N(Z_0)). \tag{2}$$

Moreover, the functions λ and u are ∞ times differentiable on $N(Z_0)$, and the differentials at Z_0 are

$$\mathrm{d}\lambda = \frac{v_0^*(\mathrm{d}Z)u_0}{v_0^* u_0} \tag{3}$$

and

$$\mathrm{d}u = (\lambda_0 I - Z_0)^+ \left(I - \frac{u_0 v_0^*}{v_0^* u_0}\right)(\mathrm{d}Z)u_0, \tag{4}$$

where v_0 is an eigenvector associated with the eigenvalue $\bar{\lambda}_0$ of Z_0^*, so that $Z_0^* v_0 = \bar{\lambda}_0 v_0$.

Note. It seems natural to normalize u by $v_0^* u = 1$ instead of $u_0^* u = 1$. Such a normalization does not, however, lead to a Moore–Penrose inverse in (4). Another possible normalization, $u^* u = 1$, also leads to trouble, as the proof shows.

Proof. The fact that the functions λ and u exist and are ∞ times differentiable (i.e. analytic) in a neighbourhood of Z_0 is proved in the same way as in Theorem 7, using the complex analogue of Theorem 3.3 and Theorem 3.4, instead of Theorem 3.5. To find $\mathrm{d}\lambda$ we differentiate both sides of $Zu = \lambda u$, and obtain

$$(\mathrm{d}Z)u_0 + Z_0 \mathrm{d}u = (\mathrm{d}\lambda)u_0 + \lambda_0 \mathrm{d}u, \tag{5}$$

where $\mathrm{d}u$ and $\mathrm{d}\lambda$ are defined at Z_0. We now pre-multiply by v_0^*, and since $v_0^* Z_0 = \lambda_0 v_0^*$ and $v_0^* u_0 \neq 0$ (Why?), we obtain

$$\mathrm{d}\lambda = \frac{v_0^*(\mathrm{d}Z)u_0}{v_0^* u_0}. \tag{6}$$

To find $\mathrm{d}u$ we again define $Y_0 = \lambda_0 I - Z_0$, and rewrite (5) as

$$\begin{aligned} Y_0 \mathrm{d}u &= (\mathrm{d}Z)u_0 - (\mathrm{d}\lambda)u_0 \\ &= (\mathrm{d}Z)u_0 - \left(\frac{v_0^*(\mathrm{d}Z)u_0}{v_0^* u_0}\right)u_0 \\ &= \left(I - \frac{u_0 v_0^*}{v_0^* u_0}\right)(\mathrm{d}Z)u_0. \end{aligned} \tag{7}$$

Pre-multiplying by Y_0^+ we obtain

$$Y_0^+ Y_0 \mathrm{d}u = Y_0^+ \left(I - \frac{u_0 v_0^*}{v_0^* u_0}\right)(\mathrm{d}Z)u_0. \tag{8}$$

(Note that $Y_0^+ u_0 \neq 0$ in general.) To complete the proof we must again show that

$$Y_0^+ Y_0 \mathrm{d}u = \mathrm{d}u. \tag{9}$$

From $Y_0 u_0 = 0$ we have $u_0^* Y_0^* = 0'$ and hence $u_0^* Y_0^+ = 0'$. Also, since u is normalized by $u_0^* u = 1$, we have $u_0^* \mathrm{d}u = 0$. (Note that $u^* u = 1$ does *not* imply $u_0^* \mathrm{d}u = 0$.) Hence

$$u_0^*(Y_0^+ : \mathrm{d}u) = 0'. \tag{10}$$

It follows that

$$r(Y_0^+ : \mathrm{d}u) = r(Y_0^+) \tag{11}$$

which implies (9). From (8) and (9), (4) follows. □

Exercises

1. Show that $v_0^* u_0 \neq 0$.
2. Given the conditions of Theorem 8, show that

$$\mathrm{d}\bar{\lambda} = \frac{u_0^*(\mathrm{d}Z)^* v_0}{u_0^* v_0}$$

and

$$\mathrm{d}v^* = v_0^*(\mathrm{d}Z)\left(I - \frac{u_0 v_0^*}{v_0^* u_0}\right)(\lambda_0 I - Z_0)^+.$$

3. Show that

$$\mathrm{d}u = (\lambda_0 I - Z_0)^+ (\mathrm{d}Z) u_0$$

if, and only if, $\mathrm{d}\lambda = 0$ or v_0 is a multiple of u_0.

10. TWO ALTERNATIVE EXPRESSIONS FOR dλ

As we have seen, the differential (9.3) of the eigenvalue function associated with a simple eigenvalue λ_0 of a (complex) matrix Z_0 can be expressed as

$$\mathrm{d}\lambda = \operatorname{tr} P_0 \mathrm{d}Z, \qquad P_0 = \frac{u_0 v_0^*}{v_0^* u_0}, \tag{1}$$

where u_0 and v_0 are (right and left) eigenvectors of Z_0 associated with λ_0:

$$Z_0 u_0 = \lambda_0 u_0, \qquad v_0^* Z_0 = \lambda_0 v_0^*, \qquad u_0^* u_0 = v_0^* v_0 = 1. \tag{2}$$

The matrix P_0 is idempotent with $r(P_0) = 1$.

Let us now express P_0 in two other ways: first as a product of $n-1$ matrices, and then as a weighted sum of the matrices $I, Z_0, \ldots, Z_0^{n-1}$.

Theorem 9

Let $\lambda_1, \lambda_2, \ldots, \lambda_n$ be the eigenvalues of a matrix $Z_0 \in \mathbb{C}^{n\times n}$, and assume that λ_i is simple. Then a scalar function $\lambda_{(i)}$ exists, defined in a neighbourhood $N(Z_0) \subset \mathbb{C}^{n\times n}$ of Z_0, such that $\lambda_{(i)}(Z_0) = \lambda_i$ and $\lambda_{(i)}(Z)$ is a (simple) eigenvalue of Z for every $Z \in N(Z_0)$. Moreover, $\lambda_{(i)}$ is ∞ times differentiable on $N(Z_0)$, and

$$\mathrm{d}\lambda_{(i)} = \operatorname{tr}\left\{\left[\prod_{\substack{j=1\\ j\neq i}}^{n}\left(\frac{\lambda_j I - Z_0}{\lambda_j - \lambda_i}\right)\right]\mathrm{d}Z\right\}. \tag{3}$$

If we assume, in addition, that *all* eigenvalues of Z_0 are simple, we may express $d\lambda_{(i)}$ also as

$$d\lambda_{(i)} = \operatorname{tr}\left(\sum_{j=1}^{n} v^{ij} Z_0^{j-1} dZ\right) \qquad (i=1,\ldots,n), \tag{4}$$

where v^{ij} is the typical element of the inverse of the Vandermonde matrix

$$V = \begin{bmatrix} 1 & 1 & \ldots & 1 \\ \lambda_1 & \lambda_2 & \ldots & \lambda_n \\ \vdots & \vdots & & \vdots \\ \lambda_1^{n-1} & \lambda_2^{n-1} & \ldots & \lambda_n^{n-1} \end{bmatrix}. \tag{5}$$

Note. In expression (3) it is not demanded that the eigenvalues are all distinct, neither that they are all non-zero. In (4), however, the eigenvalues are assumed to be distinct. Still, one (but only one) eigenvalue may be zero.

Proof. Consider the following two matrices of order $n \times n$:

$$A = \lambda_i I - Z_0 \qquad \text{and} \qquad B = \prod_{j \neq i} (\lambda_j I - Z_0). \tag{6}$$

The Cayley–Hamilton theorem (Theorem 1.10) asserts that

$$AB = BA = 0. \tag{7}$$

Further, since λ_i is a simple eigenvalue of Z_0 and using the corollary to Theorem 1.19, we find that $r(A) = n-1$. Hence application of Theorem 3.6 shows that

$$B = \mu u_0 v_0^*, \tag{8}$$

where u_0 and v_0^* are defined in (2), and μ is an arbitrary scalar.

To determine the scalar μ, we use Schur's decomposition theorem (Theorem 1.12) and write

$$S^* Z_0 S = \Lambda + R, \qquad S^* S = I, \tag{9}$$

where Λ is a diagonal matrix containing $\lambda_1, \lambda_2, \ldots, \lambda_n$ on its diagonal, and R is *strictly* upper triangular. Then,

$$\begin{aligned} \operatorname{tr} B &= \operatorname{tr} \prod_{j \neq i} (\lambda_j I - Z_0) = \operatorname{tr} \prod_{j \neq i} (\lambda_j I - \Lambda - R) \\ &= \operatorname{tr} \prod_{j \neq i} (\lambda_j I - \Lambda) = \prod_{j \neq i} (\lambda_j - \lambda_i). \end{aligned} \tag{10}$$

From (8) we also have

$$\operatorname{tr} B = \mu v_0^* u_0, \tag{11}$$

and since $v_0^* u_0$ is non-zero, we find

$$\mu = \frac{\prod_{j \neq i} (\lambda_j - \lambda_i)}{v_0^* u_0}. \tag{12}$$

Hence,

$$\prod_{j\neq i}\left(\frac{\lambda_j I - Z_0}{\lambda_j - \lambda_i}\right) = \frac{u_0 v_0^*}{v_0^* u_0}, \tag{13}$$

which by (1) is what we wanted to show.

Let us next prove (4). Since all eigenvalues of Z_0 are now assumed to be distinct, there exists by Theorem 1.15 a non-singular matrix T such that

$$T^{-1}Z_0 T = \Lambda. \tag{14}$$

Therefore,

$$\sum_j v^{ij} Z_0^{j-1} = T\left(\sum_j v^{ij}\Lambda^{j-1}\right)T^{-1}. \tag{15}$$

If we denote by E_{ii} the $n\times n$ matrix with a one in its i-th diagonal position and zeros elsewhere, and by δ_{ik} the Kronecker delta, then

$$\sum_j v^{ij}\Lambda^{j-1} = \sum_j v^{ij}\left(\sum_k \lambda_k^{j-1}E_{kk}\right) = \sum_k\left(\sum_j v^{ij}\lambda_k^{j-1}\right)E_{kk}$$

$$= \sum_k \delta_{ik}E_{kk} = E_{ii}, \tag{16}$$

because the scalar expression $\Sigma_j v^{ij}\lambda_k^{j-1}$ is the inner product of the i-th row of V^{-1} and the k-th column of V, that is

$$\sum_j v^{ij}\lambda_k^{j-1} = \delta_{ik}. \tag{17}$$

Inserting (16) in (15) yields

$$\sum_j v^{ij}Z_0^{j-1} = TE_{ii}T^{-1} = (Te_i)(e_i' T^{-1}), \tag{18}$$

where e_i is the i-th unit vector. Since λ_i is a simple eigenvalue of Z_0, we have

$$Te_i = \gamma u_0 \qquad \text{and} \qquad e_i' T^{-1} = \delta v_0^* \tag{19}$$

for some scalars γ and δ. Further,

$$1 = e_i' T^{-1} Te_i = \gamma\delta v_0^* u_0. \tag{20}$$

Hence,

$$\sum_j v^{ij} Z_0^{j-1} = (Te_i)(e_i' T^{-1}) = \gamma\delta u_0 v_0^* = \frac{u_0 v_0^*}{v_0^* u_0}. \tag{21}$$

This concludes the proof, by (1). □

Exercise

1. Show that the elements in the first column of V^{-1} sum to one, and that the elements in any other column of V^{-1} sum to zero.

11. THE SECOND DIFFERENTIAL OF THE EIGENVALUE FUNCTION

One application of the differential of the eigenvector du is to obtain the second differential of the eigenvalue $d^2\lambda$. We consider first the case where X_0 is a real symmetric matrix.

Theorem 10

Under the same conditions as in Theorem 7, we have

$$d^2\lambda = 2u_0'(dX)(\lambda_0 I - X_0)^+(dX)u_0. \tag{1}$$

Proof. Twice differentiating both sides of $Xu = \lambda u$, we obtain

$$2(dX)(du) + X_0 d^2u = (d^2\lambda)u_0 + 2(d\lambda)(du) + \lambda_0 d^2u, \tag{2}$$

where all differentials are evaluated at X_0. Pre-multiplying by u_0' gives

$$d^2\lambda = 2u_0'(dX)(du), \tag{3}$$

since $u_0'u_0 = 1$, $u_0'du = 0$ and $u_0'X_0 = \lambda_0 u_0'$. From Theorem 7 we have $du = (\lambda_0 I - X_0)^+(dX)u_0$. Inserting this in (3) gives (1).

The case where Z_0 is a complex $n \times n$ matrix is proved in a similar way.

Theorem 11

Under the same conditions as in Theorem 8, we have

$$d^2\lambda = \frac{2v_0^*(dZ)K_0(\lambda_0 I - Z_0)^+ K_0(dZ)u_0}{v_0^* u_0}, \tag{4}$$

where

$$K_0 = I - \frac{u_0 v_0^*}{v_0^* u_0}. \tag{5}$$

Exercises

1. Show that (1) can be written as

$$d^2\lambda = 2(d\,\mathrm{vec}\,X)'[(\lambda_0 I - X_0)^+ \otimes u_0 u_0']\,d\,\mathrm{vec}\,X$$

and also as

$$d^2\lambda = 2(d\,\mathrm{vec}\,X)'[u_0 u_0' \otimes (\lambda_0 I - X_0)^+]\,d\,\mathrm{vec}\,X.$$

2. Show that if λ_0 is the largest eigenvalue of X_0, then $d^2\lambda \geq 0$. Relate this to the fact that the largest eigenvalue is convex on the space of real symmetric matrices. (Compare Theorem 11.5.)
3. Similarly, if λ_0 is the smallest eigenvalue of X_0, show that $d^2\lambda \leq 0$.

12. MULTIPLE EIGENVALUES

The case of multiple eigenvalues is more difficult. In section 7 we considered the matrix function

$$A(\varepsilon, \delta) = \begin{pmatrix} 1+\varepsilon & \delta \\ \delta & 1-\varepsilon \end{pmatrix} \tag{1}$$

whose eigenvalues are not differentiable at (0, 0), and we concluded that it would be wise to restrict the study of multiple eigenvalues to matrix functions of one parameter only.

In this section we briefly summarize some of Lancaster's (1964) results. We consider the eigenvalues of $n \times n$ matrices A whose elements are functions of *one* parameter ξ, and we assume that (i) the elements of $A(\xi)$ are analytic functions in some neighbourhood of ξ_0, (ii) the matrix $A_0 = A(\xi_0)$ has *simple structure* (i.e. all eigenvalues of A_0 have only linear elementary divisors), and (iii) if $\lambda(\xi)$ is an eigenvalue of $A(\xi)$, then $\lambda(\xi) \to \lambda(\xi_0)$ as $\xi \to \xi_0$.

We shall denote by $A^{(q)}(\xi_0)$ the q-th derivative of $A(\xi)$ evaluated at $\xi = \xi_0$.

Theorem 12

If $A^{(q)}(\xi_0)$ is the first non-vanishing derivative of $A(\xi)$ at $\xi = \xi_0$, then the n eigenvalues $\lambda(\xi)$ of $A(\xi)$ are differentiable at least q times at ξ_0 and their first $q-1$ derivatives all vanish at ξ_0.

Now let λ_0 be an eigenvalue of A_0 with multiplicity m. Let U_0 be the $n \times m$ matrix whose m columns span the subspace of eigenvectors associated with $\lambda_0 : A_0 U_0 = \lambda_0 U_0$. Also, let V_0 be the $n \times m$ matrix whose m columns span the subspace of eigenvectors associated with the eigenvalue $\bar{\lambda}_0$ of $A_0^* : A_0^* V_0 = \bar{\lambda}_0 V_0$. We can normalize the matrices U_0 and V_0 so that $V_0^* U_0 = I_m$.

Theorem 13

If $A^{(q)}(\xi_0)$ is the first non-vanishing derivative of $A(\xi)$ at $\xi = \xi_0$, then the m derivatives $\lambda^{(q)}(\xi_0)$ (of the m eigenvalues which coincide at ξ_0) are the eigenvalues of the matrix $V_0^* A^{(q)}(\xi_0) U_0$.

Note. Compare Theorem 13 with the expression for $\mathrm{d}\lambda$ in Theorem 8.

MISCELLANEOUS EXERCISES

1. In generalizing the fundamental rule $\mathrm{d}x^k = kx^{k-1}\,\mathrm{d}x$ to matrices, show that it is not true that $\mathrm{d}X^k = kX^{k-1}\,\mathrm{d}X$. It is true, however, that

$$\mathrm{d}\,\mathrm{tr}\,X^k = k\,\mathrm{tr}\,X^{k-1}\,\mathrm{d}X \qquad (k = 1, 2, \ldots).$$

Prove that this also holds for *real* $k \geqslant 1$ when X is positive semidefinite.

2. Consider a point X_0 with distinct eigenvalues $\lambda_1, \lambda_2, \ldots, \lambda_n$. From the fact that $\operatorname{tr} X^k = \Sigma_i \lambda_i^k$, deduce that at X_0

$$\mathrm{d} \operatorname{tr} X^k = k \sum_i \lambda_i^{k-1} \mathrm{d}\lambda_i.$$

3. Conclude from the foregoing that at X_0,

$$\sum_i \lambda_i^{k-1} \mathrm{d}\lambda_i = \operatorname{tr} X_0^{k-1} \mathrm{d}X \qquad (k = 1, 2, \ldots, n).$$

Write this system of n equations as

$$\begin{bmatrix} 1 & 1 & \ldots & 1 \\ \lambda_1 & \lambda_2 & \ldots & \lambda_n \\ \vdots & \vdots & & \vdots \\ \lambda_1^{n-1} & \lambda_2^{n-1} & \ldots & \lambda_n^{n-1} \end{bmatrix} \begin{bmatrix} \mathrm{d}\lambda_1 \\ \mathrm{d}\lambda_2 \\ \vdots \\ \mathrm{d}\lambda_n \end{bmatrix} = \begin{bmatrix} \operatorname{tr} \mathrm{d}X \\ \operatorname{tr} X_0 \, \mathrm{d}X \\ \vdots \\ \operatorname{tr} X_0^{n-1} \, \mathrm{d}X \end{bmatrix}.$$

Solve $\mathrm{d}\lambda_i$. This provides an alternative proof of the second part of Theorem 9.

4. At points X where the eigenvalues $\lambda_1, \lambda_2, \ldots, \lambda_n$ of X are distinct, show that

$$\mathrm{d}|X| = \sum_i \left(\prod_{j \neq i} \lambda_j \right) \mathrm{d}\lambda_i.$$

In particular, at points where one of the eigenvalues is zero,

$$\mathrm{d}|X| = \left(\prod_{j=1}^{n-1} \lambda_j \right) \mathrm{d}\lambda_n$$

where λ_n is the (simple) zero eigenvalue.

5. Use this result and the fact that $\mathrm{d}|X| = \operatorname{tr} X^{\#} \mathrm{d}X$ and $\mathrm{d}\lambda_n = v'(\mathrm{d}X)u/v'u$, where $X^{\#}$ is the adjoint matrix of X, and $Xu = X'v = 0$, to show that

$$X^{\#} = \left(\prod_{j=1}^{n-1} \lambda_j \right) \frac{uv'}{v'u}$$

at points where $\lambda_n = 0$ is a simple eigenvalue. (Compare Theorem 3.3.)

6. Let $F: S \to \mathbb{R}^{m \times m}$ $(m \geqslant 2)$ be a matrix function, defined on a set S in $\mathbb{R}^{n \times q}$ and differentiable at a point $X_0 \in S$. Assume that $F(X)$ has a simple eigenvalue 0 at X_0 and in a neighbourhood $N(X_0) \subset S$ of X_0. (This implies that $r(F(X)) = m - 1$ for every $X \in N(X_0)$.) Then

$$(\mathrm{d}F^{\#})(X_0) = (\operatorname{tr} R_0 \mathrm{d}F) F_0^{\#} - F_0^{+} (\mathrm{d}F) F_0^{\#} - F_0^{\#} (\mathrm{d}F) F_0^{+},$$

where $F_0^{\#} = (F(X_0))^{\#}$ and $F_0^{+} = (F(X_0))^{+}$. Show that $R_0 = F_0^{+}$ if $F(X_0)$ is symmetric. What is R_0 if $F(X_0)$ is not symmetric?

7. Let $F : S \to \mathbb{R}^{m\times m}$ $(m \geqslant 2)$ be a *symmetric* matrix function, defined on a set S in $\mathbb{R}^{n\times q}$ and differentiable at a point $X_0 \in S$. Assume that $F(X)$ has a simple eigenvalue 0 at X_0 and in a neighbourhood of X_0. Let $F_0 = F(X_0)$. Then,

$$\mathrm{d}F^+(X_0) = -F_0^+(\mathrm{d}F)F_0^+.$$

8. Define the matrix function

$$\exp(X) = \sum_{k=0}^{\infty} \frac{1}{k!} X^k$$

which is well defined for every square matrix X, real or complex. Show that

$$\mathrm{d}\exp(X) = \sum_{k=0}^{\infty} \frac{1}{(k+1)!} \sum_{j=0}^{k} X^j(\mathrm{d}X)X^{k-j}$$

and, in particular,

$$\mathrm{tr}(\mathrm{d}\exp(X)) = \mathrm{tr}\,(\exp(X)(\mathrm{d}X)).$$

9. Let S_n denote the set of $n \times n$ symmetric matrices whose eigenvalues are smaller than one in absolute value. For X in S_n show that

$$(I_n - X)^{-1} = \sum_{k=0}^{\infty} X^k.$$

10. For X in S_n define

$$\log(I_n - X) = -\sum_{k=0}^{\infty} \frac{1}{k} X^k.$$

Show that

$$\mathrm{d}\log(I_n - X) = -\sum_{k=0}^{\infty} \frac{1}{k+1} \sum_{j=0}^{k} X^j(\mathrm{d}X)\, X^{k-j}$$

and, in particular,

$$\mathrm{tr}(\mathrm{d}\log(I_n - X)) = -\,\mathrm{tr}((I_n - X)^{-1}\,\mathrm{d}X).$$

BIBLIOGRAPHICAL NOTES

§5. Lemma 1 is due to Penrose (1955, p. 408) and Lemma 2 to Hearon and Evans (1968). See also Stewart (1969). Theorem 5 is due to Golub and Pereyra (1973).

§7–§11. The development follows Magnus (1985). Figure 1 was suggested to us by Roald Ramer. See also Bargmann and Nel (1974), Lancaster (1964), Sugiura (1973) and Kalaba *et al.* (1980, 1981a, 1981b).

§12. See Lancaster (1964).

CHAPTER 9

First-order differentials and Jacobian matrices

1. INTRODUCTION

We begin this chapter with some notational issues. We shall argue very strongly for a particular way of displaying the partial derivatives $\partial f_{st}(X)/\partial x_{ij}$ of a matrix function $F(X)$, one which generalizes the notion of a Jacobian matrix of a vector function to a Jacobian matrix of a matrix function.

The main tool in this chapter will be the first identification theorem (Theorem 5.11), which tells us how to obtain the derivative (Jacobian matrix) from the differential. Given a matrix function $F(X)$ we then proceed as follows: (i) compute the differential of $F(X)$, (ii) vectorize to obtain $\mathrm{d}\,\mathrm{vec}\,F(X) = A(X)\,\mathrm{d}\,\mathrm{vec}\,X$, and (iii) conclude that $\mathsf{D}F(X) = A(X)$.

The simplicity and elegance of this approach will be demonstrated by many examples.

2. CLASSIFICATION

We shall consider *scalar* functions ϕ, *vector* functions f and *matrix* functions F. Each of these may depend on *one* real variable ξ, a *vector* of real variables x, or a *matrix* of real variables X. We thus obtain the classification of functions and variables shown in Table 1.

***Table* 1** Classification of functions and variables.

	Scalar variable	Vector variable	Matrix variable
Scalar function	$\phi(\xi)$	$\phi(x)$	$\phi(X)$
Vector function	$f(\xi)$	$f(x)$	$f(X)$
Matrix function	$F(\xi)$	$F(x)$	$F(X)$

Examples

$\phi(\xi)$:	ξ^2
$\phi(x)$:	$a'x$, $x'Ax$
$\phi(X)$:	$a'Xb$, $\operatorname{tr} X'X$, $\lvert X\rvert$, $\lambda(X)$ (eigenvalue)
$f(\xi)$:	$(\xi, \xi^2)'$
$f(x)$:	Ax
$f(X)$:	Xa, $u(X)$ (eigenvector)
$F(\xi)$:	$\begin{pmatrix} 1 & \xi \\ \xi & \xi^2 \end{pmatrix}$
$F(x)$:	xx'
$F(X)$:	AXB, X^2, X^+

3. BAD NOTATION

If F is a differentiable $m \times p$ matrix function of an $n \times q$ matrix X of variables then the question naturally arises how to order the $mnpq$ partial derivatives of F. Obviously, this can be done in many ways. The purpose of this section is to convince the reader *not* to use the following notation, which, for reasons unknown, has earned itself an undeserved popularity.

Definition 1

Let ϕ be a differentiable real-valued function of an $n \times q$ matrix $X = (x_{ij})$ of real variables. Then the symbol $\partial\phi(X)/\partial X$ denotes the $n \times q$ matrix

$$\frac{\partial\phi(X)}{\partial X} = \begin{bmatrix} \partial\phi/\partial x_{11} & \dots & \partial\phi/\partial x_{1q} \\ \vdots & & \vdots \\ \partial\phi/\partial x_{n1} & \dots & \partial\phi/\partial x_{nq} \end{bmatrix}. \tag{1}$$

Definition 2

Let $F = (f_{st})$ be a differentiable $m \times p$ real matrix function of an $n \times q$ matrix X of real variables. Then the symbol $\partial F(X)/\partial X$ denotes the $mn \times pq$ matrix

$$\frac{\partial F(X)}{\partial X} = \begin{bmatrix} \partial f_{11}/\partial X & \dots & \partial f_{1p}/\partial X \\ \vdots & & \vdots \\ \partial f_{m1}/\partial X & \dots & \partial f_{mp}/\partial X \end{bmatrix}. \tag{2}$$

Before we criticize Definition 2, let us list some of its good points. Two very pleasant properties are: (i) if F is a matrix function of just one variable ξ, then $\partial F(\xi)/\partial\xi$ has the same order as $F(\xi)$; and (ii) if ϕ is a scalar function of a matrix of variables X, then $\partial\phi(X)/\partial X$ has the same order as X. In particular, if ϕ is a scalar function of a column vector x, then $\partial\phi/\partial x$ is a column vector and $\partial\phi/\partial x'$ a row vector. Another consequence of the definition is that it allows us to order the mn partial derivatives of an $m\times 1$ vector function $f(x)$, where x is an $n\times 1$ vector of variables, in four ways: namely as $\partial f/\partial x'$ (an $m\times n$ matrix), as $\partial f'/\partial x$ (an $n\times m$ matrix), as $\partial f/\partial x$ (an $mn\times 1$ vector), or as $\partial f'/\partial x'$ (a $1\times mn$ vector).

To see what is wrong with the definition, let us consider the identity function $F(X)=X$, where X is an $n\times q$ matrix of real variables. We obtain from Definition 2

$$\frac{\partial F(X)}{\partial X}=(\operatorname{vec} I_n)(\operatorname{vec} I_q)', \tag{3}$$

a matrix of rank 1. The Jacobian matrix of the identity function is, of course, I_{nq}, the $nq\times nq$ identity matrix. Hence Definition 2 does not give us the Jacobian matrix of the function F, and, indeed, the rank of the Jacobian matrix is not given by the rank of $\partial F(X)/\partial X$. This implies—and this cannot be stressed enough—that the matrix (2) displays the partial derivatives, *but nothing more*. In particular, the determinant of $\partial F(X)/\partial X$ has no interpretation, and (very important for practical work) a useful chain rule does not exist.

There exists another definition, equally unsuitable, which is based not on $\partial\phi(X)/\partial X$, but on

$$\frac{\partial F(X)}{\partial x_{ij}}=\begin{bmatrix}\partial f_{11}(X)/\partial x_{ij} & \dots & \partial f_{1p}(X)/\partial x_{ij}\\ \vdots & & \vdots\\ \partial f_{m1}(X)/\partial x_{ij} & \dots & \partial f_{mp}(X)/\partial x_{ij}\end{bmatrix}. \tag{4}$$

Definition 3

Let F be a differentiable $m\times p$ matrix function of an $n\times q$ matrix $X=(x_{ij})$ of real variables. Then the symbol $\partial F(X)//\partial X$ denotes the $mn\times pq$ matrix

$$\frac{\partial F(X)}{\partial\!\!\!=\!\! X}=\begin{bmatrix}\partial F(X)/\partial x_{11} & \dots & \partial F(X)/\partial x_{1q}\\ \vdots & & \vdots\\ \partial F(X)/\partial x_{n1} & \dots & \partial F(X)/\partial x_{nq}\end{bmatrix}. \tag{5}$$

Definition 3 is equally as bad as Definition 2, except for one point in which it has an advantage over Definition 2, namely that the expressions $\partial F(X)/\partial x_{ij}$ are much easier to evaluate than $\partial f_{st}(X)/\partial X$, because the latter expressions require us to disentangle the matrix function $F(X)$.

After these critical remarks, let us turn quickly to the only natural and viable generalization of the notion of a Jacobian matrix of a vector function to a Jacobian matrix of a matrix function.

Exercises

1. Show that

$$\frac{\partial X}{\partial\!\!\!=\!\! X}=\frac{\partial X}{\partial X}=(\operatorname{vec} I_n)(\operatorname{vec} I_q)'.$$

2. Let $f: \mathbb{R}^n \to \mathbb{R}^m$ be a differentiable vector function. Then show that

$$\partial f(x) // \partial x' = \frac{\partial f(x)}{\partial x'} = \mathsf{D}f(x),$$

is an $m \times n$ matrix of partial derivatives.

3. Show that $\partial F/\partial X$ and $\partial F//\partial X$ stand in one-to-one relationship,

$$\partial F // \partial X = K_{nm} \frac{\partial F}{\partial X} K_{pq}$$

and

$$\frac{\partial F}{\partial X} = K_{mn} \, \partial F // \partial X \, K_{qp},$$

where K is the commutation matrix (Neudecker 1982).

4. GOOD NOTATION

Let ϕ be a scalar function of an $n \times 1$ vector x. We have already encountered the *derivative* of ϕ,

$$\mathsf{D}\phi(x) = (\mathsf{D}_1\phi(x), \ldots, \mathsf{D}_n\phi(x)) = \frac{\partial \phi(x)}{\partial x'}. \tag{1}$$

If f is an $m \times 1$ vector function of x, then the derivative (or *Jacobian matrix*) of f is the $m \times n$ matrix

$$\mathsf{D}f(x) = \frac{\partial f(x)}{\partial x'}. \tag{2}$$

Since (1) is just a special case of (2), the double use of the D symbol is permitted. Generalizing these concepts to matrix functions of matrices, we arrive at the following definition.

Definition 1

Let F be a differentiable $m \times p$ real matrix function of an $n \times q$ matrix of real variables X. The *Jacobian matrix* of F at X is the $mp \times nq$ matrix

$$\mathsf{D}F(X) = \frac{\partial \operatorname{vec} F(X)}{\partial (\operatorname{vec} X)'}. \tag{3}$$

Thus $\mathsf{D}F$, $\mathsf{D}f$ and $\mathsf{D}\phi$ are all defined. The reader should compare (3) with the equivalent expression in (5.15.9).

It is worthwhile noticing that $\mathsf{D}F(X)$ and $\partial F(X)/\partial X$ contain the same $mnpq$ partial derivatives, but in a different pattern. Indeed, the orders of the two matrices are different ($\mathsf{D}F(X)$ is of the order $mp \times nq$, while $\partial F(X)/\partial X$ is of the order $mn \times pq$), and, more important, their ranks are in general different.

Since $\mathsf{D}F(X)$ is a straightforward matrix generalization of the traditional definition of the Jacobian matrix $\partial f(x)/\partial x'$, all properties of Jacobian matrices are preserved. In particular, questions relating to functions with non-zero Jacobian determinant at certain points, remain meaningful.

Definition 1 reduces the study of matrix functions of matrices to the study of vector functions of vectors, since it allows $F(X)$ and X only in their vectorized forms $\operatorname{vec} F$ and $\operatorname{vec} X$. As a result, the unattractive expressions

$$\frac{\partial F(X)}{\partial X}, \qquad \frac{\partial F(x)}{\partial x} \qquad \text{and} \qquad \frac{\partial f(X)}{\partial X} \tag{4}$$

are not needed. The same is, in principle, true for the expressions

$$\frac{\partial \phi(X)}{\partial X} \qquad \text{and} \qquad \frac{\partial F(\xi)}{\partial \xi}, \tag{5}$$

since these can be replaced by

$$\mathsf{D}\phi(X)=\frac{\partial \phi(X)}{\partial(\operatorname{vec} X)'} \qquad \text{and} \qquad \mathsf{D}F(\xi)=\frac{\partial \operatorname{vec} F(\xi)}{\partial \xi}. \tag{6}$$

However, the idea of arranging the partial derivatives of $\phi(X)$ and $F(\xi)$ into a matrix (rather than a vector) is rather appealing and sometimes useful, so we retain the expressions (5).

Exercises

1. Let F be a differentiable matrix function of an $n \times q$ matrix of variables $X=(x_{ij})$. Then
$$\mathsf{D}F(X)=\sum_{i=1}^{n}\sum_{j=1}^{q}\left(\operatorname{vec}\frac{\partial F(X)}{\partial x_{ij}}\right)(\operatorname{vec} E_{ij})',$$
where E_{ij} denotes an $n \times q$ matrix with a one in the ij-th position and zeros elsewhere.
2. Show that
$$\mathsf{D}\phi(X)=\left(\operatorname{vec}\frac{\partial \phi(X)}{\partial X}\right)'$$
and
$$\mathsf{D}F(\xi)=\operatorname{vec}\frac{\partial F(\xi)}{\partial \xi}.$$

5. IDENTIFICATION OF JACOBIAN MATRICES

Our strategy to find the Jacobian matrix of a function will *not* be to evaluate each of its partial derivatives, but rather to find the differential. In the case of a differentiable vector function $f(x)$, the first identification theorem (Theorem 5.6) tells us that there exists a one-to-one correspondence between the differential of f and its Jacobian matrix. More specifically, it states that

$$\mathrm{d}f(x) = A(x)\mathrm{d}x \tag{1}$$

implies and is implied by

$$\mathsf{D}f(x) = A(x). \tag{2}$$

Thus, once we know the differential, the Jacobian matrix is identified.

The extension to matrix functions is straightforward. The *identification theorem for matrix functions* (Theorem 5.11) states that

$$\mathrm{d}\,\mathrm{vec}\,F(X) = A(X)\,\mathrm{d}\,\mathrm{vec}\,X \tag{3}$$

implies and is implied by

$$\mathsf{D}F(X) = A(X). \tag{4}$$

Since computations with differentials are relatively easy, this identification result is extremely useful. Given a matrix function $F(X)$ we may thus proceed as follows: (i) compute the differential of $F(X)$, (ii) vectorize to obtain $\mathrm{d}\,\mathrm{vec}\,F(X) = A(X)\,\mathrm{d}\,\mathrm{vec}\,X$, and (iii) conclude that $\mathsf{D}F(X) = A(X)$.

Many examples in this chapter will demonstrate the simplicity and elegance of this approach. Let us consider one now. Let $F(X) = AXB$, where A and B are matrices of constants. Then

$$\mathrm{d}F(X) = A(\mathrm{d}X)B, \tag{5}$$

and

$$\mathrm{d}\,\mathrm{vec}\,F(X) = (B' \otimes A)\,\mathrm{d}\,\mathrm{vec}\,X, \tag{6}$$

so that

$$\mathsf{D}F(X) = B' \otimes A. \tag{7}$$

6. THE FIRST IDENTIFICATION TABLE

The identification theorem for matrix functions of matrix variables encompasses, of course, identification for matrix, vector and scalar functions of matrix, vector and scalar variables. Table 2 lists these results.

In the first identification table, ϕ is a scalar function, f an $m \times 1$ vector function and F an $m \times p$ matrix function; ξ is a scalar, x an $n \times 1$ vector and X an $n \times q$ matrix; α is a scalar, a is a column vector and A is a matrix, each of which may be a function of X, x or ξ.

7. PARTITIONING OF THE DERIVATIVE

Before the workings of the identification table are exemplified, we have to settle one further question of notation. Let ϕ be a differentiable scalar function of an

***Table* 2** The first identification table.

Function	Differential	Derivative/Jacobian	Order of D	Other notation
$\phi(\xi)$	$d\phi=\alpha\, d\xi$	$D\phi(\xi)=\alpha$	1×1	
$\phi(x)$	$d\phi=a'dx$	$D\phi(x)=a'$	$1\times n$	
$\phi(X)$	$d\phi=\operatorname{tr} A'dX$ $=(\operatorname{vec} A)'\, d\operatorname{vec} X$	$D\phi(X)=(\operatorname{vec} A)'$	$1\times nq$	$\dfrac{\partial\phi(X)}{\partial X}=A$
$f(\xi)$	$df=a\, d\xi$	$Df(\xi)=a$	$m\times 1$	
$f(x)$	$df=A\, dx$	$Df(x)=A$	$m\times n$	
$f(X)$	$df=A\, d\operatorname{vec} X$	$Df(X)=A$	$m\times nq$	
$F(\xi)$	$dF=A\, d\xi$	$DF(\xi)=\operatorname{vec} A$	$mp\times 1$	$\dfrac{\partial F(\xi)}{\partial\xi}=A$
$F(x)$	$d\operatorname{vec} F=A\, dx$	$DF(x)=A$	$mp\times n$	
$F(X)$	$d\operatorname{vec} F=A\, d\operatorname{vec} X$	$DF(X)=A$	$mp\times nq$	

$n\times 1$ vector x. Suppose that x is partitioned as

$$x'=(x_1', x_2'). \tag{1}$$

Then the derivative $D\phi(x)$ is partitioned in the same way, and we write

$$D\phi(x)=(D_1\phi(x), D_2\phi(x)), \tag{2}$$

where $D_1\phi(x)$ contains the partial derivatives of ϕ with respect to x_1, and $D_2\phi(x)$ contains the partial derivatives of ϕ with respect to x_2. As a result, if

$$d\phi(x)=a_1'(x)dx_1+a_2'(x)dx_2, \tag{3}$$

then

$$D_1\phi(x)=a_1'(x), \qquad D_2\phi(x)=a_2'(x), \tag{4}$$

and so

$$D\phi(x)=(a_1'(x), a_2'(x)). \tag{5}$$

8. SCALAR FUNCTIONS OF A VECTOR

Let us now give some examples. The two most important cases of a scalar function of a vector x are the *linear form* $a'x$ and the *quadratic form* $x'Ax$.

Let $\phi(x)=a'x$, where a is a vector of constants. Then $d\phi(x)=a'dx$, so $D\phi(x)=a'$. Next, let $\phi(x)=x'Ax$, where A is a square matrix of constants. Then

$$\begin{aligned} d\phi(x) &= d(x'Ax)=(dx)'Ax+x'Adx \\ &= ((dx)'Ax)'+x'Adx=x'A'dx+x'Adx \\ &= x'(A+A')dx, \end{aligned} \tag{1}$$

so that $D\phi(x)=x'(A+A')$. Thus we obtain Table 3.

Table 3

$\phi(x)$	$\mathrm{d}\phi(x)$	$\mathrm{D}\phi(x)$
$a'x$	$a'\mathrm{d}x$	a'
$x'Ax$	$x'(A+A')\mathrm{d}x$	$x'(A+A')$

Note: If A is symmetric, then $\mathrm{D}\phi(x)=2x'A$.

Exercises

1. If $\phi(x)=a'f(x)$, then $\mathrm{D}\phi(x)=a'\mathrm{D}f(x)$.
2. If $\phi(x)=(f(x))'g(x)$, then $\mathrm{D}\phi(x)=(g(x))'\mathrm{D}f(x)+(f(x))'\mathrm{D}g(x)$.
3. If $\phi(x)=x'Af(x)$, then $\mathrm{D}\phi(x)=(f(x))'A'+x'A\mathrm{D}f(x)$.
4. If $\phi(x)=(f(x))'Af(x)$, then $\mathrm{D}\phi(x)=(f(x))'(A+A')\mathrm{D}f(x)$.
5. If $\phi(x)=(f(x))'Ag(x)$, then $\mathrm{D}\phi(x)=(g(x))'A'\mathrm{D}f(x)+(f(x))'A\mathrm{D}g(x)$.
6. If $\phi(x)=x_1'Ax_2$, where $x=(x_1':x_2')'$, then $\mathrm{D}_1\phi(x)=x_2'A'$, $\mathrm{D}_2\phi(x)=x_1'A$ and

$$\mathrm{D}\phi(x)=x'\begin{pmatrix}0 & A\\ A' & 0\end{pmatrix}.$$

9. SCALAR FUNCTIONS OF A MATRIX, I: TRACE

There is certainly no lack of interesting examples of scalar functions of matrices. In this section we shall investigate differentials of traces of some matrix functions. Section 10 is devoted to determinants, and section 11 to eigenvalues.

The simplest case is

$$\mathrm{d}\,\mathrm{tr}\,X=\mathrm{tr}\,\mathrm{d}X=\mathrm{tr}\,I\,\mathrm{d}X, \tag{1}$$

implying

$$\frac{\partial\,\mathrm{tr}\,X}{\partial X}=I. \tag{2}$$

More interesting is the trace of the (positive semidefinite) matrix function $X'X$. We have

$$\begin{aligned}\mathrm{d}\,\mathrm{tr}\,X'X&=\mathrm{tr}\,\mathrm{d}(X'X)=\mathrm{tr}((\mathrm{d}X)'X+X'\mathrm{d}X)\\&=\mathrm{tr}(\mathrm{d}X)'X+\mathrm{tr}\,X'\mathrm{d}X=2\,\mathrm{tr}\,X'\mathrm{d}X.\end{aligned} \tag{3}$$

Hence,

$$\frac{\partial\,\mathrm{tr}\,X'X}{\partial X}=2X. \tag{4}$$

Next consider the trace of X^2, where X is square. This gives

$$\mathrm{d}\,\mathrm{tr}\,X^2=\mathrm{tr}\,\mathrm{d}X^2=\mathrm{tr}((\mathrm{d}X)X+X\mathrm{d}X)=2\,\mathrm{tr}\,X\mathrm{d}X, \tag{5}$$

Table 4

$\phi(X)$	$\mathrm{d}\phi(X)$	$\mathsf{D}\phi(X)$	$\partial\phi(X)/\partial X$
$\operatorname{tr} AX$	$\operatorname{tr} A\mathrm{d}X$	$(\operatorname{vec} A')'$	A'
$\operatorname{tr} XAX'B$	$\operatorname{tr}(AX'B+A'X'B')\mathrm{d}X$	$(\operatorname{vec}(B'XA'+BXA))'$	$B'XA'+BXA$
$\operatorname{tr} XAXB$	$\operatorname{tr}(AXB+BXA)\mathrm{d}X$	$(\operatorname{vec}(B'X'A'+A'X'B'))'$	$B'X'A'+A'X'B'$

and thus

$$\frac{\partial \operatorname{tr} X^2}{\partial X}=2X'. \tag{6}$$

In Table 4 we present straightforward generalizations of the three cases just considered. The proofs are easy and are left to the reader.

Exercises

1. Show that $\operatorname{tr} BX'$, $\operatorname{tr} XB$, $\operatorname{tr} X'B$, $\operatorname{tr} BXC$ and $\operatorname{tr} BX'C$ can all be written as $\operatorname{tr} AX$ and determine their derivatives.
2. Show that
$$\partial \operatorname{tr} X'AX/\partial X=(A+A')X;$$
$$\partial \operatorname{tr} XAX'/\partial X=X(A+A');$$
$$\partial \operatorname{tr} XAX/\partial X=X'A'+A'X'.$$
3. Show that
$$\partial \operatorname{tr} AX^{-1}/\partial X=-(X^{-1}AX^{-1})'.$$
4. Use the previous results to find the derivatives of $a'Xb$, $a'XX'a$ and $a'X^{-1}a$.
5. Show that for square X,
$$\partial \operatorname{tr} X^p/\partial X=p(X')^{p-1} \qquad (p=1, 2, \ldots).$$
6. If $\phi(X)=\operatorname{tr} F(X)$, then $\mathsf{D}\phi(X)=(\operatorname{vec} I)'\mathsf{D}F(X)$.
7. Determine the derivative of $\phi(X)=\operatorname{tr} F(X)AG(X)B$.
8. Determine the derivative of $\phi(X, Z)=\operatorname{tr} AXBZ$.

10. SCALAR FUNCTIONS OF A MATRIX, II: DETERMINANT

Recall that the differential of a determinant is given by

$$\mathrm{d}|X|=|X|\operatorname{tr} X^{-1}\mathrm{d}X, \tag{1}$$

if X is a square non-singular matrix (Theorem 8.1). As a result the derivative is

$$|X|(\operatorname{vec}(X^{-1})')', \tag{2}$$

and

$$\frac{\partial |X|}{\partial X} = |X|(X')^{-1}. \tag{3}$$

This is easily verified from the identification table.

Let us now employ equation (1) to find the differential and derivative of the determinant of some simple matrix functions of X. The first of these is $|XX'|$, where X is not necessarily a square matrix, but must have full row rank in order to ensure that the determinant is non-zero (in fact, positive). The differential is

$$\begin{aligned} \mathrm{d}|XX'| &= |XX'| \operatorname{tr}(XX')^{-1}\mathrm{d}(XX') \\ &= |XX'| \operatorname{tr}(XX')^{-1}((\mathrm{d}X)X' + X(\mathrm{d}X)') \\ &= |XX'| [\operatorname{tr}(XX')^{-1}(\mathrm{d}X)X' + \operatorname{tr}(XX')^{-1}X(\mathrm{d}X)'] \\ &= |XX'| [\operatorname{tr} X'(XX')^{-1}\mathrm{d}X + \operatorname{tr}(\mathrm{d}X)X'(XX')^{-1}] \\ &= 2|XX'| \operatorname{tr} X'(XX')^{-1}\mathrm{d}X. \end{aligned} \tag{4}$$

As a result

$$\frac{\partial |XX'|}{\partial X} = 2|XX'|(XX')^{-1}X. \tag{5}$$

Similarly we find for $|X'X| \neq 0$,

$$\mathrm{d}|X'X| = 2|X'X| \operatorname{tr}(X'X)^{-1}X'\mathrm{d}X, \tag{6}$$

so that

$$\frac{\partial |X'X|}{\partial X} = 2|X'X|X(X'X)^{-1}. \tag{7}$$

Finally, let us consider the determinant of X^2, where X is non-singular. Since $|X^2| = |X|^2$, we have

$$\mathrm{d}|X^2| = \mathrm{d}|X|^2 = 2|X|\mathrm{d}|X| = 2|X|^2 \operatorname{tr} X^{-1}\mathrm{d}X. \tag{8}$$

These results are summarized in Table 5, where each determinant is assumed to be non-zero.

Table 5

$\phi(X)$	$\mathrm{d}\phi(X)$	$\mathrm{D}\phi(X)$	$\partial\phi(X)/\partial X$
$\lvert X\rvert$	$\lvert X\rvert \operatorname{tr} X^{-1}\mathrm{d}X$	$\lvert X\rvert(\operatorname{vec}(X^{-1})')'$	$\lvert X\rvert(X^{-1})'$
$\lvert XX'\rvert$	$2\lvert XX'\rvert \operatorname{tr} X'(XX')^{-1}\mathrm{d}X$	$2\lvert XX'\rvert(\operatorname{vec}(XX')^{-1}X)'$	$2\lvert XX'\rvert(XX')^{-1}X$
$\lvert X'X\rvert$	$2\lvert X'X\rvert \operatorname{tr}(X'X)^{-1}X'\mathrm{d}X$	$2\lvert X'X\rvert(\operatorname{vec} X(X'X)^{-1})'$	$2\lvert X'X\rvert X(X'X)^{-1}$
$\lvert X^2\rvert$	$2\lvert X\rvert^2 \operatorname{tr} X^{-1}\mathrm{d}X$	$2\lvert X\rvert^2(\operatorname{vec}(X^{-1})')'$	$2\lvert X\rvert^2(X^{-1})'$

Exercises

1. Show that $\partial|AXB|/\partial X=|AXB|A'(B'X'A')^{-1}B'$, if the inverse exists.
2. Let $F(X)$ be a square non-singular matrix function of X, and $G(X)=C(F(X))^{-1}A$. Then

$$\partial|F(X)|/\partial X=\begin{cases} |F(X)|(GXB+G'XB'), & \text{if } F(X)=AXBX'C, \\ |F(X)|(BXG+B'XG'), & \text{if } F(X)=AX'BXC, \\ |F(X)|(GXB+BXG)', & \text{if } F(X)=AXBXC. \end{cases}$$

3. Generalize (3) and (8) for non-singular X to

$$\partial|X^p|/\partial X=p|X|^p(X')^{-1},$$

a formula that holds for positive and negative integers.
4. Determine the derivative of $\phi(X)=\log|X'AX|$, where A is positive definite and $X'AX$ non-singular.
5. Determine the derivative of $\phi(X)=|AF(X)BG(X)C|$, and verify (3), (5), (7) and (8) as special cases.

11. SCALAR FUNCTIONS OF A MATRIX, III: EIGENVALUE

Let X_0 be a real symmetric $n\times n$ matrix, and let u_0 be a normalized eigenvector associated with a simple eigenvalue λ_0 of X_0. Then we know from section 8.8 that unique and differentiable functions $\lambda=\lambda(X)$ and $u=u(X)$ exist for all X in a neighbourhood $N(X_0)$ of X_0 satisfying

$$\lambda(X_0)=\lambda_0, \qquad u(X_0)=u_0 \tag{1}$$

and

$$Xu(X)=\lambda(X)u(X), \qquad u(X)'u(X)=1 \qquad (X\in N(X_0)). \tag{2}$$

The differential of λ at X_0 is then

$$\mathrm{d}\lambda=u_0'(\mathrm{d}X)u_0. \tag{3}$$

Hence we obtain the derivative

$$\mathrm{D}\lambda(X)=\frac{\partial\lambda}{\partial(\mathrm{vec}\,X)'}=u_0'\otimes u_0' \tag{4}$$

and the gradient (a column vector!)

$$\nabla\lambda(X)=u_0\otimes u_0. \tag{5}$$

We can also display the partial derivatives in a matrix:

$$\frac{\partial\lambda}{\partial X}=u_0u_0'. \tag{6}$$

12. TWO EXAMPLES OF VECTOR FUNCTIONS

Let us consider a set of variables $y_1, \ldots, y_m$, and suppose that these are known linear combinations of another set of variables $x_1, \ldots, x_n$, so that

$$y_i = \sum_j a_{ij} x_j \qquad (i = 1, \ldots, m). \tag{1}$$

Then

$$y = f(x) = Ax, \tag{2}$$

and since $\mathrm{d}f(x) = A\mathrm{d}x$, we have for the Jacobian matrix

$$\mathsf{D}f(x) = A. \tag{3}$$

If, on the other hand, the y_i are linearly related to variables x_{ij} such that

$$y_i = \sum_j a_j x_{ij} \qquad (i = 1, \ldots, m), \tag{4}$$

then this defines a vector function

$$y = f(X) = Xa. \tag{5}$$

The differential in this case is

$$\mathrm{d}f(X) = (\mathrm{d}X)a = \mathrm{vec}(\mathrm{d}X)a = (a' \otimes I)\ \mathrm{d}\,\mathrm{vec}\,X \tag{6}$$

and we find for the Jacobian matrix

$$\mathsf{D}f(X) = a' \otimes I. \tag{7}$$

Exercises

1. Show that the Jacobian matrix of the vector function $f(x) = Ag(x)$ is $\mathsf{D}f(x) = A\mathsf{D}g(x)$, and generalize this to the case where A is a matrix function of x.
2. Show that the Jacobian matrix of the vector function $f(x) = (x'x)a$ is $\mathsf{D}f(x) = 2ax'$, and generalize this to the case where a is a vector function of x.
3. Determine the Jacobian matrix of the vector function $f(x) = \nabla\phi(x)$. This matrix is of course the *Hessian matrix* of ϕ.
4. Show that the Jacobian matrix of the vector function $f(X) = X'a$ is $\mathsf{D}f(X) = I \otimes a'$.
5. Under the conditions of section 11, show that the derivative at X_0 of the eigenvector function $u(X)$ is given by

$$\mathsf{D}u(X) = \frac{\partial u(X)}{\partial(\mathrm{vec}\,X)'} = u_0' \otimes (\lambda_0 I_n - X_0)^+.$$

13. MATRIX FUNCTIONS

An example of a matrix function of a *vector* of variables x is

$$F(x) = xx'. \tag{1}$$

The differential is

$$\mathrm{d}xx' = (\mathrm{d}x)x' + x(\mathrm{d}x)', \tag{2}$$

so that

$$\begin{aligned}\mathrm{d}\,\mathrm{vec}\,xx' &= (x \otimes I)\,\mathrm{d}\,\mathrm{vec}\,x + (I \otimes x)\,\mathrm{d}\,\mathrm{vec}\,x' \\ &= (I \otimes x + x \otimes I)\mathrm{d}x. \end{aligned} \tag{3}$$

Hence,

$$\mathsf{D}F(x) = I \otimes x + x \otimes I. \tag{4}$$

Next we consider four simple examples of matrix functions of a *matrix* of variables X, where the order of X is $n \times q$. First the identity function

$$F(X) = X. \tag{5}$$

Clearly, $\mathrm{d}\,\mathrm{vec}\,F(X) = \mathrm{d}\,\mathrm{vec}\,X$, so that

$$\mathsf{D}F(X) = I_{nq}. \tag{6}$$

More interesting is the transpose function

$$F(X) = X'. \tag{7}$$

We obtain

$$\mathrm{d}\,\mathrm{vec}\,F(X) = \mathrm{d}\,\mathrm{vec}\,X' = K_{nq}\,\mathrm{d}\,\mathrm{vec}\,X. \tag{8}$$

Hence,

$$\mathsf{D}F(X) = K_{nq}. \tag{9}$$

The commutation matrix K is likely to play a role whenever the transpose of a matrix of variables occurs. For example, when

$$F(X) = XX', \tag{10}$$

then

$$\mathrm{d}F(X) = (\mathrm{d}X)X' + X(\mathrm{d}X)'$$

and

$$\begin{aligned}\mathrm{d}\,\mathrm{vec}\,F(X) &= (X \otimes I_n)\,\mathrm{d}\,\mathrm{vec}\,X + (I_n \otimes X)\,\mathrm{d}\,\mathrm{vec}\,X' \\ &= (X \otimes I_n)\,\mathrm{d}\,\mathrm{vec}\,X + (I_n \otimes X)K_{nq}\,\mathrm{d}\,\mathrm{vec}\,X \\ &= [(X \otimes I_n) + K_{nn}(X \otimes I_n)]\mathrm{d}\,\mathrm{vec}\,X \\ &= (I_{n^2} + K_{nn})(X \otimes I_n)\mathrm{d}\,\mathrm{vec}\,X. \end{aligned} \tag{11}$$

Hence,

$$\mathsf{D}F(X) = 2N_n(X \otimes I_n), \tag{12}$$

where $N_n = \frac{1}{2}(I_{n^2} + K_{nn})$ is a symmetric idempotent matrix with rank $\frac{1}{2}n(n+1)$ (see Theorem 3.11).

In a similar fashion we obtain from

$$F(X)=X'X, \tag{13}$$

$$\mathrm{d}\,\mathrm{vec}\,F(X) = (I_{q^2} + K_{qq})(I_q \otimes X')\mathrm{d}\,\mathrm{vec}\,X, \tag{14}$$

so that

$$\mathsf{D}F(X)=2N_q(I_q \otimes X'). \tag{15}$$

These results are summarized in Table 6, where X is an $n \times q$ matrix of variables.

Table 6

$F(X)$	$\mathrm{d}F(X)$	$\mathsf{D}F(X)$
X	$\mathrm{d}X$	I_{nq}
X'	$(\mathrm{d}X)'$	K_{nq}
XX'	$(\mathrm{d}X)X' + X(\mathrm{d}X)'$	$2N_n(X \otimes I_n)$
$X'X$	$(\mathrm{d}X)'X + X'\mathrm{d}X$	$2N_q(I_q \otimes X')$

If X is a non-singular $n \times n$ matrix, then the matrix function

$$F(X)=X^{-1} \tag{16}$$

has differential

$$\mathrm{d}F(X)= -X^{-1}(\mathrm{d}X)X^{-1}. \tag{17}$$

Taking vecs we obtain

$$\mathrm{d}\,\mathrm{vec}\,F(X)= -((X')^{-1} \otimes X^{-1})\,\mathrm{d}\,\mathrm{vec}\,X. \tag{18}$$

Hence,

$$\mathsf{D}F(X)= -(X')^{-1} \otimes X^{-1}. \tag{19}$$

Finally, if X is a square matrix of variables, then we can consider

$$F(X)=X^p \qquad (p=1, 2, \ldots). \tag{20}$$

We take differentials,

$$\mathrm{d}F(X)=(\mathrm{d}X)X^{p-1} + X(\mathrm{d}X)X^{p-2} + \ldots + X^{p-1}(\mathrm{d}X)$$

$$= \sum_{j=1}^{p} X^{j-1}(\mathrm{d}X)X^{p-j}, \tag{21}$$

and vecs,

$$\mathrm{d}\,\mathrm{vec}\,F(X)=\left(\sum_{j=1}^{p}(X')^{p-j} \otimes X^{j-1}\right)\mathrm{d}\,\mathrm{vec}\,X. \tag{22}$$

Hence

$$\mathsf{D}F(X)= \sum_{j=1}^{p}(X')^{p-j} \otimes X^{j-1}. \tag{23}$$

The last two examples are summarized in Table 7.

Table 7

$F(X)$	$\mathrm{d}F(X)$	$\mathrm{D}F(X)$	Conditions
X^{-1}	$-X^{-1}(\mathrm{d}X)X^{-1}$	$-(X')^{-1}\otimes X^{-1}$	X non-singular
X^p	$\sum_{j=1}^{p} X^{j-1}(\mathrm{d}X)X^{p-j}$	$\sum_{j=1}^{p} (X')^{p-j}\otimes X^{j-1}$	X square, $p=1, 2, \ldots$

Exercises

1. Find the Jacobian matrix of the matrix functions AXB and $AX^{-1}B$.
2. Find the Jacobian matrix of the matrix functions XAX', $X'AX$, XAX and $X'AX'$.
3. What is the Jacobian matrix of the Moore–Penrose inverse $F(X)=X^{+}$ (see section 8.5).
4. What is the Jacobian matrix of the adjoint matrix $F(X)=X^{\#}$ (see section 8.6).
5. Find the Jacobian matrix of the matrix function $F(X)=AG(X)BH(X)C$, where A, B and C are constant matrices.

14. KRONECKER PRODUCTS

An interesting problem arises in the treatment of Kronecker products. Consider the matrix function

$$F(X,Y)=X\otimes Y. \tag{1}$$

The differential is easily found as

$$\mathrm{d}F(X,Y)=(\mathrm{d}X)\otimes Y+X\otimes \mathrm{d}Y, \tag{2}$$

and, upon taking vecs, we obtain

$$\mathrm{d}\,\mathrm{vec}\,F(X,Y)=\mathrm{vec}(\mathrm{d}X\otimes Y)+\mathrm{vec}(X\otimes \mathrm{d}Y). \tag{3}$$

In order to find the Jacobian of F we must find matrices $A(Y)$ and $B(X)$ such that

$$\mathrm{vec}(\mathrm{d}X\otimes Y)=A(Y)\,\mathrm{d}\,\mathrm{vec}\,X \tag{4}$$

and

$$\mathrm{vec}(X\otimes \mathrm{d}Y)=B(X)\,\mathrm{d}\,\mathrm{vec}\,Y, \tag{5}$$

in which case the Jacobian matrix of $F(X, Y)$ takes the partitioned form

$$\mathrm{D}F(X, Y)=(A(Y):B(X)). \tag{6}$$

The crucial step here is to realize that we can express the vec of a Kronecker product of two matrices in terms of the Kronecker product of their vecs, that is

$$\mathrm{vec}(X\otimes Y)=(I_q\otimes K_{rm}\otimes I_p)(\mathrm{vec}\,X\otimes \mathrm{vec}\,Y), \tag{7}$$

where it is assumed that X is an $n \times q$ matrix and Y is a $p \times r$ matrix (see Theorem 3.10).

Using (7) we now write

$$\begin{aligned}\operatorname{vec}(\mathrm{d}X \otimes Y) &= (I_q \otimes K_{rn} \otimes I_p)(\mathrm{d} \operatorname{vec} X \otimes \operatorname{vec} Y) \\ &= (I_q \otimes K_{rn} \otimes I_p)(I_{nq} \otimes \operatorname{vec} Y)\, \mathrm{d} \operatorname{vec} X. \end{aligned} \tag{8}$$

Hence,

$$\begin{aligned}A(Y) &= (I_q \otimes K_{rn} \otimes I_p)(I_{nq} \otimes \operatorname{vec} Y) \\ &= I_q \otimes ((K_{rn} \otimes I_p)(I_n \otimes \operatorname{vec} Y)). \end{aligned} \tag{9}$$

In a similar fashion we find

$$\begin{aligned}B(X) &= (I_q \otimes K_{rn} \otimes I_p)(\operatorname{vec} X \otimes I_{pr}) \\ &= ((I_q \otimes K_{rn})(\operatorname{vec} X \otimes I_r)) \otimes I_p. \end{aligned} \tag{10}$$

We thus obtain the useful formula

$$\begin{aligned}\mathrm{d} \operatorname{vec}(X \otimes Y) = (I_q \otimes K_{rn} \otimes I_p)[&(I_{nq} \otimes \operatorname{vec} Y)\mathrm{d} \operatorname{vec} X \\ &+ (\operatorname{vec} X \otimes I_{pr})\mathrm{d} \operatorname{vec} Y], \end{aligned} \tag{11}$$

from which the Jacobian matrix of the matrix function $F(X, Y) = X \otimes Y$ follows:

$$\mathsf{D}F(X, Y) = (I_q \otimes K_{rn} \otimes I_p)(I_{nq} \otimes \operatorname{vec} Y : \operatorname{vec} X \otimes I_{pr}). \tag{12}$$

Exercises

1. Let $F(X,Y) = XX' \otimes YY'$, where X has n rows and Y has p rows (the number of columns of X and Y is irrelevant). Show that

$$\begin{aligned}\mathrm{d} \operatorname{vec} F(X, Y) = (I_n \otimes K_{pn} \otimes I_p)[&(G_n(X) \otimes \operatorname{vec} YY')\, \mathrm{d} \operatorname{vec} X \\ &+ (\operatorname{vec} XX' \otimes G_p(Y))\mathrm{d} \operatorname{vec} Y],\end{aligned}$$

where

$$G_m(A) = (I_{m^2} + K_{mm})(A \otimes I_m)$$

for any matrix A having m rows. Compute $\mathsf{D}F(X,Y)$.

2. Find the differential and the derivative of the matrix function $F(X,Y) = X \odot Y$ (Hadamard product).

15. SOME OTHER PROBLEMS

Suppose we want to find the Jacobian matrix of the real-valued function $\phi: \mathbb{R}^{n \times q} \to \mathbb{R}$ given by

$$\phi(X) = \sum_{i=1}^{n} \sum_{j=1}^{q} x_{ij}^2. \tag{1}$$

We can, of course, obtain the Jacobian matrix by first calculating (easy, in this case) the partial derivatives. More appealing, however, is to note that

$$\phi(X)=\operatorname{tr} XX', \tag{2}$$

from which we obtain

$$\mathrm{d}\phi(X)=2\operatorname{tr} X'\mathrm{d}X \tag{3}$$

and

$$\frac{\partial\phi(X)}{\partial X}=2X. \tag{4}$$

This example is, of course, very simple. But the idea of expressing a function of X in terms of the *matrix* X rather than in terms of the *elements* x_{ij} is often important. Some more examples should clarify this.

Let $\phi(X)$ be defined as the sum of the n^2 elements of X^{-1}. Then we let $\imath$ be the $n\times 1$ sum vector $(1, 1, \ldots, 1)'$ and write

$$\phi(X)=\imath'X^{-1}\imath \tag{5}$$

from which we obtain easily

$$\mathrm{d}\phi(X)=-\operatorname{tr} X^{-1}\imath\imath'X^{-1}\mathrm{d}X \tag{6}$$

and hence

$$\frac{\partial\phi(X)}{\partial X}=-(X')^{-1}\imath\imath'(X')^{-1}. \tag{7}$$

Consider another example. Let $F(X)$ be the $n\times(n-1)$ matrix function of an $n\times n$ matrix of variables X defined as X^{-1} without its last column. Then let E_n be the $n\times(n-1)$ matrix obtained from the identity matrix I_n by deleting its last column, i.e.

$$E_n=\begin{pmatrix} I_{n-1} \\ 0' \end{pmatrix}. \tag{8}$$

With E_n so defined, we can express $F(X)$ as

$$F(X)=X^{-1}E_n. \tag{9}$$

It is then simple to find

$$\mathrm{d}F(X)=-X^{-1}(\mathrm{d}X)X^{-1}E_n=-X^{-1}(\mathrm{d}X)F(X), \tag{10}$$

and hence

$$\mathsf{D}F(X)=-(F'(X)\otimes X^{-1}). \tag{11}$$

As a final example consider the real-valued function $\phi(X)$ defined as the ij-th element of X^2. In this case we can write

$$\phi(X)=e_i'X^2e_j, \tag{12}$$

where e_i and e_j are unit vectors. Hence

$$\begin{aligned}\mathrm{d}\phi(X)&=e_i'(\mathrm{d}X)Xe_j+e_i'X(\mathrm{d}X)e_j\\&=\operatorname{tr}(Xe_je_i'+e_je_i'X)\mathrm{d}X,\end{aligned} \tag{13}$$

so that

$$\frac{\partial\phi(X)}{\partial X} = e_i e_j' X' + X' e_i e_j'. \tag{14}$$

BIBLIOGRAPHICAL NOTES

§3. See also Magnus and Neudecker (1985) and Pollock (1985).

CHAPTER 10

Second-order differentials and Hessian matrices

1. INTRODUCTION

While in Chapter 9 the main tool was the first identification theorem, in the present chapter it is the *second* identification theorem (Theorem 6.13) which plays the central role. The second identification theorem tells us how to obtain the Hessian matrix from the second differential, and the purpose of this chapter is to demonstrate its workings in practice.

2. THE HESSIAN MATRIX OF A MATRIX FUNCTION

For a scalar function ϕ of an $n \times 1$ vector x, the Hessian matrix of ϕ at x was introduced in section 6.3; it is the $n \times n$ matrix of second-order partial derivatives $\mathsf{D}^2_{ji}\phi(x)$ denoted by

$$\mathsf{H}\phi(x) \qquad \text{or} \qquad \frac{\partial^2 \phi(x)}{\partial x\, \partial x'}. \tag{1}$$

We note that

$$\mathsf{H}\phi(x) = \frac{\partial}{\partial x'}\left(\frac{\partial \phi(x)}{\partial x'}\right)' = \mathsf{D}(\mathsf{D}\phi(x))'. \tag{2}$$

For a vector function $f : \mathbb{R}^n \to \mathbb{R}^m$ we defined the Hessian matrix as the stacked matrix

$$\mathsf{H}f(x) = \begin{bmatrix} \mathsf{H}f_1(x) \\ \mathsf{H}f_2(x) \\ \vdots \\ \mathsf{H}f_m(x) \end{bmatrix}. \tag{3}$$

Without much difficulty one verifies that

$$\mathsf{H}f(x) = \frac{\partial}{\partial x'}\operatorname{vec}\left(\frac{\partial f(x)}{\partial x'}\right)' = \mathsf{D}(\mathsf{D}f(x))'. \tag{4}$$

This suggests the following definition of the Hessian matrix of a matrix function (compare section 6.14).

Definition

Let F be a twice differentiable $m \times p$ matrix function of an $n \times q$ matrix X. The *Hessian matrix* of F at X is the $mnpq \times nq$ matrix

$$\mathsf{H}F(X) = \mathsf{D}(\mathsf{D}F(X))'. \tag{5}$$

Exercises

1. Show that
$$\mathsf{H}F(X) = \frac{\partial}{\partial(\operatorname{vec} X)'}\operatorname{vec}\left(\frac{\partial \operatorname{vec} F(X)}{\partial(\operatorname{vec} X)'}\right)'.$$
2. Write out $\mathsf{H}F(X)$ in terms of the Hessian matrices $\mathsf{H}F_{ij}(X)$ of its component functions.
3. Evaluate $\mathsf{D}^2 f(x) = \mathsf{D}(\mathsf{D}f(x))$. Compare $\mathsf{D}^2 f(x)$ with $\mathsf{D}(\mathsf{D}f(x))'$, and conclude that the latter expression is more practicable as a definition for the Hessian matrix than the former.

3. IDENTIFICATION OF HESSIAN MATRICES

The second identification theorem (Theorem 6.6) allows us to identify the Hessian matrix of a *scalar* function from its second differential. More precisely, it tells us that

$$\mathrm{d}^2\phi(x) = (\mathrm{d}x)'B\mathrm{d}x \tag{1}$$

implies and is implied by

$$\mathsf{H}\phi(x) = \tfrac{1}{2}(B + B'), \tag{2}$$

where B may depend on x, but not on $\mathrm{d}x$.

The second identification theorem for vector functions (Theorem 6.7) allows us to identify the Hessian matrix of an $m \times 1$ *vector* function $f(x)$. If $B_1, B_2, \ldots, B_m$ are square matrices and

$$B = \begin{bmatrix} B_1 \\ B_2 \\ \vdots \\ B_m \end{bmatrix}, \tag{3}$$

then

$$\mathrm{d}^2 f(x) = (I_m \otimes \mathrm{d}x)'B\,\mathrm{d}x \tag{4}$$

implies and is implied by

$$\mathsf{H}f(x)=\tfrac{1}{2}(B+(B')_v), \tag{5}$$

where B may depend on x, and

$$(B')_v=\begin{bmatrix} B'_1 \\ B'_2 \\ \vdots \\ B'_m \end{bmatrix}. \tag{6}$$

The extension to *matrix* functions is straightforward. The *second identification theorem for matrix functions* (Theorem 6.13) states that

$$\mathrm{d}^2 \operatorname{vec} F(X)=(I_{mp}\otimes \mathrm{d} \operatorname{vec} X)'\, B \,\mathrm{d} \operatorname{vec} X \tag{7}$$

implies and is implied by

$$\mathsf{H}\, F(X)=\tfrac{1}{2}(B+(B')_v) \tag{8}$$

where $F(X)$ is an $m\times p$ matrix function of an $n\times q$ matrix of variables X,

$$B=\begin{bmatrix} B_{11} \\ \vdots \\ B_{m1} \\ \vdots \\ B_{1p} \\ \vdots \\ B_{mp} \end{bmatrix}, \qquad (B')_v=\begin{bmatrix} B'_{11} \\ \vdots \\ B'_{m1} \\ \vdots \\ B'_{1p} \\ \vdots \\ B'_{mp} \end{bmatrix}, \tag{9}$$

and the B_{ij} are square $nq\times nq$ matrices.

4. THE SECOND IDENTIFICATION TABLE

These considerations lead to Table 1.

***Table* 1** The second identification table.

Function	Second differential	Hessian matrix	Order of H
$\phi(\xi)$	$\mathrm{d}^2\phi=\beta(\mathrm{d}\xi)^2$	$\mathsf{H}\phi(\xi)=\beta$	1×1
$\phi(x)$	$\mathrm{d}^2\phi=(\mathrm{d}x)'B\,\mathrm{d}x$	$\mathsf{H}\phi(x)=\frac{1}{2}(B+B')$	$n\times n$
$\phi(X)$	$\mathrm{d}^2\phi=(\mathrm{d}\operatorname{vec}X)'B\,\mathrm{d}\operatorname{vec}X$	$\mathsf{H}\phi(X)=\frac{1}{2}(B+B')$	$nq\times nq$
$f(\xi)$	$\mathrm{d}^2f=b(\mathrm{d}\xi)^2$	$\mathsf{H}f(\xi)=b$	$m\times 1$
$f(x)$	$\mathrm{d}^2f=(I_m\otimes \mathrm{d}x)'B\,\mathrm{d}x$	$\mathsf{H}f(x)=\frac{1}{2}(B+(B')_v)$	$mn\times n$
$f(X)$	$\mathrm{d}^2f=(I_m\otimes \mathrm{d}\operatorname{vec}X)'B\,\mathrm{d}\operatorname{vec}X$	$\mathsf{H}f(X)=\frac{1}{2}(B+(B')_v)$	$mnq\times nq$
$F(\xi)$	$\mathrm{d}^2F=B(\mathrm{d}\xi)^2$	$\mathsf{H}F(\xi)=\operatorname{vec}B$	$mp\times 1$
$F(x)$	$\mathrm{d}^2\operatorname{vec}F=(I_{mp}\otimes \mathrm{d}x)'B\,\mathrm{d}x$	$\mathsf{H}F(x)=\frac{1}{2}(B+(B')_v)$	$mnp\times n$
$F(X)$	$\mathrm{d}^2\operatorname{vec}F=(I_{mp}\otimes \mathrm{d}\operatorname{vec}X)'B\,\mathrm{d}\operatorname{vec}X$	$\mathsf{H}F(X)=\frac{1}{2}(B+(B')_v)$	$mnpq\times nq$

In the second identification table, ϕ is a scalar function, f an $m \times 1$ vector function and F an $m \times p$ matrix function; ξ is a scalar, x an $n \times 1$ vector and X an $n \times q$ matrix; β is a scalar, b is a column vector and B is a matrix, each of which may be a function of X, x or ξ. In the case of a vector function f, we have

$$B = \begin{bmatrix} B_1 \\ B_2 \\ \vdots \\ B_m \end{bmatrix} \quad \text{and} \quad (B')_v = \begin{bmatrix} B'_1 \\ B'_2 \\ \vdots \\ B'_m \end{bmatrix}. \tag{1}$$

In the case of a matrix function F, we have

$$B = \begin{bmatrix} B_{11} \\ \vdots \\ B_{m1} \\ \vdots \\ B_{1p} \\ \vdots \\ B_{mp} \end{bmatrix} \quad \text{and} \quad (B')_v = \begin{bmatrix} B'_{11} \\ \vdots \\ B'_{m1} \\ \vdots \\ B'_{1p} \\ \vdots \\ B'_{mp} \end{bmatrix}. \tag{2}$$

The matrices $B_1, B_2, \ldots, B_m$ (respectively, $B_{11}, \ldots, B_{mp}$) are square matrices of order $n \times n$ if f (or F) is a function of an $n \times 1$ vector x; the order of these matrices is $nq \times nq$ if f (or F) is a function of an $n \times q$ matrix X.

Exercises

1. Evaluate the Hessian matrix of $\phi(x) = a'x$ and $\phi(x) = x'Ax$.
2. At every point where the $n \times n$ matrix X is non-singular, show that the Hessian matrix of the real-valued function $\phi(X) = |X|$ is

$$\mathsf{H}\phi(X) = |X| K_n(X^{-1} \otimes I_n)'[(\operatorname{vec} I_n)(\operatorname{vec} I_n)' - I_{n^2}](I_n \otimes X^{-1}).$$

Show that $\mathsf{H}\phi(X)$ is non-singular for every $n \geq 2$.

5. AN EXPLICIT FORMULA FOR THE HESSIAN MATRIX

It is sometimes difficult to find the Jacobian matrix or Hessian matrix of a matrix function from the identification tables. In such cases it is convenient to have an expression which gives explicitly the Jacobian matrix or Hessian matrix in terms of the partial derivatives.

Let F be an $m \times p$ matrix function of an $n \times q$ matrix of variables X. If $q = 1$, we write x instead of X. Let e_i and e_s be $n \times 1$ unit vectors with a one in the i-th (s-th) place and zeros elsewhere, and let E_{ij} and E_{st} be $n \times q$ matrices with a one in the ij-th (st-th) position and zeros elsewhere. The Jacobian matrix of $F(x)$ can be expressed as

$$\mathsf{D}F(x) = \sum_{i=1}^{n} \left(\operatorname{vec} \frac{\partial F}{\partial x_i} \right) e_i' \tag{1}$$

and, as noted in section 9.4 (exercise 1), the Jacobian matrix of $F(X)$ can be expressed as

$$\mathsf{D}F(X) = \sum_{i=1}^{n} \sum_{j=1}^{q} \left(\operatorname{vec} \frac{\partial F}{\partial x_{ij}} \right) (\operatorname{vec} E_{ij})'. \tag{2}$$

Similar expressions can be found for the Hessian matrix of $F(x)$ and $F(X)$. We have in fact

$$\mathsf{H}F(x) = \sum_{i=1}^{n} \sum_{s=1}^{n} \operatorname{vec} \left(\frac{\partial^2 F}{\partial x_s \, \partial x_i} \right) \otimes e_i e_s' \tag{3}$$

and

$$\mathsf{H}F(X) = \sum_{i=1}^{n} \sum_{j=1}^{q} \sum_{s=1}^{n} \sum_{t=1}^{q} \left(\operatorname{vec} \frac{\partial^2 F}{\partial x_{st} \, \partial x_{ij}} \right) \otimes (\operatorname{vec} E_{ij})(\operatorname{vec} E_{st})'. \tag{4}$$

The verification of the results is left to the reader.

6. SCALAR FUNCTIONS

In many cases the second differential of a real-valued function $\phi(X)$ takes one of the two forms

$$\operatorname{tr} B(\mathrm{d}X)'C(\mathrm{d}X) \qquad \text{or} \qquad \operatorname{tr} B(\mathrm{d}X)C(\mathrm{d}X). \tag{1}$$

The following result will then prove useful.

Theorem 1

Let ϕ be a twice differentiable real-valued function of an $n \times q$ matrix X. Then the following two relationships hold between the second differential and the Hessian matrix of ϕ at X:

$$\mathrm{d}^2\phi(X) = \operatorname{tr} B(\mathrm{d}X)'C\mathrm{d}X \;\leftrightarrow\; \mathsf{H}\phi(X) = \tfrac{1}{2}(B' \otimes C + B \otimes C')$$

and

$$\mathrm{d}^2\phi(X) = \operatorname{tr} B(\mathrm{d}X)C\mathrm{d}X \;\leftrightarrow\; \mathsf{H}\phi(X) = \tfrac{1}{2}K_{qn}(B' \otimes C + C' \otimes B).$$

Proof. Using the fact, established in Theorem 2.3, that

$$\operatorname{tr} ABCD = (\operatorname{vec} B')'(A' \otimes C) \operatorname{vec} D, \tag{2}$$

we obtain

$$\operatorname{tr} B(\mathrm{d}X)' C\mathrm{d}X = (\mathrm{d} \operatorname{vec} X)'(B' \otimes C) \, \mathrm{d} \operatorname{vec} X \tag{3}$$

and

$$\begin{aligned}\operatorname{tr} B(\mathrm{d}X)C\mathrm{d}X &= (\mathrm{d}\operatorname{vec} X')'(B' \otimes C)\,\mathrm{d}\operatorname{vec} X \\ &= (\mathrm{d}\operatorname{vec} X)'K_{qn}(B' \otimes C)\,\mathrm{d}\operatorname{vec} X. \end{aligned} \tag{4}$$

The result now follows from the second identification table. □

Let us give three examples. First, consider the quadratic function

$$\phi(X) = \operatorname{tr} X'AX. \tag{5}$$

Twice taking differentials, we obtain

$$\mathrm{d}^2\phi(X) = 2\operatorname{tr}(\mathrm{d}X)'A\mathrm{d}X, \tag{6}$$

so that

$$\mathsf{H}\phi(X) = I \otimes (A + A'). \tag{7}$$

As a second example, consider the real-valued function

$$\phi(X) = \operatorname{tr} X^{-1}, \tag{8}$$

defined for every non-singular $n \times n$ matrix X. We have

$$\mathrm{d}\phi(X) = -\operatorname{tr} X^{-1}(\mathrm{d}X)X^{-1}, \tag{9}$$

and therefore

$$\begin{aligned}\mathrm{d}^2\phi(X) &= -\operatorname{tr}(\mathrm{d}X^{-1})(\mathrm{d}X)X^{-1} - \operatorname{tr} X^{-1}(\mathrm{d}X)(\mathrm{d}X^{-1}) \\ &= 2\operatorname{tr} X^{-1}(\mathrm{d}X)X^{-1}(\mathrm{d}X)X^{-1} = 2\operatorname{tr} X^{-2}(\mathrm{d}X)X^{-1}\mathrm{d}X, \end{aligned} \tag{10}$$

so that the Hessian matrix becomes

$$\mathsf{H}\phi(X) = K_n(X'^{-2} \otimes X^{-1} + X'^{-1} \otimes X^{-2}). \tag{11}$$

Finally, if λ_0 is a simple eigenvalue of a real symmetric $n \times n$ matrix X_0 with associated eigenvector u_0, then there exists a twice differentiable 'eigenvalue function' λ such that $\lambda(X_0) = \lambda_0$ (see Theorem 8.7). The second differential at X_0 is given in Theorem 8.10; it is

$$\begin{aligned}\mathrm{d}^2\lambda &= 2u_0'(\mathrm{d}X)(\lambda_0 I - X_0)^+(\mathrm{d}X)u_0 \\ &= 2\operatorname{tr} u_0u_0'(\mathrm{d}X)(\lambda_0 I - X_0)^+\mathrm{d}X. \end{aligned} \tag{12}$$

Hence the Hessian matrix is

$$\mathsf{H}\lambda(X) = K_n[u_0u_0' \otimes (\lambda_0 I - X_0)^+ + (\lambda_0 I - X_0)^+ \otimes u_0u_0']. \tag{13}$$

Exercises

1. Show that the Hessian matrix of $\phi(X) = \operatorname{tr} AXBX'$ is $\mathsf{H}\phi(X) = B' \otimes A + B \otimes A'$.
2. Show that the Hessian matrix of $\phi(X) = \operatorname{tr} X^2$ is $\mathsf{H}\phi(X) = 2K_n$ if X is an $n \times n$ matrix.

3. Determine the Hessian matrix of $\phi(X)=a'XX'a$.
4. At points where the $n\times n$ matrix X has a positive determinant, show that the Hessian matrix of $\phi(X)=\log|X|$ is

$$\mathsf{H}\phi(X)=-K_n[(X')^{-1}\otimes X^{-1}].$$

7. VECTOR FUNCTIONS

Let us consider one example of a *vector* function, namely

$$f(x)=\phi(x)a, \tag{1}$$

where ϕ is a real-valued function of an $n\times 1$ vector of variables x, and a is an $m\times 1$ vector of constants. The second differential is

$$\begin{aligned}\mathrm{d}^2f(x)&=\mathrm{d}^2\phi(x)a=[(\mathrm{d}x)'(\mathsf{H}\phi(x))(\mathrm{d}x)]a\\&=a(\mathrm{d}x)'(\mathsf{H}\phi(x))\mathrm{d}x=(a\otimes(\mathrm{d}x)')(\mathsf{H}\phi(x))\mathrm{d}x\\&=(I_m\otimes\mathrm{d}x)'(a\otimes I_n)(\mathsf{H}\phi(x))\mathrm{d}x,\end{aligned} \tag{2}$$

so that

$$\mathsf{H}f(x)=(a\otimes I_n)\mathsf{H}\phi(x) \tag{3}$$

according to the second identification table.

8. MATRIX FUNCTIONS, I

We shall consider two examples of Hessian matrices of a *matrix* function. The first is a matrix function of an $n\times 1$ *vector* x:

$$F(x)=\tfrac{1}{2}xx'. \tag{1}$$

It is easy to obtain

$$\mathrm{d}^2F(x)=(\mathrm{d}x)(\mathrm{d}x)', \tag{2}$$

from which we find

$$\mathrm{d}^2\operatorname{vec}F(x)=\operatorname{vec}(\mathrm{d}x)(\mathrm{d}x)'=(I_n\otimes\mathrm{d}x)\mathrm{d}x. \tag{3}$$

We now use the fact that

$$\mathrm{d}x=[I_n\otimes(\mathrm{d}x)']\operatorname{vec}I_n \tag{4}$$

to obtain

$$\begin{aligned}I_n\otimes\mathrm{d}x&=I_n\otimes[(I_n\otimes(\mathrm{d}x)')\operatorname{vec}I_n]\\&=[I_n\otimes I_n\otimes(\mathrm{d}x)'][I_n\otimes\operatorname{vec}I_n].\end{aligned} \tag{5}$$

Substituting (5) in (3) yields

$$\mathrm{d}^2 \operatorname{vec} F(x) = (I_{n^2} \otimes \mathrm{d}x)'(I_n \otimes \operatorname{vec} I_n)\mathrm{d}x. \tag{6}$$

The Hessian matrix then follows from the second identification table; it is

$$\mathsf{H}F(x) = \tfrac{1}{2}[I_n \otimes \operatorname{vec} I_n + (I_n \otimes \operatorname{vec} I_n)'_v]. \tag{7}$$

Alternatively we can use the expression (3) of section 5. We find

$$\frac{\partial^2 F(x)}{\partial x_s \, \partial x_i} = \tfrac{1}{2}(e_i e_s' + e_s e_i') \tag{8}$$

and thus

$$\begin{aligned}
\mathsf{H}F(x) &= \tfrac{1}{2} \sum_{i=1}^{n} \sum_{s=1}^{n} [\operatorname{vec}(e_i e_s' + e_s e_i')] \otimes e_i e_s' \\
&= \tfrac{1}{2} \sum_{i=1}^{n} \sum_{s=1}^{n} (e_s \otimes e_i \otimes e_s' \otimes e_i + e_i \otimes e_s \otimes e_s' \otimes e_i) \\
&= \tfrac{1}{2} \sum_{i=1}^{n} \sum_{s=1}^{n} [e_s \otimes e_s' \otimes e_i \otimes e_i + (K_n \otimes I_n)(e_s \otimes e_s' \otimes e_i \otimes e_i)] \\
&= [\tfrac{1}{2}(I_{n^2} + K_n) \otimes I_n](I_n \otimes \operatorname{vec} I_n).
\end{aligned} \tag{9}$$

In this case, the second derivation is more straightforward than the first; moreover, it leads to a more appealing (although, of course, equivalent) expression, namely (9) rather than (7).

Exercise

1. Show that $(I_n \otimes \operatorname{vec} I_n)'_v = (K_n \otimes I_n)(I_n \otimes \operatorname{vec} I_n)$.

9. MATRIX FUNCTIONS, II

The second example is a matrix function of an $n \times q$ *matrix* X:

$$F(X) = \tfrac{1}{2} X X'. \tag{1}$$

We find

$$\frac{\partial F(X)}{\partial x_{ij}} = \tfrac{1}{2}(E_{ij} X' + X E_{ij}') \tag{2}$$

and thus

$$\begin{aligned}
\frac{\partial^2 F(X)}{\partial x_{st} \partial x_{ij}} &= \tfrac{1}{2}(E_{ij} E_{st}' + E_{st} E_{ij}') \\
&= \tfrac{1}{2}\delta_{jt}(e_i e_s' + e_s e_i'),
\end{aligned} \tag{3}$$

where δ_{jt} denotes the Kronecker delta. Using formula (4) of section 5 we obtain the Hessian matrix

$$\begin{aligned}
\mathsf{H}F(X) &= \tfrac{1}{2}\sum_{i=1}^{n}\sum_{j=1}^{q}\sum_{s=1}^{n}\sum_{t=1}^{q}\delta_{jt}[\operatorname{vec}(e_i e_s' + e_s e_i')] \otimes (\operatorname{vec} E_{ij})(\operatorname{vec} E_{st})' \\
&= \tfrac{1}{2}\sum_{i=1}^{n}\sum_{s=1}^{n}[\operatorname{vec}(e_i e_s' + e_s e_i')] \otimes \left(\sum_{t=1}^{q}(\operatorname{vec} E_{it})(\operatorname{vec} E_{st})'\right) \\
&= \tfrac{1}{2}\sum_{i=1}^{n}\sum_{s=1}^{n}[\operatorname{vec}(e_i e_s' + e_s e_i')] \otimes I_q \otimes e_i e_s' \\
&= \tfrac{1}{2}\sum_{i=1}^{n}\sum_{s=1}^{n}(K_{n^2,q} \otimes I_n)[I_q \otimes (\operatorname{vec}(e_i e_s' + e_s e_i')) \otimes e_i e_s'] \\
&= (K_{n^2,q} \otimes I_n)(I_q \otimes A), \qquad (4)
\end{aligned}$$

where A is the Hessian matrix derived in the previous section:

$$A = [\tfrac{1}{2}(I_{n^2} + K_n) \otimes I_n][I_n \otimes \operatorname{vec} I_n). \qquad (5)$$

Part Four—

Inequalities

CHAPTER 11

Inequalities

1. INTRODUCTION

Inequalities occur in many disciplines. In economics they occur primarily because economics is concerned with optimizing behaviour. That is, we often want to find an x^* such that $\phi(x^*) \geq \phi(x)$ for all x in some set. The equivalence of the inequality

$$\phi(x) \geq 0 \qquad \text{for all } x \text{ in } S \tag{1}$$

and the minimization problem

$$\min_{x \in S} \phi(x) = 0 \tag{2}$$

suggests that inequalities can often be tackled using differential calculus. We shall see in this chapter that this method does not always lead to success, but if it does we shall use it.

The chapter falls naturally into several parts. In sections 1–4 we discuss (matrix analogues of) the Cauchy–Schwarz inequality and the arithmetic–geometric means inequality. Sections 5–14 are devoted to inequalities concerning eigenvalues and contain *inter alia* Fischer's min–max theorem and Poincaré's separation theorem. In section 15 we prove Hadamard's inequality. In sections 16–23 we use Karamata's inequality to prove a representation theorem for $(\operatorname{tr} A^p)^{1/p}$, $p > 1$, A positive semidefinite, which in turn is used to establish matrix analogues of the inequalities of Hölder and Minkowski. Sections 24 and 25 contain Minkowski's determinant theorem. In sections 26–28 several inequalities concerning the weighted means of order p are discussed. Finally, in sections 29–32, we turn to least-squares inequalities.

2. THE CAUCHY–SCHWARZ INEQUALITY

We begin our discussion of inequalities with the following fundamental result.

Theorem 1

For any two vectors a and b of the same order we have

$$(a'b)^2 \leqslant (a'a)(b'b) \tag{1}$$

with equality if and only if a and b are linearly dependent.

Let us give two proofs.

First proof. For any matrix A, $\operatorname{tr} A'A \geqslant 0$ with equality if and only if $A=0$, see (1.10.8). Now define

$$A = ab' - ba'. \tag{2}$$

Then,

$$\operatorname{tr} A'A = 2(a'a)(b'b) - 2(a'b)^2 \geqslant 0 \tag{3}$$

with equality if and only if $ab' = ba'$, that is, if and only if a and b are linearly dependent. □

Second proof. If $b=0$ the result is trivial. Assume therefore that $b \neq 0$, and consider the matrix

$$M = I - (1/b'b)bb'. \tag{4}$$

The matrix M is symmetric idempotent, and therefore positive semidefinite. Hence,

$$(a'a)(b'b) - (a'b)^2 = (b'b)a'Ma \geqslant 0. \tag{5}$$

Equality in (5) implies $a'Ma = 0$, and hence $Ma = 0$, that is, $a = \alpha b$ with $\alpha = a'b/b'b$. The result follows. □

Exercises

1. If A is positive semidefinite, show that

$$(x'Ay)^2 \leqslant (x'Ax)(y'Ay)$$

with equality if and only if Ax and Ay are linearly dependent.

2. Hence show that, for $A = (a_{ij})$ positive semidefinite,

$$|a_{ij}| \leqslant \max_i |a_{ii}|.$$

3. Show that

$$(x'y)^2 \leqslant (x'Ax)(y'A^{-1}y)$$

for every positive definite matrix A, with equality if and only if x and $A^{-1}y$ are linearly dependent.

4. Given $x \neq 0$, define $\psi(A) = (x'A^{-1}x)^{-1}$ for A positive definite. Show that

$$\psi(A) = \min_y \frac{y'Ay}{(y'x)^2}.$$

5. Prove Bergstrom's inequality:

$$x'(A+B)^{-1}x \leqslant \frac{(x'A^{-1}x)(x'B^{-1}x)}{x'(A^{-1}+B^{-1})x}$$

for any positive definite matrices A and B. [*Hint*: Use the fact that $\psi(A+B) \geqslant \psi(A) + \psi(B)$ where ψ is defined in Exercise 4.]

6. Show that

$$|(1/n)\sum x_i| \leqslant \left((1/n)\sum x_i^2\right)^{1/2}$$

with equality if and only if $x_1 = x_2 = \ldots = x_n$.

7. If all eigenvalues of A are real, show that

$$|(1/n)\operatorname{tr} A| \leqslant [(1/n)\operatorname{tr} A^2]^{1/2}$$

with equality if and only if the eigenvalues of the $n \times n$ matrix A are all equal.

8. Prove the triangle inequality: $\|x+y\| \leqslant \|x\| + \|y\|$.

3. MATRIX ANALOGUES OF THE CAUCHY–SCHWARZ INEQUALITY

The Cauchy–Schwarz inequality can be extended to matrices in several ways.

Theorem 2

For any two real matrices A and B of the same order, we have

$$(\operatorname{tr} A'B)^2 \leqslant (\operatorname{tr} A'A)(\operatorname{tr} B'B) \tag{1}$$

with equality if and only if one of the matrices A and B is a multiple of the other; also

$$\operatorname{tr}(A'B)^2 \leqslant \operatorname{tr}(A'A)(B'B) \tag{2}$$

with equality if and only if $AB' = BA'$; and

$$|A'B|^2 \leqslant |A'A|\,|B'B| \tag{3}$$

with equality if and only if $A'A$ or $B'B$ is singular, or $B = AQ$ for some non-singular matrix Q.

Proof. The first inequality follows from Theorem 1 by letting $a = \operatorname{vec} A$ and $b = \operatorname{vec} B$. To prove the second inequality, let $X = AB'$ and $Y = BA'$ and apply (1) to the matrices X and Y. This gives

$$(\operatorname{tr} BA'BA')^2 \leqslant (\operatorname{tr} BA'AB')(\operatorname{tr} AB'BA'), \tag{4}$$

from which (2) follows. The condition for equality in (2) is easily established.

Finally, to prove (3), assume that $|A'B| \neq 0$. (If $|A'B| = 0$, the result is trivial.) Then both A and B have full column rank, so that $A'A$ and $B'B$ are non-singular.

Now define

$$G = B'A(A'A)^{-1}A'B, \qquad H = B'(I - A(A'A)^{-1}A')B, \tag{5}$$

and notice that G is positive definite and H positive semidefinite (because $I - A(A'A)^{-1}A'$ is idempotent). Since $|G+H| \geq |G|$ by Theorem 1.22, with equality if and only if $H=0$, we obtain

$$|B'B| \geq |B'A(A'A)^{-1}A'B| = |A'B|^2|A'A|^{-1} \tag{6}$$

with equality if and only if $B'(I - A(A'A)^{-1}A')B = 0$, that is, if and only if $(I - A(A'A)^{-1}A')B = 0$. This concludes the proof. □

Exercises

1. Show that $\operatorname{tr}(A'B)^2 \leq \operatorname{tr}(AA')(BB')$ with equality if and only if $A'B$ is symmetric.
2. Prove Schur's inequality $\operatorname{tr} A^2 \leq \operatorname{tr} A'A$ with equality if and only if A is symmetric. [*Hint*: Use the commutation matrix.]

4. THE THEOREM OF THE ARITHMETIC AND GEOMETRIC MEANS

The most famous of all inequalities is the arithmetic–geometric mean inequality which was first proved (assuming equal weights) by Euclid. In its simplest form it asserts that

$$x^{\alpha}y^{1-\alpha} \leq \alpha x + (1-\alpha)y \qquad (0 < \alpha < 1) \tag{1}$$

for every non-negative x and y, with equality if and only if $x=y$. Let us demonstrate the general theorem.

Theorem 3

For any two $n \times 1$ vectors $x = (x_1, x_2, \ldots, x_n)'$ and $a = (\alpha_1, \alpha_2, \ldots, \alpha_n)'$ satisfying $x_i \geq 0$, $\alpha_i > 0$, $\sum_{i=1}^n \alpha_i = 1$, we have

$$\prod_{i=1}^n x_i^{\alpha_i} \leq \sum_{i=1}^n \alpha_i x_i \tag{2}$$

with equality if and only if $x_1 = x_2 = \ldots = x_n$.

Proof. Assume that $x_i > 0$, $i = 1, \ldots, n$ (if at least one x_i is zero the result is trivially true), and define

$$\phi(x) = \sum_{i=1}^n \alpha_i x_i - \prod_{i=1}^n x_i^{\alpha_i}. \tag{3}$$

We wish to show that $\phi(x) \geq 0$ for all positive x. Differentiating ϕ, we obtain

$$\begin{aligned} \mathrm{d}\phi &= \sum_{i=1}^n \alpha_i \mathrm{d}x_i - \sum_{i=1}^n \alpha_i x_i^{\alpha_i - 1}(\mathrm{d}x_i)\prod_{j \neq i} x_j^{\alpha_j} \\ &= \sum_{i=1}^n \left(\alpha_i - (\alpha_i/x_i)\prod_{j=1}^n x_j^{\alpha_j}\right)\mathrm{d}x_i. \end{aligned} \tag{4}$$

The first-order conditions are therefore

$$(\alpha_i/x_i)\prod_{j=1}^{n} x_j^{\alpha_j}=\alpha_i \qquad (i=1,\ldots,n), \tag{5}$$

that is,

$$x_1=x_2=\ldots=x_n. \tag{6}$$

At such points $\phi(x)=0$. Since $\prod_{i=1}^{n} x_i^{\alpha_i}$ is concave, $\phi(x)$ is convex. Hence by Theorem 7.8, ϕ has an absolute minimum (namely zero) at every point where $x_1=x_2=\ldots=x_n$. □

Exercises

1. Prove (1) by using the fact that the log function is concave on $(0, \infty)$.
2. Use Theorem 3 to show that
$$|A|^{1/n}\leqslant(1/n)\operatorname{tr}A \tag{7}$$
for every $n \times n$ positive semidefinite A. Also show that equality occurs if and only if $A = \mu I$ for some $\mu \geqslant 0$.
3. Prove (7) directly for positive definite A by letting $A = X'X$ (X square) and defining
$$\phi(X)=(1/n)\operatorname{tr}X'X-|X|^{2/n}. \tag{8}$$
Show that
$$\mathrm{d}\phi=(2/n)\operatorname{tr}(X'-|X|^{2/n}X^{-1})\mathrm{d}X \tag{9}$$
and
$$\mathrm{d}^2\phi=(2/n)\operatorname{tr}(\mathrm{d}X)'(\mathrm{d}X)+(2/n)|X|^{2/n}[\operatorname{tr}(X^{-1}\mathrm{d}X)^2-(2/n)(\operatorname{tr}X^{-1}\mathrm{d}X)^2]. \tag{10}$$

5. THE RAYLEIGH QUOTIENT

In the next few sections we shall investigate inequalities concerning eigenvalues of real symmetric matrices. We shall adopt the convention to arrange the eigenvalues $\lambda_1, \lambda_2, \ldots, \lambda_n$ of a real symmetric matrix A in *increasing* order, so that

$$\lambda_1\leqslant\lambda_2\leqslant\ldots\leqslant\lambda_n. \tag{1}$$

Our first result concerns the bounds of the Rayleigh quotient, $x'Ax/x'x$.

Theorem 4

For any real symmetric $n\times n$ matrix A,

$$\lambda_1\leqslant\frac{x'Ax}{x'x}\leqslant\lambda_n. \tag{2}$$

Proof. Let S be an orthogonal $n\times n$ matrix such that

$$S'AS=\Lambda=\operatorname{diag}(\lambda_1, \lambda_2, \ldots, \lambda_n) \tag{3}$$

and let $y = S'x$. Since

$$\lambda_1 y'y \leqslant y'\Lambda y \leqslant \lambda_n y'y, \tag{4}$$

we obtain

$$\lambda_1 x'x \leqslant x'Ax \leqslant \lambda_n x'x, \tag{5}$$

because $x'Ax = y'\Lambda y$ and $x'x = y'y$. The result follows. □

Since the extrema of $x'Ax/x'x$ can be achieved (by choosing x to be an eigenvector associated with λ_1 or λ_n), Theorem 4 implies that we may *define* λ_1 and λ_n as follows:

$$\lambda_1 = \min_x \frac{x'Ax}{x'x}, \tag{6}$$

$$\lambda_n = \max_x \frac{x'Ax}{x'x}. \tag{7}$$

The representations (6) and (7) show that we can express λ_1 and λ_n (two *nonlinear* functions of A) as an envelope of *linear* functions of A. This technique is called *quasilinearization*: the right-hand sides of (6) and (7) are quasilinear representations of λ_1 and λ_n. We shall encounter some useful applications of this technique in the next few sections.

Exercises

1. Use the quasilinear representations (6) and (7) to show that

   $$\lambda_1(A+B) \geqslant \lambda_1(A),\ \lambda_n(A+B) \geqslant \lambda_n(A),$$

   $$\lambda_1(A) \operatorname{tr} B \leqslant \operatorname{tr} AB \leqslant \lambda_n(A) \operatorname{tr} B$$

   for any $n \times n$ symmetric matrix A and positive semidefinite $n \times n$ matrix B.
2. If A is a symmetric $n \times n$ matrix and A_k a $k \times k$ principal submatrix of A, then prove

   $$\lambda_1(A) \leqslant \lambda_1(A_k) \leqslant \lambda_k(A_k) \leqslant \lambda_n(A).$$

   (A generalization of this result is given in Theorem 12.)
3. Show that

   $$\lambda_1(A+B) \geqslant \lambda_1(A) + \lambda_1(B),$$
   $$\lambda_n(A+B) \leqslant \lambda_n(A) + \lambda_n(B)$$

   for any two symmetric $n \times n$ matrices A and B. (See also Theorem 5.)

6. CONCAVITY OF λ_1, CONVEXITY OF λ_n

As an immediate consequence of the definitions (5.6) and (5.7), let us prove Theorem 5, thus illustrating the usefulness of quasilinear representations.

Theorem 5

For any two real symmetric matrices A and B of order n, and $0 \leqslant \alpha \leqslant 1$,

$$\lambda_1[\alpha A+(1-\alpha)B] \geqslant \alpha\lambda_1(A)+(1-\alpha)\lambda_1(B),$$
$$\lambda_n[\alpha A+(1-\alpha)B] \leqslant \alpha\lambda_n(A)+(1-\alpha)\lambda_n(B).$$

Hence, λ_1 is concave and λ_n convex on the space of real symmetric matrices.

Proof. Using the representation (5.6), we obtain

$$\begin{aligned}\lambda_1[\alpha A+(1-\alpha)B] &= \min_x \frac{x'[\alpha A+(1-\alpha)B]x}{x'x} \\ &\geqslant \alpha \min_x \frac{x'Ax}{x'x} + (1-\alpha)\min_x \frac{x'Bx}{x'x} \\ &= \alpha\lambda_1(A)+(1-\alpha)\lambda_1(B).\end{aligned}$$

The analogue for λ_n is proved similarly. □

7. VARIATIONAL DESCRIPTION OF EIGENVALUES

The representation of λ_1 and λ_n given in (5.6) and (5.7) can be extended in the following way.

Theorem 6

Let A be a real symmetric $n \times n$ matrix with eigenvalues $\lambda_1 \leqslant \lambda_2 \leqslant \ldots \leqslant \lambda_n$. Let $S=(s_1, s_2, \ldots, s_n)$ be an orthogonal $n \times n$ matrix which diagonalizes A, so that

$$S'AS = \operatorname{diag}(\lambda_1, \lambda_2, \ldots, \lambda_n). \tag{1}$$

Then, for $k=1, 2, \ldots, n$,

$$\lambda_k = \min_{R'_{k-1}x=0} \frac{x'Ax}{x'x} = \max_{T'_{k+1}x=0} \frac{x'Ax}{x'x}, \tag{2}$$

where

$$R_k=(s_1, s_2, \ldots, s_k), \qquad T_k=(s_k, s_{k+1}, \ldots, s_n). \tag{3}$$

Moreover, if $\lambda_1 = \lambda_2 = \ldots = \lambda_k$, then

$$\frac{x'Ax}{x'x} = \lambda_1 \qquad \text{if and only if } x = \sum_{i=1}^{k} \alpha_i s_i \tag{4}$$

for some set of real numbers $\alpha_1,\ldots,\alpha_k$ not all zero. Similarly, if $\lambda_l=\lambda_{l+1}=\ldots=\lambda_n$, then

$$\frac{x'Ax}{x'x}=\lambda_n \quad \text{if and only if } x=\sum_{j=l}^{n}\alpha_j s_j \tag{5}$$

for some set of real numbers $\alpha_l,\ldots,\alpha_n$ not all zero.

Proof. Let us prove the first representation of λ_k in (2), the second being proved in the same way.

As in the proof of Theorem 4, let $y=S'x$. Partitioning S and y as

$$S=(R_{k-1},T_k), \qquad y=\begin{pmatrix} y_1 \\ y_2 \end{pmatrix}, \tag{6}$$

we may express x as

$$x=Sy=R_{k-1}y_1+T_k y_2. \tag{7}$$

Hence,

$$R'_{k-1}x=0 \leftrightarrow y_1=0 \leftrightarrow x=T_k y_2. \tag{8}$$

It follows that

$$\min_{R'_{k-1}x=0}\frac{x'Ax}{x'x}=\min_{x=T_k y_2}\frac{x'Ax}{x'x}=\min_{y_2}\frac{y'_2(T'_kAT_k)y_2}{y'_2y_2}=\lambda_k, \tag{9}$$

using Theorem 4 and the fact that $T'_kAT_k=\operatorname{diag}(\lambda_k,\lambda_{k+1},\ldots,\lambda_n)$. The case of equality is easily proved and is left to the reader. □

Useful as the representations in (2) may be, there is one problem in using them, namely that the representations are *not* quasilinear, because R_{k-1} and T_{k+1} also depend on A. A quasilinear representation of the eigenvalues was first obtained by Fischer (1905).

8. FISCHER'S MIN–MAX THEOREM

We shall obtain Fischer's result by use of the following theorem, of interest in itself.

Theorem 7

Let A be a real symmetric $n\times n$ matrix with eigenvalues $\lambda_1\leqslant\lambda_2\leqslant\ldots\leqslant\lambda_n$. Let $1\leqslant k\leqslant n$. Then,

$$\min_{B'x=0}\frac{x'Ax}{x'x}\leqslant\lambda_k \tag{1}$$

for *every* $n\times(k-1)$ matrix B, and

$$\max_{C'x=0}\frac{x'Ax}{x'x}\geqslant\lambda_k \tag{2}$$

for *every* $n\times(n-k)$ matrix C.

Proof. Let B be an arbitrary $n \times (k-1)$ matrix, and denote (normalized) eigenvectors associated with the eigenvalues $\lambda_1, \ldots, \lambda_n$ of A by $s_1, s_2, \ldots, s_n$.

Let $R = (s_1, s_2, \ldots, s_k)$, so that

$$R'AR = \text{diag}(\lambda_1, \lambda_2, \ldots, \lambda_k), \qquad R'R = I_k. \tag{3}$$

Now consider the $(k-1) \times k$ matrix $B'R$. Since the rank of $B'R$ cannot exceed $k-1$, its k columns are linearly dependent. Thus

$$B'Rp = 0 \tag{4}$$

for some $k \times 1$ vector $p \neq 0$. Then, choosing $x = Rp$, we obtain

$$\min_{B'x=0} \frac{x'Ax}{x'x} \leqslant \frac{p'(R'AR)p}{p'p} \leqslant \lambda_k, \tag{5}$$

using (3) and Theorem 4. This proves (1). The proof of (2) is similar. □

Let us now demonstrate Fischer's famous min–max theorem.

Theorem 8 (Fischer)

Let A be a real symmetric $n \times n$ matrix with eigenvalues $\lambda_1 \leqslant \lambda_2 \leqslant \ldots \leqslant \lambda_n$. Then λ_k $(1 \leqslant k \leqslant n)$ may be defined as

$$\lambda_k = \max_{B'B = I_{k-1}} \min_{B'x=0} \frac{x'Ax}{x'x}, \tag{6}$$

or equivalently as

$$\lambda_k = \min_{C'C = I_{n-k}} \max_{C'x=0} \frac{x'Ax}{x'x}, \tag{7}$$

where, as the notation indicates, B is an $n \times (k-1)$ matrix and C is an $n \times (n-k)$ matrix.

Proof. Again we shall prove only the first representation, leaving the proof of (7) as an exercise for the reader.

As in the proof of Theorem 6, let R_{k-1} be a semi-orthogonal $n \times (k-1)$ matrix, satisfying

$$AR_{k-1} = R_{k-1}\Lambda_{k-1}, \qquad R'_{k-1}R_{k-1} = I_{k-1}, \tag{8}$$

where

$$\Lambda_{k-1} = \text{diag}(\lambda_1, \lambda_2, \ldots, \lambda_{k-1}). \tag{9}$$

Then, defining

$$\phi(B) = \min_{B'x=0} \frac{x'Ax}{x'x}, \tag{10}$$

we obtain

$$\lambda_k = \phi(R_{k-1}) = \max_{B = R_{k-1}} \phi(B) \leqslant \max_{B'B = I_{k-1}} \phi(B) \leqslant \lambda_k, \tag{11}$$

where the first equality follows from Theorem 6, and the last inequality from Theorem 7. Hence,

$$\lambda_k = \max_{B'B=I_{k-1}} \phi(B) = \max_{B'B=I_{k-1}} \min_{B'x=0} \frac{x'Ax}{x'x}. \tag{12}$$

□

Exercises

1. Let A be a square $n \times n$ matrix (not necessarily symmetric). Show that for every $n \times 1$ vector x

$$(x'Ax)^2 \leqslant (x'AA'x)(x'x)$$

and hence

$$\frac{1}{2}\left|\frac{x'(A+A')x}{x'x}\right| \leqslant \left(\frac{x'AA'x}{x'x}\right)^{1/2}.$$

2. Use exercise 1 and Theorems 6 and 7 to prove that

$$\tfrac{1}{2}|\lambda_k(A+A')| \leqslant [\lambda_k(AA')]^{1/2} \qquad (k = 1, \ldots, n)$$

for every $n \times n$ matrix A. (This was first proved by Fan and Hoffman (1955). Related inequalities are given in Amir-Moéz and Fass (1962).)

9. MONOTONICITY OF THE EIGENVALUES

The usefulness of the quasilinear representation of the eigenvalues in Theorem 8, as opposed to the representation in Theorem 6, is clearly brought out in the proof of Theorem 9.

Theorem 9

For any symmetric matrix A and positive semidefinite matrix B,

$$\lambda_k(A+B) \geqslant \lambda_k(A) \qquad (k = 1, 2, \ldots, n). \tag{1}$$

If B is positive definite, then the inequality is strict.

Proof. For any $n \times (k-1)$ matrix P, we have

$$\begin{aligned}\min_{P'x=0} \frac{x'(A+B)x}{x'x} &= \min_{P'x=0}\left(\frac{x'Ax}{x'x} + \frac{x'Bx}{x'x}\right)\\ &\geqslant \min_{P'x=0}\frac{x'Ax}{x'x} + \min_{P'x=0}\frac{x'Bx}{x'x}\\ &\geqslant \min_{P'x=0}\frac{x'Ax}{x'x} + \min_{x}\frac{x'Bx}{x'x} \geqslant \min_{P'x=0}\frac{x'Ax}{x'x},\end{aligned} \tag{2}$$

and hence, by Theorem 8,

$$\lambda_k(A+B) = \max_{P'P=I_{k-1}} \min_{P'x=0} \frac{x'(A+B)x}{x'x}$$

$$\geq \max_{P'P=I_{k-1}} \min_{P'x=0} \frac{x'Ax}{x'x} = \lambda_k(A). \qquad (3)$$

If B is positive definite, the last inequality in (2) is strict, and so the inequality in (3) is strict too. □

Exercises

1. Prove Theorem 9 by means of the representation (8.7) rather than (8.6).
2. Show how an application of Theorem 6 fails to prove Theorem 9.

10. THE POINCARÉ SEPARATION THEOREM

In section 8 we employed Theorems 6 and 7 to prove Fischer's min–max theorem. Let us now demonstrate another consequence of Theorems 6 and 7: Poincaré's separation theorem.

Theorem 10

Let A be a real symmetric $n \times n$ matrix with eigenvalues $\lambda_1 \leq \lambda_2 \leq \ldots \leq \lambda_n$, and let G be a semi-orthogonal $n \times k$ matrix $(1 \leq k \leq n)$, so that $G'G = I_k$. Then the eigenvalues $\mu_1 \leq \mu_2 \leq \ldots \leq \mu_k$ of $G'AG$ satisfy

$$\lambda_i \leq \mu_i \leq \lambda_{n-k+i} \qquad (i = 1, 2, \ldots, k). \qquad (1)$$

Note. For $k = 1$, Theorem 10 reduces to Theorem 4. For $k = n$, we obtain the well-known result that the symmetric matrices A and $G'AG$ have the same set of eigenvalues, if G is orthogonal (see Theorem 1.5).

Proof. Let $1 \leq i \leq k$, and let R be a semi-orthogonal $n \times (i-1)$ matrix whose columns are eigenvectors of A associated with $\lambda_1, \lambda_2, \ldots, \lambda_{i-1}$. Then,

$$\lambda_i = \min_{R'x=0} \frac{x'Ax}{x'x} \leq \min_{\substack{R'x=0 \\ x=Gy}} \frac{x'Ax}{x'x} = \min_{R'Gy=0} \frac{y'G'AGy}{y'y} \leq \mu_i, \qquad (2)$$

using Theorems 6 and 7.

Next, let $n-k+1 \leq j \leq n$, and let T be a semi-orthogonal $n \times (n-j)$ matrix whose columns are eigenvectors of A associated with $\lambda_{j+1}, \ldots, \lambda_n$. Then we obtain in the same way

$$\lambda_j = \max_{T'x=0} \frac{x'Ax}{x'x} \geq \max_{\substack{T'x=0 \\ x=Gy}} \frac{x'Ax}{x'x} = \max_{T'Gy=0} \frac{y'G'AGy}{y'y} \geq \mu_{k-n+j}. \qquad (3)$$

Choosing $j=n-k+i$ $(1\leqslant i\leqslant k)$ in (3) thus yields $\mu_i\leqslant\lambda_{n-k+i}$, and the result follows. □

11. TWO COROLLARIES OF POINCARÉ'S THEOREM

The Poincaré theorem is of such fundamental importance that we shall present in this and the next two sections a number of special cases. The first of these is not merely a special case, but an equivalent formulation of the same result: see exercise 2.

Theorem 11

Let A be a real symmetric $n\times n$ matrix with eigenvalues $\lambda_1\leqslant\lambda_2\leqslant\ldots\leqslant\lambda_n$, and let M be an idempotent symmetric $n\times n$ matrix of rank k $(1\leqslant k\leqslant n)$. Denoting the eigenvalues of the $n\times n$ matrix MAM, apart from $n-k$ zeros, by $\mu_1\leqslant\mu_2\leqslant\ldots\leqslant\mu_k$, we have

$$\lambda_i\leqslant\mu_i\leqslant\lambda_{n-k+i}\qquad(i=1,2,\ldots,k).$$

Proof. Immediate from Theorem 10 by writing $M=GG'$, $G'G=I_k$ (see (1.17.13)), and noting that $GG'AGG'$ and $G'AG$ have the same eigenvalues, apart from $n-k$ zeros. □

Another special case of Theorem 10 is Theorem 12.

Theorem 12

If A is a real symmetric $n\times n$ matrix and A_k a $k\times k$ principal submatrix of A, then

$$\lambda_i(A)\leqslant\lambda_i(A_k)\leqslant\lambda_{n-k+i}(A)\qquad(i=1,\ldots,k).$$

Proof. Let G be the $n\times k$ matrix

$$G=\begin{pmatrix}I_k\\0\end{pmatrix}$$

or a row permutation thereof. Then $G'G=I_k$ and $G'AG$ is a $k\times k$ principal submatrix of A. The result now follows from Theorem 10. □

Exercises

1. Let A be a real symmetric $n\times n$ matrix with eigenvalues $\lambda_1\leqslant\lambda_2\leqslant\ldots\leqslant\lambda_n$, and let B be the real symmetric $(n+1)\times(n+1)$ matrix

$$B=\begin{pmatrix}A&b\\b'&\alpha\end{pmatrix}.$$

Then the eigenvalues $\mu_1 \leqslant \mu_2 \leqslant \ldots \leqslant \mu_{n+1}$ of B satisfy

$$\mu_1 \leqslant \lambda_1 \leqslant \mu_2 \leqslant \lambda_2 \leqslant \ldots \leqslant \lambda_n \leqslant \mu_{n+1}.$$

[*Hint*: Use Theorem 12.]

2. Obtain Theorem 10 as a special case of Theorem 11.

12. FURTHER CONSEQUENCES OF THE POINCARÉ THEOREM

An immediate consequence of Poincaré's inequality (Theorem 10) is the following theorem.

Theorem 13

For any real symmetric $n \times n$ matrix A with eigenvalues $\lambda_1 \leqslant \lambda_2 \leqslant \ldots \leqslant \lambda_n$,

$$\min_{X'X=I_k} \operatorname{tr} X'AX = \sum_{i=1}^{k} \lambda_i, \tag{1}$$

$$\max_{X'X=I_k} \operatorname{tr} X'AX = \sum_{i=1}^{k} \lambda_{n-k+i}. \tag{2}$$

Proof. Denoting the k eigenvalues of $X'AX$ by $\mu_1 \leqslant \mu_2 \leqslant \ldots \leqslant \mu_k$, we have from Theorem 10,

$$\sum_{i=1}^{k} \lambda_i \leqslant \sum_{i=1}^{k} \mu_i \leqslant \sum_{i=1}^{k} \lambda_{n-k+i}. \tag{3}$$

Noting that

$$\sum_{i=1}^{k} \mu_i = \operatorname{tr} X'AX,$$

and that the bounds in (3) can be attained by suitable choices of X, the result follows. □

An important special case of Theorem 13, which we shall use in section 17 is the following.

Theorem 14

Let $A=(a_{ij})$ be a real symmetric $n \times n$ matrix with eigenvalues $\lambda_1 \leqslant \lambda_2 \leqslant \ldots \leqslant \lambda_n$. Then,

$$\lambda_1 \leqslant a_{ii} \leqslant \lambda_n \qquad (i=1, \ldots, n), \tag{4}$$

$$\lambda_1 + \lambda_2 \leqslant a_{ii} + a_{jj} \leqslant \lambda_{n-1} + \lambda_n \qquad (i \neq j = 1, \ldots, n), \tag{5}$$

and so on. In particular, for $k=1, 2, \ldots, n$,

$$\sum_{i=1}^{k} \lambda_i \leqslant \sum_{i=1}^{k} a_{ii} \leqslant \sum_{i=1}^{k} \lambda_{n-k+i}. \tag{6}$$

Proof. Theorem 13 implies that the inequality

$$\sum_{i=1}^{k} \lambda_i \leqslant \operatorname{tr} X'AX \leqslant \sum_{i=1}^{k} \lambda_{n-k+i} \tag{7}$$

is valid for every $n \times k$ matrix X satisfying $X'X = I_k$. Taking $X = (I_k, 0)'$ or a row permutation thereof, the result follows. □

Exercise

1. Prove Theorem 13 directly from Theorem 6 without using Poincaré's theorem. [*Hint*: Write $\operatorname{tr} X'AX = \sum_{i=1}^{k} x_i' A x_i$ where $X = (x_1, \ldots, x_k)$.]

13. MULTIPLICATIVE VERSION

Let us now obtain the multiplicative versions of Theorems 13 and 14 for the positive definite case.

Theorem 15

For any positive definite $n \times n$ matrix A with eigenvalues $\lambda_1 \leqslant \lambda_2 \leqslant \ldots \leqslant \lambda_n$,

$$\min_{X'X = I_k} |X'AX| = \prod_{i=1}^{k} \lambda_i, \tag{1}$$

$$\max_{X'X = I_k} |X'AX| = \prod_{i=1}^{k} \lambda_{n-k+i}. \tag{2}$$

Proof. As in the proof of Theorem 13, let $\mu_1 \leqslant \mu_2 \leqslant \ldots \leqslant \mu_k$ be the eigenvalues of $X'AX$. Then Theorem 10 implies

$$\prod_{i=1}^{k} \lambda_i \leqslant \prod_{i=1}^{k} \mu_i \leqslant \prod_{i=1}^{k} \lambda_{n-k+i}. \tag{3}$$

Since

$$\prod_{i=1}^{k} \mu_i = |X'AX|,$$

and the bounds in (3) can be attained by suitable choices of X, the result follows. □

Theorem 16

Let $A = (a_{ij})$ be a positive definite $n \times n$ matrix with eigenvalues $\lambda_1 \leqslant \lambda_2 \leqslant \ldots \leqslant \lambda_n$, and define

$$A_k = \begin{bmatrix} a_{11} & \ldots & a_{1k} \\ \vdots & & \vdots \\ a_{k1} & \ldots & a_{kk} \end{bmatrix} \qquad (k = 1, \ldots, n). \tag{4}$$

Then, for $k=1, 2, \dots, n$,

$$\prod_{i=1}^{k} \lambda_i \leqslant |A_k| \leqslant \prod_{i=1}^{k} \lambda_{n-k+i}. \tag{5}$$

Proof. Theorem 15 implies that the inequality

$$\prod_{i=1}^{k} \lambda_i \leqslant |X'AX| \leqslant \prod_{i=1}^{k} \lambda_{n-k+i} \tag{6}$$

is valid for every $n \times k$ matrix X satisfying $X'X=I_k$. Taking $X=(I_k, 0)'$, the result follows. □

Exercises

1. Prove Theorem 16 using Theorem 12 rather than Theorem 15.
2. Use Theorem 16 to show that a symmetric $n \times n$ matrix A is positive definite if and only if $|A_k|>0$, $k=1, \dots, n$. This gives an alternative proof of Theorem 1.29.

14. THE MAXIMUM OF A BILINEAR FORM

Theorem 6 together with the Cauchy–Schwarz inequality allows a generalization from quadratic to bilinear forms.

Theorem 17

Let A be an $m \times n$ matrix with rank $r \geqslant 1$. Let $\lambda_1 \leqslant \lambda_2 \leqslant \dots \leqslant \lambda_r$ denote the positive eigenvalues of AA' and let $S=(s_1, \dots, s_r)$ be a semi-orthogonal $m \times r$ matrix such that

$$AA'S=S\Lambda, \qquad S'S=I_r, \qquad \Lambda=\text{diag}(\lambda_1, \dots, \lambda_r). \tag{1}$$

Then, for $k=1, 2, \dots, r$,

$$(x'Ay)^2 \leqslant \lambda_k \tag{2}$$

for every $x \in \mathbb{R}^m$ and $y \in \mathbb{R}^n$ satisfying

$$x'x=1, \qquad y'y=1, \qquad s_i'x=0 \qquad (i=k+1, \dots, r). \tag{3}$$

Moreover, if $\lambda_j=\lambda_{j+1}= \dots =\lambda_k$, and either $\lambda_{j-1}<\lambda_j$ or $j=1$, then equality in (2) occurs if and only if $x=x_*$ and $y=y_*$, where

$$x_* = \sum_{i=j}^{k} \alpha_i s_i, \qquad y_* = \pm \lambda_k^{-1/2} A'x_* \tag{4}$$

for some set of real numbers $\alpha_j, \dots, \alpha_k$ satisfying $\sum_{i=j}^{k} \alpha_i^2=1$. (If λ_k is a simple eigenvalue of AA', then $j=k$ and x_* and y_* are unique, apart from sign.) Moreover, y_* is an eigenvector of $A'A$ associated with the eigenvalue λ_k, and

$$x_* = \pm \lambda_k^{-1/2} Ay_*, \qquad s_i'Ay_*=0 \qquad (i=k+1, \dots, r). \tag{5}$$

Proof. Let x and y be arbitrary vectors in $\mathbb{R}^m$ and $\mathbb{R}^n$, respectively, satisfying (3). Then

$$(x'Ay)^2 \leqslant x'AA'x \leqslant \lambda_k, \tag{6}$$

where the first inequality is Cauchy–Schwarz and the second follows from Theorem 6.

Equality occurs if and only if $y = \gamma A'x$ for some $\gamma \neq 0$ (to make the first inequality of (6) into an equality) and

$$x = \sum_{i=j}^{k} \alpha_i s_i \tag{7}$$

for some $\alpha_j, \ldots, \alpha_k$ satisfying $\sum_{i=j}^{k} \alpha_i^2 = 1$ (because of the requirement that $x'x = 1$). From (7) it follows that $AA'x = \lambda_k x$, so that

$$1 = y'y = \gamma^2 x'AA'x = \gamma^2 \lambda_k. \tag{8}$$

Hence $\gamma = \pm \lambda_k^{-1/2}$ and $y = \pm \lambda_k^{-1/2} A'x$.

Furthermore,

$$Ay = \pm \lambda_k^{-1/2} AA'x = \pm \lambda_k^{-1/2} \lambda_k x = \pm \lambda_k^{1/2} x \tag{9}$$

implying

$$A'Ay = \pm \lambda_k^{1/2} A'x = \lambda_k y, \tag{10}$$

and also

$$s_i' Ay = \pm \lambda_k^{1/2} s_i' x = 0 \qquad (i = k+1, \ldots, r). \tag{11}$$

□

15. HADAMARD'S INEQUALITY

The following inequality is a very famous one, and is due to Hadamard.

Theorem 18

For any real $n \times n$ matrix $A = (a_{ij})$,

$$|A|^2 \leqslant \prod_{i=1}^{n} \left(\sum_{j=1}^{n} a_{ij}^2 \right) \tag{1}$$

with equality if and only if AA' is a diagonal matrix or A has a zero row.

Proof. Assume that A is non-singular. (If $|A| = 0$, the result is trivial.) Then AA' is positive definite and hence, by Theorem 1.28,

$$|A|^2 = |AA'| \leqslant \prod_{i=1}^{n} (AA')_{ii} = \prod_{i=1}^{n} \left(\sum_{j=1}^{n} a_{ij}^2 \right) \tag{2}$$

with equality if and only if AA' is diagonal. □

16. AN INTERLUDE: KARAMATA'S INEQUALITY

Let $x=(x_1,x_2,\ldots,x_n)'$ and $y=(y_1,y_2,\ldots,y_n)'$ be two $n\times 1$ vectors. We say that y is *majorized* by x, and write

$$(y_1,\ldots,y_n)\prec(x_1,\ldots,x_n), \tag{1}$$

when the following three conditions are satisfied:

$$x_1+x_2+\ldots+x_n=y_1+y_2+\ldots+y_n, \tag{2}$$

$$x_1\leqslant x_2\leqslant\ldots\leqslant x_n, \qquad y_1\leqslant y_2\leqslant\ldots\leqslant y_n, \tag{3}$$

$$x_1+x_2+\ldots+x_k\leqslant y_1+y_2+\ldots+y_k \qquad (1\leqslant k\leqslant n-1). \tag{4}$$

Theorem 19

Let ϕ be a real-valued convex function defined on an interval S in $\mathbb{R}$. If $(y_1,\ldots,y_n)\prec(x_1,\ldots,x_n)$, then

$$\sum_{i=1}^{n}\phi(x_i)\geqslant\sum_{i=1}^{n}\phi(y_i). \tag{5}$$

If, in addition, ϕ is strictly convex on S, then equality in (5) occurs if and only if $x_i=y_i$ $(i=1,\ldots,n)$.

Proof. The first part of the theorem is a well known result (see Hardy *et al.* 1952, Beckenbach and Bellman 1961). Let us prove the second part, which investigates when equality in (5) can occur. Clearly, if $x_i=y_i$ for all i, then $\sum_{i=1}^{n}\phi(x_i)=\sum_{i=1}^{n}\phi(y_i)$. To prove the converse, assume that ϕ is strictly convex. We must then demonstrate the truth of the following statement: if $(y_1,\ldots,y_n)\prec(x_1,\ldots,x_n)$ and $\sum_{i=1}^{n}\phi(x_i)=\sum_{i=1}^{n}\phi(y_i)$, then $x_i=y_i$ $(i=1,\ldots,n)$.

Let us proceed by induction. For $n=1$ the statement is trivially true. Assume the statement to be true for $n=1,2,\ldots,N-1$. Assume also that $(y_1,\ldots,y_N)\prec(x_1,\ldots,x_N)$ and $\sum_{i=1}^{N}\phi(x_i)=\sum_{i=1}^{N}\phi(y_i)$. We shall show that $x_i=y_i$ $(i=1,\ldots,N)$.

Assume first that $\sum_{i=1}^{k}x_i<\sum_{i=1}^{k}y_i$ (strict inequality) for $k=1,\ldots,N-1$. Replace y_i by z_i where

$$z_1=y_1-\varepsilon, \qquad z_i=y_i \quad (i=2,\ldots,N-1), \qquad z_N=y_N+\varepsilon. \tag{6}$$

Then as we can choose $\varepsilon>0$ arbitrarily small, $(z_1,\ldots,z_N)\prec(x_1,\ldots x_N)$. Hence, by the first part of the theorem,

$$\sum_{i=1}^{N}\phi(x_i)\geqslant\sum_{i=1}^{N}\phi(z_i). \tag{7}$$

On the other hand,

$$\begin{aligned}\sum_{i=1}^{N} \phi(x_i) &= \sum_{i=1}^{N} \phi(y_i)\\ &= \sum_{i=1}^{N} \phi(z_i) + \varepsilon\left(\frac{\phi(y_1)-\phi(y_1-\varepsilon)}{\varepsilon} - \frac{\phi(y_N+\varepsilon)-\phi(y_N)}{\varepsilon}\right)\\ &< \sum_{i=1}^{N} \phi(z_i), \qquad (8)\end{aligned}$$

which contradicts (7). (See Figure 1 to see how the inequality in (8) is obtained.)

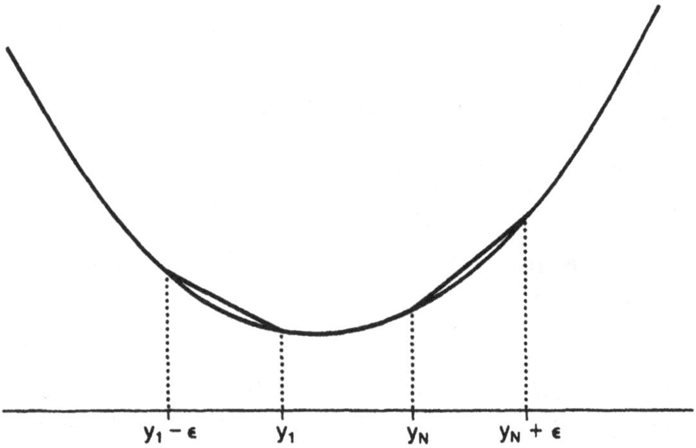

Figure **1** Diagram showing that

$$\frac{\phi(y_1)-\phi(y_1-\varepsilon)}{\varepsilon} < \frac{\phi(y_N+\varepsilon)-\phi(y_N)}{\varepsilon}$$

Next assume that $\sum_{i=1}^{m} x_i = \sum_{i=1}^{m} y_i$ for some m $(1 \leqslant m \leqslant N-1)$. Then $(y_1, \ldots, y_m) \prec (x_1, \ldots, x_m)$ and $(y_{m+1}, \ldots, y_N) \prec (x_{m+1}, \ldots, x_N)$. The first part of the theorem implies

$$\sum_{i=1}^{m} \phi(x_i) \geqslant \sum_{i=1}^{m} \phi(y_i), \qquad \sum_{i=m+1}^{N} \phi(x_i) \geqslant \sum_{i=m+1}^{N} \phi(y_i), \qquad (9)$$

and since $\sum_{i=1}^{N} \phi(x_i) = \sum_{i=1}^{N} \phi(y_i)$ by assumption, it follows that the $\geqslant$ signs in (9) can be replaced by = signs. The induction hypothesis applied to the sets $(x_1, \ldots, x_m)$ and $(y_1, \ldots, y_m)$ then yields $x_i = y_i$ $(i = 1, \ldots, m)$; the induction hypothesis applied to the sets $(x_{m+1}, \ldots, x_N)$ and $(y_{m+1}, \ldots, y_N)$ yields $x_i = y_i$ $(i = m+1, \ldots, N)$. This completes the proof. □

17. KARAMATA'S INEQUALITY APPLIED TO EIGENVALUES

An important application of Karamata's inequality is the next result, which provides the basis of the analysis in the next few sections.

Theorem 20

Let $A=(a_{ij})$ be a real symmetric $n\times n$ matrix with eigenvalues $\lambda_1, \lambda_2, \ldots, \lambda_n$. Then for any convex function ϕ we have

$$\sum_{i=1}^{n} \phi(\lambda_i) \geqslant \sum_{i=1}^{n} \phi(a_{ii}). \tag{1}$$

Moreover, if ϕ is strictly convex, then equality in (1) occurs if and only if A is diagonal.

Proof. Without loss of generality we may assume that

$$\lambda_1 \leqslant \lambda_2 \leqslant \ldots \leqslant \lambda_n, \qquad a_{11} \leqslant a_{22} \leqslant \ldots \leqslant a_{nn}. \tag{2}$$

Theorem 14 implies that $(a_{11}, \ldots, a_{nn})$ is majorized by $(\lambda_1, \ldots, \lambda_n)$ in the sense of section 16. Karamata's inequality (Theorem 19) then yields (1). For strictly convex ϕ, Theorem 19 implies that equality in (1) holds if and only if $\lambda_i = a_{ii}$ for $i=1, \ldots, n$, and, by Theorem 1.30, this is the case if and only if A is diagonal. □

Exercise

1. Prove Theorem 1.28 as a special case of Theorem 20. [*Hint*: Choose $\phi(x) = -\log x, x>0$.]

18. AN INEQUALITY CONCERNING POSITIVE SEMIDEFINITE MATRICES

If A is positive semidefinite then, by Theorem 1.13, we can write $A = S\Lambda S'$, where Λ is diagonal with non-negative diagonal elements. We now define the p-th power of A as $A^p = S\Lambda^p S'$. In particular, $A^{1/2}$ (for positive semidefinite A) is the *unique* positive semidefinite matrix $S\Lambda^{1/2}S'$.

Theorem 21

Let $A=(a_{ij})$ be a positive semidefinite $n\times n$ matrix. Then

$$\operatorname{tr} A^p \geqslant \sum a_{ii}^p \qquad (p>1) \tag{1}$$

and

$$\operatorname{tr} A^p \leqslant \sum a_{ii}^p \qquad (0<p<1), \tag{2}$$

with equality if and only if A is diagonal.

Proof. Let $p>1$, and define $\phi(x)=x^p$ $(x\geq 0)$. The function ϕ is continuous and strictly convex. Hence Theorem 20 implies that

$$\operatorname{tr} A^p = \sum_{i=1}^{n} \lambda_i^p(A) = \sum_{i=1}^{n} \phi(\lambda_i(A)) \geq \sum_{i=1}^{n} \phi(a_{ii}) = \sum_{i=1}^{n} a_{ii}^p \tag{3}$$

with equality if and only if A is diagonal. Next, let $0<p<1$, define $\psi(x)=-x^p$ $(x\geq 0)$, and proceed in the same way to make the proof complete. □

19. A REPRESENTATION THEOREM FOR $(\Sigma a_i^p)^{1/p}$

As a preliminary to Theorem 23, let us prove Theorem 22.

Theorem 22

Let $p>1$, $q=p/(p-1)$ and $a_i \geq 0$ $(i=1,\ldots,n)$. Then

$$\sum_{i=1}^{n} a_i x_i \leq \left(\sum_{i=1}^{n} a_i^p\right)^{1/p} \tag{1}$$

for every set of non-negative numbers $x_1, x_2, \ldots, x_n$ satisfying $\sum_{i=1}^{n} x_i^q = 1$. Equality in (1) occurs if and only if $a_1 = a_2 = \ldots. = a_n = 0$ or

$$x_i^q = a_i^p \Big/ \sum_{j=1}^{n} a_j^p \qquad (i=1,\ldots,n). \tag{2}$$

Note. We call this theorem a *representation* theorem because (1) can be alternatively written as

$$\max_{x\in S} \sum_{i=1}^{n} a_i x_i = \left(\sum_{i=1}^{n} a_i^p\right)^{1/p} \tag{3}$$

where

$$S = \left\{x = (x_1,\ldots,x_n) : x_i \geq 0, \sum_{i=1}^{n} x_i^q = 1\right\}. \tag{4}$$

Proof. Let us consider the maximization problem

$$\text{maximize} \quad \sum_{i=1}^{n} a_i x_i \tag{5}$$

$$\text{subject to} \quad \sum_{i=1}^{n} x_i^q = 1. \tag{6}$$

We form the Lagrangian function

$$\psi(x) = \sum_{i=1}^{n} a_i x_i - \lambda q^{-1}\left(\sum_{i=1}^{n} x_i^q - 1\right), \tag{7}$$

and differentiate. This yields

$$d\psi(x) = \sum_{i=1}^{n} (a_i - \lambda x_i^{q-1})dx_i. \tag{8}$$

From (8) we obtain the first-order conditions:

$$\lambda x_i^{q-1} = a_i \qquad (i = 1, \ldots, n), \tag{9}$$

$$\sum_{i=1}^{n} x_i^q = 1. \tag{10}$$

Solving for x_i and λ, we obtain

$$x_i = \left(a_i^p \Big/ \sum_i a_i^p\right)^{1/q} \qquad (i = 1, \ldots, n), \tag{11}$$

$$\lambda = \left(\sum_i a_i^p\right)^{1/p}. \tag{12}$$

Since $q > 1$, $\psi(x)$ is concave; it follows from Theorem 7.13 that $\sum a_i x_i$ has an absolute maximum under the constraint (6) at every point where (11) is satisfied. The constrained maximum is

$$\sum_i a_i \left(a_i^p \Big/ \sum_i a_i^p\right)^{1/q} = \sum_i a_i^p \Big/ \left(\sum_i a_i^p\right)^{1/q} = \left(\sum_i a_i^p\right)^{1/p}. \tag{13}$$

□

20. A REPRESENTATION THEOREM FOR (tr $A^p)^{1/p}$

An important generalization of Theorem 22, which provides the basis for proving matrix analogues of the fundamental inequalities of Hölder and Minkowski (Theorems 24 and 26), is given in Theorem 23.

Theorem 23

Let $p > 1$, $q = p/(p-1)$ and let $A \neq 0$ be a positive semidefinite $n \times n$ matrix. Then

$$\operatorname{tr} AX \leqslant (\operatorname{tr} A^p)^{1/p} \tag{1}$$

for every positive semidefinite $n \times n$ matrix X satisfying $\operatorname{tr} X^q = 1$. Equality in (1) occurs if and only if

$$X^q = (1/\operatorname{tr} A^p)A^p. \tag{2}$$

Proof. Let X be an arbitrary positive semidefinite $n \times n$ matrix satisfying $\operatorname{tr} X^q = 1$. Let S be an orthogonal matrix such that $S'XS = \Lambda$, where Λ is diagonal and has the eigenvalues of X as its diagonal elements. Define $B = (b_{ij}) = S'AS$.

Then

$$\operatorname{tr} AX = \operatorname{tr} B\Lambda = \sum_i b_{ii}\lambda_i \tag{3}$$

and

$$\operatorname{tr} X^q = \operatorname{tr} \Lambda^q = \sum_i \lambda_i^q. \tag{4}$$

Hence, by Theorem 22,

$$\operatorname{tr} AX \leqslant \left(\sum_i b_{ii}^p\right)^{1/p}. \tag{5}$$

Since A is positive semidefinite, so is B, and Theorem 21 thus implies that

$$\sum_i b_{ii}^p \leqslant \operatorname{tr} B^p. \tag{6}$$

Combining (5) and (6) we obtain

$$\operatorname{tr} AX \leqslant (\operatorname{tr} B^p)^{1/p} = (\operatorname{tr} A^p)^{1/p}. \tag{7}$$

Equality in (5) occurs iff

$$\lambda_i^q = b_{ii}^p \Big/ \sum_i b_{ii}^p \qquad (i=1, \ldots, n) \tag{8}$$

and equality in (6) iff B is diagonal. Hence, equality in (1) occurs iff

$$\Lambda^q = B^p/\operatorname{tr} B^p, \tag{9}$$

which is equivalent to (2). This concludes the proof. □

21. HÖLDER'S INEQUALITY

In its simplest form Hölder's inequality asserts that

$$x_1^\alpha y_1^{1-\alpha} + x_2^\alpha y_2^{1-\alpha} \leqslant (x_1+x_2)^\alpha (y_1+y_2)^{1-\alpha} \qquad (0<\alpha<1) \tag{1}$$

for every non-negative x_1, x_2, y_1, y_2. This inequality can be extended in two directions. We can show (by simple mathematical induction) that

$$\sum_{i=1}^m x_i^\alpha y_i^{1-\alpha} \leqslant \left(\sum_{i=1}^m x_i\right)^\alpha \left(\sum_{i=1}^m y_i\right)^{1-\alpha} \qquad (0<\alpha<1) \tag{2}$$

for every $x_i \geqslant 0$, $y_i \geqslant 0$; and also, arranging the induction differently, that

$$\prod_{j=1}^n x_j^{\alpha_j} + \prod_{j=1}^n y_j^{\alpha_j} \leqslant \prod_{j=1}^n (x_j + y_j)^{\alpha_j} \tag{3}$$

for every $x_j \geqslant 0$, $y_j \geqslant 0$, $\alpha_j > 0$, $\sum_{j=1}^n \alpha_j = 1$.

Combining (2) and (3), we obtain the following result.

Hölder's inequality

Let $X = (x_{ij})$ be a non-negative $m \times n$ matrix (that is, a matrix all of whose elements are non-negative), and let $\alpha_j > 0$ $(j = 1,\ldots, n)$, $\sum_{j=1}^{n} \alpha_j = 1$. Then

$$\sum_{i=1}^{m} \prod_{j=1}^{n} x_{ij}^{\alpha_j} \leqslant \prod_{j=1}^{n} \left(\sum_{i=1}^{m} x_{ij} \right)^{\alpha_j} \tag{4}$$

with equality if and only if either $r(X) = 1$ or one of the columns of X is the null vector.

In this section we want to show how Theorem 23 can be used to obtain the matrix analogue of (2).

Theorem 24

For any two positive semidefinite matrices A and B of the same order, $A \neq 0$, $B \neq 0$, and $0 < \alpha < 1$, we have

$$\operatorname{tr} A^{\alpha} B^{1-\alpha} \leqslant (\operatorname{tr} A)^{\alpha} (\operatorname{tr} B)^{1-\alpha}, \tag{5}$$

with equality if and only if $B = \mu A$ for some scalar $\mu > 0$.

Proof. Let $p = 1/\alpha$, $q = 1/(1-\alpha)$ and assume $B \neq 0$. Now define

$$X = \frac{B^{1/q}}{(\operatorname{tr} B)^{1/q}}. \tag{6}$$

Then $\operatorname{tr} X^q = 1$, and hence Theorem 23 applied to $A^{1/p}$ yields

$$\operatorname{tr} A^{1/p} B^{1/q} \leqslant (\operatorname{tr} A)^{1/p} (\operatorname{tr} B)^{1/q}, \tag{7}$$

which is (5). According to Theorem 23, equality in (5) can occur only if $X^q = (1/\operatorname{tr} A)A$, that is, if $B = \mu A$ for some $\mu > 0$. □

Exercises

1. Let A and B be positive semidefinite, and $0 < \alpha < 1$. Define the symmetric matrix
$$C = \alpha A + (1-\alpha)B - A^{\alpha/2} B^{1-\alpha} A^{\alpha/2}.$$
Show that $\operatorname{tr} C \geqslant 0$ with equality if and only if $A = B$.
2. For every $x > 0$
$$x^{\alpha} \leqslant \alpha x + 1 - \alpha \qquad (0 < \alpha < 1)$$
$$x^{\alpha} \geqslant \alpha x + 1 - \alpha \qquad (\alpha > 1 \text{ or } \alpha < 0),$$
with equality if and only if $x = 1$.
3. If A and B are positive semidefinite *and commute* (that is, $AB = BA$), then the matrix C of exercise 1 is positive semidefinite. Moreover, C is non-singular (hence positive definite) if and only if $A - B$ is non-singular.

4. Let $p>1$ and $q=p/(p-1)$. Show that for any two positive semidefinite matrices $A\neq 0$ and $B\neq 0$ of the same order,

$$\operatorname{tr} AB \leqslant (\operatorname{tr} A^p)^{1/p}(\operatorname{tr} B^q)^{1/q}$$

with equality if and only if $B=\mu A^{p-1}$ for some $\mu>0$.

22. CONCAVITY OF $\log|A|$

In exercise 1 of the previous section we saw that

$$\operatorname{tr} A^\alpha B^{1-\alpha} \leqslant \operatorname{tr}(\alpha A+(1-\alpha)B) \qquad (0<\alpha<1) \tag{1}$$

for any pair of positive semidefinite matrices A and B. Let us now demonstrate the multiplicative analogue of (1).

Theorem 25

For any two positive semidefinite matrices A and B of the same order, and $0<\alpha<1$, we have

$$|A|^\alpha |B|^{1-\alpha} \leqslant |\alpha A+(1-\alpha)B| \tag{2}$$

with equality if and only if $A=B$ or $|\alpha A+(1-\alpha)B|=0$.

Proof. If either A or B is singular, the result is obvious. Assume therefore that A and B are both positive definite. Applying exercise 2 of section 21 to the eigenvalues $\lambda_1,\ldots,\lambda_n$ of the positive definite matrix $B^{-1/2}AB^{-1/2}$ yields

$$\lambda_i^\alpha \leqslant \alpha\lambda_i+(1-\alpha) \qquad (i=1,\ldots,n), \tag{3}$$

and hence, multiplying both sides of (3) over $i=1,\ldots,n$,

$$|B^{-1/2}AB^{-1/2}|^\alpha \leqslant |\alpha B^{-1/2}AB^{-1/2}+(1-\alpha)I|. \tag{4}$$

From (4) we obtain

$$\begin{aligned} |A|^\alpha|B|^{1-\alpha} &= |B||B^{-1/2}AB^{-1/2}|^\alpha \leqslant |B||\alpha B^{-1/2}AB^{-1/2}+(1-\alpha)I| \\ &= |B^{1/2}(\alpha B^{-1/2}AB^{-1/2}+(1-\alpha)I)B^{1/2}| = |\alpha A+(1-\alpha)B|. \end{aligned} \tag{5}$$

There is equality in (5) iff there is equality in (3), which occurs iff every eigenvalue of $B^{-1/2}AB^{-1/2}$ equals one, that is, iff $A=B$. □

Another way of expressing the result of Theorem 25 is to say that the real-valued function ϕ defined by $\phi(A)=\log|A|$ is *concave* on the set of positive definite matrices. This is seen by taking logarithms on both sides of (2). We note, however, that the function ψ given by $\psi(A)=|A|$ is neither convex nor concave on the set of positive definite matrices. This is easily seen by taking

$$A=\begin{pmatrix}1 & 0\\ 0 & 1\end{pmatrix}, \qquad B=\begin{pmatrix}1+\delta & 0\\ 0 & 1+\varepsilon\end{pmatrix} \qquad (\delta>-1,\ \varepsilon>-1). \tag{6}$$

Then, for $\alpha=\frac{1}{2}$,

$$\alpha|A|+(1-\alpha)|B|-|\alpha A+(1-\alpha)B| = \delta\varepsilon/4, \tag{7}$$

which can take positive or negative values depending on whether δ and ε have the same or opposite signs.

Exercises

1. Show that, for A positive definite,
$$\mathrm{d}^2 \log|A| = -\operatorname{tr} A^{-1}(\mathrm{d}A)A^{-1}(\mathrm{d}A) < 0$$
for all $\mathrm{d}A \neq 0$. (Compare Theorem 25.)
2. Show that the matrix inverse is 'matrix convex' on the set of positive definite matrices. That is, show that the matrix
$$C(\lambda) = \lambda A^{-1} + (1-\lambda)B^{-1} - (\lambda A + (1-\lambda)B)^{-1}$$
is positive semidefinite for all positive definite A and B and $0<\lambda<1$ (Moore 1973).
3. Furthermore, show that $C(\lambda)$ is positive definite for all $\lambda \in (0, 1)$ if and only if $|A-B| \neq 0$ (Moore 1973).
4. Show that
$$x'(A+B)^{-1}x \leqslant \frac{(x'A^{-1}x)(x'B^{-1}x)}{x'(A^{-1}+B^{-1})x} \leqslant \frac{x'A^{-1}x + x'B^{-1}x}{4}$$
[*Hint*: Use Exercise 2 and Bergstrom's inequality, section 11.2.]

23. MINKOWSKI'S INEQUALITY

Minkowski's inequality, in its most rudimentary form, states that
$$[(x_1+y_1)^p + (x_2+y_2)^p]^{1/p} \leqslant (x_1^p + x_2^p)^{1/p} + (y_1^p + y_2^p)^{1/p} \tag{1}$$
for every non-negative x_1, x_2, y_1, y_2 and $p>1$. As in Hölder's inequality, (1) can be extended in two directions. We have
$$\left(\sum_{i=1}^{m}(x_i+y_i)^p\right)^{1/p} \leqslant \left(\sum_{i=1}^{m} x_i^p\right)^{1/p} + \left(\sum_{i=1}^{m} y_i^p\right)^{1/p} \tag{2}$$
for every $x_i \geqslant 0$, $y_i \geqslant 0$ and $p>1$; and also
$$\left[\left(\sum_{j=1}^{n} x_j\right)^p + \left(\sum_{j=1}^{n} y_j\right)^p\right]^{1/p} \leqslant \sum_{j=1}^{n}(x_j^p + y_j^p)^{1/p} \tag{3}$$
for every $x_j \geqslant 0$, $y_j \geqslant 0$ and $p>1$. Notice that if in (3) we replace x_j by $x_j^{1/p}$, y_j by $y_j^{1/p}$ and then p by $1/p$, we obtain
$$\left(\sum_{j=1}^{n}(x_j+y_j)^p\right)^{1/p} \geqslant \left(\sum_{j=1}^{n} x_j^p\right)^{1/p} + \left(\sum_{j=1}^{n} y_j^p\right)^{1/p} \tag{4}$$
for every $x_j \geqslant 0$, $y_j \geqslant 0$ and $0<p<1$.

All these cases are contained in the following inequality.

Minkowski's inequality

Let $X=(x_{ij})$ be a non-negative $m\times n$ matrix (that is, $x_{ij}\geqslant 0$ for $i=1,\ldots,m$ and $j=1,\ldots,n$), and let $p>1$. Then

$$\left[\sum_{i=1}^{m}\left(\sum_{j=1}^{n}x_{ij}\right)^{p}\right]^{1/p}\leqslant\sum_{j=1}^{n}\left(\sum_{i=1}^{m}x_{ij}^{p}\right)^{1/p} \tag{5}$$

with equality if and only if $r(X)=1$.

Let us now obtain, again by using Theorem 23, the matrix analogue of (2).

Theorem 26

For any two positive semidefinite matrices A and B of the same order ($A\neq 0$, $B\neq 0$), and $p>1$, we have

$$[\operatorname{tr}(A+B)^{p}]^{1/p}\leqslant(\operatorname{tr}A^{p})^{1/p}+(\operatorname{tr}B^{p})^{1/p}, \tag{6}$$

with equality if and only if $A=\mu B$ for some $\mu>0$.

Proof. Let $p>1$, $q=p/(p-1)$ and let

$$R=\{X: X\in\mathbb{R}^{n\times n},\ X \text{ positive semidefinite, } \operatorname{tr}X^{q}=1\}. \tag{7}$$

An equivalent version of Theorem 23 then states that

$$\max_{R}\operatorname{tr}AX=(\operatorname{tr}A^{p})^{1/p} \tag{8}$$

for every positive semidefinite $n\times n$ matrix A. Using this representation, we obtain

$$\begin{aligned}[\operatorname{tr}(A+B)^{p}]^{1/p}&=\max_{R}\operatorname{tr}(A+B)X\\ &\leqslant\max_{R}\operatorname{tr}AX+\max_{R}\operatorname{tr}BX\\ &=(\operatorname{tr}A^{p})^{1/p}+(\operatorname{tr}B^{p})^{1/p}.\end{aligned} \tag{9}$$

Equality in (9) can occur only if the same X maximizes $\operatorname{tr}AX$, $\operatorname{tr}BX$ and $\operatorname{tr}(A+B)X$, which implies, by Theorem 23, that A^{p}, B^{p} and $(A+B)^{p}$ are proportional, and hence that A and B must be proportional. □

24. QUASILINEAR REPRESENTATION OF $|A|^{1/n}$

In section 4 we established (exercise 2) that

$$(1/n)\operatorname{tr}A\geqslant|A|^{1/n} \tag{1}$$

for every positive semidefinite $n \times n$ matrix A. The following theorem generalizes this result.

Theorem 27

Let $A \neq 0$ be a positive semidefinite $n \times n$ matrix. Then

$$(1/n) \operatorname{tr} AX \geqslant |A|^{1/n} \tag{2}$$

for every positive definite $n \times n$ matrix X satisfying $|X| = 1$, with equality if and only if

$$X = |A|^{1/n} A^{-1}. \tag{3}$$

Let us give two proofs.

First proof. Let $A \neq 0$ be positive semidefinite and X positive definite with $|X| = 1$. Denote the eigenvalues of $X^{1/2}AX^{1/2}$ by $\lambda_1, \ldots, \lambda_n$. Then $\lambda_i \geqslant 0$ $(i = 1, \ldots, n)$, and Theorem 3 implies that

$$\prod_{i=1}^{n} \lambda_i^{1/n} \leqslant (1/n) \sum_{i=1}^{n} \lambda_i \tag{4}$$

with equality if and only if $\lambda_1 = \lambda_2 = \ldots = \lambda_n$. Rewriting (4) in terms of the matrices A and X we obtain

$$|X^{1/2}AX^{1/2}|^{1/n} \leqslant (1/n) \operatorname{tr} X^{1/2}AX^{1/2} \tag{5}$$

and hence, since $|X| = 1$,

$$(1/n) \operatorname{tr} AX \geqslant |A|^{1/n}. \tag{6}$$

Equality in (6) occurs if and only if all eigenvalues of $X^{1/2}AX^{1/2}$ are equal, that is, iff

$$X^{1/2}AX^{1/2} = \mu I_n \tag{7}$$

for some $\mu > 0$. (Notice that $\mu = 0$ cannot occur, because it would imply $A = 0$, which we have excluded.) From (7) we obtain $A = \mu X^{-1}$ and hence $X = \mu A^{-1}$. Taking determinants on both sides we find $\mu = |A|^{1/n}$ since $|X| = 1$. □

Second proof. In this proof we view the inequality (2) as the solution of the following constrained minimization problem:

$$\underset{X}{\text{minimize}} \quad (1/n) \operatorname{tr} AX \tag{8}$$

$$\text{subject to} \quad \log|X| = 0, \ X \text{ positive definite}, \tag{9}$$

where A is a given positive semidefinite $n \times n$ matrix. To take into account the positive definiteness of X we write $X = YY'$ where Y is a square matrix of order n;

the minimization problem then becomes

$$\underset{Y}{\text{minimize}} \quad (1/n)\operatorname{tr} Y'AY \tag{10}$$

$$\text{subject to} \quad \log|Y|^2=0. \tag{11}$$

To solve (10)–(11) we form the Lagrangian function

$$\psi(Y)=(1/n)\operatorname{tr} Y'AY-\lambda\log|Y|^2 \tag{12}$$

and differentiate. This yields

$$\begin{aligned}\mathrm{d}\psi(Y)&=(2/n)\operatorname{tr} Y'A\mathrm{d}Y-2\lambda\operatorname{tr} Y^{-1}\mathrm{d}Y\\&=2\operatorname{tr}[(1/n)Y'A-\lambda Y^{-1}]\mathrm{d}Y.\end{aligned} \tag{13}$$

From (13) we obtain the first-order conditions

$$(1/n)Y'A=\lambda Y^{-1} \tag{14}$$

$$|Y|^2=1. \tag{15}$$

Pre-multiplying both sides of (14) with $n(Y')^{-1}$ gives

$$A=n\lambda(YY')^{-1}, \tag{16}$$

which shows that $\lambda>0$ and A is non-singular. (If $\lambda=0$, then A is the null matrix; this case we have excluded from the beginning.) Taking determinants in (16) we obtain, using (15),

$$n\lambda=|A|^{1/n}. \tag{17}$$

Inserting this in (16) and rearranging yields

$$YY'=|A|^{1/n}A^{-1}. \tag{18}$$

Since tr $Y'AY$ is convex, $\log|Y|^2$ concave (Theorem 25) and $\lambda>0$, it follows that $\psi(Y)$ is convex. Hence Theorem 7.13 implies that $(1/n)\operatorname{tr} Y'AY$ has an absolute minimum under the constraint (11) at every point where (18) is satisfied. The constrained minimum is

$$(1/n)\operatorname{tr} Y'AY=(1/n)\operatorname{tr}|A|^{1/n}A^{-1}A=|A|^{1/n}. \tag{19}$$

□

Exercises

1. Use exercise 4.2 to prove that

$$(1/n)\operatorname{tr} AX\geqslant|A|^{1/n}|X|^{1/n}$$

for every two positive semidefinite $n\times n$ matrices A and X, with equality if and only if $A=0$ or $X=\mu A^{-1}$ for some $\mu\geqslant 0$.

2. Hence obtain Theorem 27 as a special case.

25. MINKOWSKI'S DETERMINANT THEOREM

Using the quasilinear representation given in Theorem 27, let us establish Minkowski's determinant theorem.

Theorem 28

For any two positive semidefinite $n\times n$ matrices $A\neq 0$ and $B\neq 0$,

$$|A+B|^{1/n}\geqslant|A|^{1/n}+|B|^{1/n} \tag{1}$$

with equality if and only if $|A+B|=0$ or $A=\mu B$ for some $\mu>0$.

Proof. Let A and B be two positive semidefinite matrices, and assume that $A\neq 0$, $B\neq 0$. If $|A|=0$, $|B|>0$, we clearly have $|A+B|>|B|$. If $|A|>0$, $|B|=0$, we have $|A+B|>|A|$. If $|A|=|B|=0$, we have $|A+B|\geqslant 0$. Hence, if A or B is singular, the inequality (1) holds, and equality occurs iff $|A+B|=0$.

Assume next that A and B are positive definite. Using the representation in Theorem 27, we then have

$$\begin{aligned}|A+B|^{1/n}&=\min_X\,(1/n)\operatorname{tr}(A+B)X\\ &\geqslant\min_X\,(1/n)\operatorname{tr}AX+\min_X\,(1/n)\operatorname{tr}BX\\ &=|A|^{1/n}+|B|^{1/n},\end{aligned} \tag{2}$$

where the minimum is taken over all positive definite X satisfying $|X|=1$. Equality occurs only if the same X minimizes $(1/n)\operatorname{tr}AX$, $(1/n)\operatorname{tr}BX$ and $(1/n)\operatorname{tr}(A+B)X$, which implies that A^{-1}, B^{-1} and $(A+B)^{-1}$ must be proportional, and hence that A and B must be proportional. □

26. WEIGHTED MEANS OF ORDER p

Definition

Let x be an $n\times 1$ vector with positive components $x_1, x_2, \ldots, x_n$, and let a be an $n\times 1$ vector of positive weights $\alpha_1, \alpha_2, \ldots, \alpha_n$, so that

$$0<\alpha_i<1, \qquad \sum_{i=1}^{n}\alpha_i=1. \tag{1}$$

Then, for any real $p\neq 0$, the expression

$$M_p(x,a)=\left(\sum_{i=1}^{n}\alpha_i x_i^p\right)^{1/p} \tag{2}$$

is called the *weighted mean of order p* of $x_1,\ldots,x_n$ with weights $\alpha_1,\ldots,\alpha_n$.

Note. This definition can be extended to non-negative x if we set $M_p(x, a) = 0$ in the case where $p < 0$ and one or more of the x_i are zero. We shall, however, confine ourselves to positive x.

The functional form defined by (2) occurs frequently in the economics literature. For example, if we multiply $M_p(x, a)$ by a constant, we obtain the CES (constant elasticity of substitution) functional form.

Theorem 29

$M_p(x, a)$ is (positively) linearly homogeneous in x, that is,

$$M_p(\lambda x, a) = \lambda M_p(x, a) \tag{3}$$

for every $\lambda > 0$.

Proof. Immediate from the definition. □

One would expect a mean of n numbers to lie between the smallest and largest of the n numbers. This is indeed the case here as we shall now demonstrate.

Theorem 30

We have

$$\min_{1 \leq i \leq n} x_i \leq M_p(x, a) \leq \max_{1 \leq i \leq n} x_i \tag{4}$$

with equality if and only if $x_1 = x_2 = \ldots = x_n$.

Proof. We first prove the theorem for $p = 1$. Since

$$\sum_{i=1}^{n} \alpha_i [x_i - M_1(x, a)] = 0, \tag{5}$$

we have either $x_1 = x_2 = \ldots = x_n$, or else $x_i < M_1(x, a)$ for at least one x_i and $M_1(x, a) < x_i$ for at least one other x_i. This proves the theorem for $p = 1$.

For $p \neq 1$, we let $y_i = x_i^p$ $(i = 1, \ldots, n)$. Then, since

$$\min_{1 \leq i \leq n} y_i \leq M_1(y, a) \leq \max_{1 \leq i \leq n} y_i, \tag{6}$$

we obtain

$$\min_{1 \leq i \leq n} x_i^p \leq [M_p(x, a)]^p \leq \max_{1 \leq i \leq n} x_i^p. \tag{7}$$

This implies that $[M_p(x, a)]^p$ lies between $(\min x_i)^p$ and $(\max x_i)^p$, and completes the proof. □

Let us next investigate the behaviour of $M_p(x, a)$ when p tends to 0 or to $\pm\infty$.

Theorem 31

$$\lim_{p \to 0} M_p(x, a) = \prod_{i=1}^{n} x_i^{\alpha_i} \tag{8}$$

$$\lim_{p\to\infty} M_p(x, a) = \max x_i \tag{9}$$

$$\lim_{p\to-\infty} M_p(x, a) = \min x_i. \tag{10}$$

Proof. To prove (8) we let

$$\phi(p) = \log\left(\sum_{i=1}^{n} \alpha_i x_i^p\right)$$

and $\psi(p) = p$, so that

$$\log M_p(x, a) = \phi(p)/\psi(p). \tag{11}$$

Then $\phi(0) = \psi(0) = 0$, and

$$\phi'(p) = \left(\sum_{i=1}^{n} \alpha_i x_i^p\right)^{-1} \sum_{j=1}^{n} \alpha_j x_j^p \log x_j, \qquad \psi'(p) = 1. \tag{12}$$

By l'Hôpital's rule,

$$\lim_{p\to 0} \frac{\phi(p)}{\psi(p)} = \lim_{p\to 0} \frac{\phi'(p)}{\psi'(p)} = \frac{\phi'(0)}{\psi'(0)} = \sum_{j=1}^{n} \alpha_j \log x_j = \log\left(\prod_{j=1}^{n} x_j^{\alpha_j}\right), \tag{13}$$

and (8) follows. To prove (9), let

$$x_k = \max_{1\leqslant i\leqslant n} x_i$$

(k is not necessarily unique). Then, for $p > 0$,

$$\alpha_k^{1/p} x_k \leqslant M_p(x, a) \leqslant x_k \tag{14}$$

which implies (9). Finally, to prove (10), let $q = -p$ and $y_i = 1/x_i$. Then

$$M_p(x, a) = [M_q(y, a)]^{-1} \tag{15}$$

and hence

$$\lim_{p\to-\infty} M_p(x, a) = \lim_{q\to\infty} [M_q(y, a)]^{-1} = \left(\max_{1\leqslant i\leqslant n} y_i\right)^{-1} = \min_{1\leqslant i\leqslant n} x_i. \tag{16}$$

□

27. SCHLÖMILCH'S INEQUALITY

The limiting result (26.8) in the previous section suggests that it is convenient to define

$$M_0(x, a) = \prod_{i=1}^{n} x_i^{\alpha_i}, \tag{1}$$

since in that case $M_p(x, a)$, regarded as a function of p, is continuous for every p in

$\mathbb{R}$. The arithmetic–geometric mean inequality (Theorem 3) then takes the form

$$M_0(x,a) \leqslant M_1(x,a), \tag{2}$$

and is a special case of the following result, due to Schlömilch.

Theorem 32

If not all x_i are equal, $M_p(x,a)$ is a monotonically increasing function of p. That is, if $p<q$, then

$$M_p(x,a) \leqslant M_q(x,a) \tag{3}$$

with equality if and only if $x_1 = x_2 = \ldots = x_n$.

Proof. Assume that not all x_i are equal. (If they are, the result is trivial.) We show first that

$$\mathrm{d}M_p(x,a)/\mathrm{d}p > 0 \qquad \text{for all } p \neq 0. \tag{4}$$

Define

$$\phi(p) = \log\left(\sum_{i=1}^{n} \alpha_i x_i^p\right).$$

Then $M_p(x,a) = \exp[\phi(p)/p]$ and

$$\begin{aligned}
\mathrm{d}M_p(x,a)/\mathrm{d}p &= p^{-2} M_p(x,a)\,[p\phi'(p) - \phi(p)] \\
&= p^{-2} M_p(x,a)\left[\left(\sum_i \alpha_i x_i^p\right)^{-1} \sum_j \alpha_j x_j^p \log x_j^p - \log \sum_i \alpha_i x_i^p\right] \\
&= \frac{M_p(x,a)}{p^2 \sum_i \alpha_i x_i^p}\left[\sum_i \alpha_i g(x_i^p) - g\left(\sum_i \alpha_i x_i^p\right)\right],
\end{aligned} \tag{5}$$

where the real-valued function g is defined for $z>0$ by $g(z) = z\log z$. Since g is strictly convex (see exercise 4.9.1), condition (4) follows. Hence $M_p(x,a)$ is strictly increasing on $(-\infty, 0)$ and $(0, \infty)$, and since $M_p(x,a)$ is continuous at $p=0$, the result follows. □

28. CURVATURE PROPERTIES OF $M_p(x,a)$

The curvature properties of the weighted means of order p follow from the sign of the Hessian matrix.

Theorem 33

$M_p(x,a)$ is a concave function of x for $p \leqslant 1$ and a convex function of x for $p \geqslant 1$. In particular,

$$M_p(x,a) + M_p(y,a) \leqslant M_p(x+y,a) \qquad (p<1) \tag{1}$$

and

$$M_p(x,a)+M_p(y,a)\geqslant M_p(x+y,a) \qquad (p>1) \tag{2}$$

with equality if and only if x and y are linearly dependent.

Proof. Let $p \neq 0$, $p \neq 1$. (If $p=1$, the result is obvious.) Let

$$\phi(x)=\sum_i \alpha_i x_i^p,$$

so that $M(x)\equiv M_p(x,a)=[\phi(x)]^{1/p}$. Then

$$\mathrm{d}M(x)=\frac{M(x)}{p\phi(x)}\,\mathrm{d}\phi(x), \tag{3}$$

$$\mathrm{d}^2M(x)=\frac{(1-p)\,M(x)}{[p\phi(x)]^2}\{[\mathrm{d}\phi(x)]^2+[p/(1-p)]\,\phi(x)\,\mathrm{d}^2\phi(x)\}. \tag{4}$$

Now, since

$$\mathrm{d}\phi(x)=p\sum_i \alpha_i x_i^{p-1}\mathrm{d}x_i, \qquad \mathrm{d}^2\phi(x)=p(p-1)\sum_i \alpha_i x_i^{p-2}(\mathrm{d}x_i)^2, \tag{5}$$

we obtain

$$\mathrm{d}^2M(x)=\frac{(1-p)\,M(x)}{[\phi(x)]^2}\left[\left(\sum_i \alpha_i x_i^{p-1}\mathrm{d}x_i\right)^2-\phi(x)\sum_i \alpha_i x_i^{p-2}(\mathrm{d}x_i)^2\right]. \tag{6}$$

Let $\lambda_i=\alpha_i x_i^{p-2}$ $(i=1,\ldots,n)$ and Λ the diagonal $n\times n$ matrix with $\lambda_1,\ldots,\lambda_n$ on its diagonal. Then

$$\begin{aligned}\mathrm{d}^2M(x)&=\frac{(1-p)M(x)}{[\phi(x)]^2}[(x'\Lambda\mathrm{d}x)^2-\phi(x)\,(\mathrm{d}x)'\Lambda\mathrm{d}x]\\&=\frac{(p-1)\,M(x)}{[\phi(x)]^2}(\mathrm{d}x)'\Lambda^{1/2}[\phi(x)\,I_n-\Lambda^{1/2}xx'\Lambda^{1/2}]\,\Lambda^{1/2}\,\mathrm{d}x.\end{aligned} \tag{7}$$

The matrix $\phi(x)I-\Lambda^{1/2}xx'\Lambda^{1/2}$ is positive semidefinite, because all but one of its eigenvalues equal $\phi(x)$, and the remaining eigenvalue is zero. (Note that $x'\Lambda x=\phi(x)$.) Hence $\mathrm{d}^2M(x)\geqslant 0$ for $p>1$ and $\mathrm{d}^2M(x)\leqslant 0$ for $p<1$. The result then follows from Theorem 7.7. (The second part of the theorem also follows from Minkowski's inequality by writing $\alpha_i^{1/p}x_i$ for x_i and $\alpha_i^{1/p}y_i$ for y_i in (23.2) and (23.4).)

Exercises

1. Show that $p\log M_p(x,a)$ is a convex function of p.
2. Hence show that the function $M_p\equiv M_p(x,a)$ satisfies

$$M_p^p\leqslant\prod_{i=1}^{n} M_{p_i}^{\delta_i p_i}$$

for every p_i, where

$$p=\sum_{i=1}^{n}\delta_i p_i, \qquad 0<\delta_i<1, \qquad \sum_{i=1}^{n}\delta_i=1.$$

29. LEAST SQUARES

The last topic in this chapter on inequalities deals with least squares problems. In Theorem 34 we wish to approximate a given vector d by a linear combination of the columns of a given matrix A.

Theorem 34 (least squares)

Let A be a given $n\times k$ matrix, and d a given $n\times 1$ vector. Then

$$(Ax-d)'(Ax-d)\geqslant d'(I-AA^+)d \tag{1}$$

for every x in $\mathbb{R}^k$, with equality if and only if

$$x=A^+d+(I-A^+A)q \tag{2}$$

for some q in $\mathbb{R}^k$.

Note. In the special case where A has full column rank k, we have $A^+=(A'A)^{-1}A'$ and hence a *unique* vector x_* exists which minimizes $(Ax-d)'(Ax-d)$ over all x, namely

$$x_*=(A'A)^{-1}A'd. \tag{3}$$

Proof. Consider the real-valued function $\phi:\mathbb{R}^k\to\mathbb{R}$ defined by

$$\phi(x)=(Ax-d)'(Ax-d). \tag{4}$$

Differentiating ϕ we obtain

$$\mathrm{d}\phi=2(Ax-d)'\mathrm{d}(Ax-d)=2(Ax-d)'A\mathrm{d}x. \tag{5}$$

Since ϕ is convex it has an absolute minimum at points x which satisfy $\mathrm{d}\phi(x)=0$, that is,

$$A'Ax=A'd. \tag{6}$$

Using Theorem 2.12 we see that equation (6) is consistent, and that its general solution is given by

$$\begin{aligned} x&=(A'A)^+A'd+[I-(A'A)^+A'A]q \\ &=A^+d+(I-A^+A)q. \end{aligned} \tag{7}$$

Hence $Ax=AA^+d$, and the absolute minimum is

$$(Ax-d)'(Ax-d)=d'(I-AA^+)d. \tag{8}$$

□

30. GENERALIZED LEAST SQUARES

As an immediate generalization of Theorem 34, let us prove Theorem 35.

Theorem 35 (generalized least squares)

Let A be a given $n\times k$ matrix, d a given $n\times 1$ vector and B a given positive semidefinite $n\times n$ matrix. Then

$$(Ax-d)'B(Ax-d)\geqslant d'Cd \tag{1}$$

with

$$C=B-BA(A'BA)^{+}A'B \tag{2}$$

for every x in $\mathbb{R}^k$, with equality if and only if

$$x=(A'BA)^{+}A'Bd+[I-(A'BA)^{+}A'BA]q \tag{3}$$

for some q in $\mathbb{R}^k$.

Proof. Let $d_0=B^{1/2}d$ and $A_0=B^{1/2}A$, and apply Theorem 34. □

Exercises

1. Consider the matrix C defined in (2). Show that (i) C is symmetric and positive semidefinite, (ii) $CA=0$, and (iii) C is idempotent if B is idempotent.
2. Consider the solution for x in (3). Show that (i) x is unique if and only if $A'BA$ is non-singular, and (ii) Ax is unique if and only if $r(A'BA)=r(A)$.

31. RESTRICTED LEAST SQUARES

The next result determines the minimum of a quadratic form when x is subject to linear restrictions.

Theorem 36 (restricted least squares)

Let A be a given $n\times k$ matrix, d a given $n\times 1$ vector and B a given positive semidefinite $n\times n$ matrix. Further let R be a given $m\times k$ matrix and r a given $m\times 1$ vector such that $RR^{+}r=r$. Then

$$(Ax-d)'B(Ax-d)\geqslant\begin{pmatrix}d\\r\end{pmatrix}'\begin{pmatrix}C_{11} & C_{12}\\ C_{12}' & C_{22}\end{pmatrix}\begin{pmatrix}d\\r\end{pmatrix} \tag{1}$$

for every x in $\mathbb{R}^k$ satisfying $Rx=r$. Here

$$C_{11}=B+BAN^{+}R'(RN^{+}R')^{+}RN^{+}A'B-BAN^{+}A'B$$

$$C_{12}=-BAN^{+}R'(RN^{+}R')^{+},\qquad C_{22}=(RN^{+}R')^{+}-I,$$

and $N = A'BA + R'R$. Equality occurs if and only if

$$x = x_0 + N^+R'(RN^+R')^+(r - Rx_0) + (I - N^+N)q, \tag{2}$$

where $x_0 = N^+A'Bd$ and q is an arbitrary $k \times 1$ vector.

Proof. Define the Lagrangian function

$$\psi(x) = \tfrac{1}{2}(Ax - d)'B(Ax - d) - l'(Rx - r), \tag{3}$$

where l is an $m \times 1$ vector of Lagrange multipliers. Differentiating ψ we obtain

$$\mathrm{d}\psi = x'A'BA\mathrm{d}x - d'BA\mathrm{d}x - l'R\mathrm{d}x. \tag{4}$$

The first-order conditions are therefore

$$A'BAx - R'l = A'Bd, \tag{5}$$

$$Rx = r, \tag{6}$$

which we can write as one equation as follows:

$$\begin{pmatrix} A'BA & R' \\ R & 0 \end{pmatrix} \begin{pmatrix} x \\ -l \end{pmatrix} = \begin{pmatrix} A'Bd \\ r \end{pmatrix}. \tag{7}$$

According to Theorem 3.23, equation (7) in x and l has a solution if and only if

$$A'Bd \in \mathcal{M}(A'BA, R') \qquad \text{and} \qquad r \in \mathcal{M}(R), \tag{8}$$

in which case the general solution for x is

$$x = [N^+ - N^+R'(RN^+R')^+RN^+]A'Bd + N^+R'(RN^+R')^+r + (I - NN^+)q \tag{9}$$

where $N = A'BA + R'R$ and q is arbitrary.

The consistency conditions (8) being satisfied, the general solution for x is given by (9) which we rewrite as

$$x = x_0 + N^+R'(RN^+R')^+(r - Rx_0) + (I - NN^+)q, \tag{10}$$

where $x_0 = N^+A'Bd$. Since ψ is convex (independent of the signs of the components of l), constrained absolute minima occur at points x satisfying (10). The value of the absolute minimum is obtained by inserting (10) in $(Ax - d)'B(Ax - d)$. □

Exercises

1. Let V be a given positive semidefinite matrix, and let A be a given matrix and b a given vector such that $b \in \mathcal{M}(A)$. The class of solutions to the problem

$$\begin{array}{ll} \text{minimize} & x'Vx \\ \text{subject to} & Ax = b \end{array}$$

is given by

$$x = V_0^+ A'(AV_0^+ A')^+ b + (I - V_0^+ V_0)q,$$

where $V_0 = V + A'A$ and q is an arbitrary vector. Moreover, if $\mathcal{M}(A') \subset \mathcal{M}(V)$, then the solution simplifies to

$$x = V^+ A'(AV^+ A')^+ b + (I - V^+ V)q.$$

2. Hence show that

$$\min_{Ax=b} x'Vx = b'Cb,$$

where

$$C = (AV_0^+ A')^+ - I.$$

Also show that, if $\mathcal{M}(A') \subset \mathcal{M}(V)$, the matrix C simplifies to

$$C = (AV^+ A')^+.$$

32. RESTRICTED LEAST SQUARES: MATRIX VERSION

Finally let us prove the following matrix version of Theorem 36 which we shall use in section 13.16.

Theorem 37

Let B be a given positive semidefinite matrix, and let W and R be given matrices such that $\mathcal{M}(W) \subset \mathcal{M}(R)$. Then

$$\operatorname{tr} X'BX \geqslant \operatorname{tr} W'CW \tag{1}$$

for every X satisfying $RX = W$. Here

$$C = (RB_0^+ R')^+ - I, \qquad B_0 = B + R'R. \tag{2}$$

Equality occurs if and only if

$$X = B_0^+ R'(RB_0^+ R')^+ W + (I - B_0 B_0^+)Q, \tag{3}$$

where Q is an arbitrary matrix of appropriate order.

Moreover, if $\mathcal{M}(R') \subset \mathcal{M}(B)$, then C simplifies to

$$C = (RB^+ R')^+, \tag{4}$$

with equality occurring if and only if

$$X = B^+ R'(RB^+ R')^+ W + (I - BB^+)Q. \tag{5}$$

Proof. Consider the Lagrangian function

$$\psi(X) = \tfrac{1}{2} \operatorname{tr} X'BX - \operatorname{tr} L'(RX - W), \tag{6}$$

where L is a matrix of Lagrange multipliers. Differentiating leads to

$$\mathrm{d}\psi(X) = \operatorname{tr} X'B\mathrm{d}X - \operatorname{tr} L'R\mathrm{d}X. \tag{7}$$

Hence we obtain the first-order conditions

$$BX = R'L, \tag{8}$$

$$RX = W, \tag{9}$$

which we write as one matrix equation

$$\begin{pmatrix} B & R' \\ R & 0 \end{pmatrix} \begin{pmatrix} X \\ -L \end{pmatrix} = \begin{pmatrix} 0 \\ W \end{pmatrix}. \tag{10}$$

According to Theorem 3.24, equation (10) is consistent, because $\mathcal{M}(W) \subset \mathcal{M}(R)$; the solution for X is

$$X = B_0^+ R'(RB_0^+ R')^+ W + (I - B_0 B_0^+)Q \tag{11}$$

in general, and

$$X = B^+ R'(RB^+ R')^+ W + (I - BB^+)Q \tag{12}$$

if $\mathcal{M}(R') \in \mathcal{M}(B)$. Since ψ is convex, the constrained absolute minima occur at points X satisfying (11) or (12). The value of the absolute minimum is obtained by inserting (11) or (12) in tr $X'BX$. □

Exercise

1. Let X be given by (3). Show that $X'a$ is unique if and only if $a \in \mathcal{M}(B:R')$.

MISCELLANEOUS EXERCISES

1. Show that $\log x \leq x - 1$ for every $x > 0$ with equality if and only if $x = 1$.
2. Hence show that $\log |A| \leq \operatorname{tr} A - n$ for every positive definite $n \times n$ matrix A, with equality if and only if $A = I_n$.
3. Show that

$$|A + B|/|A| \leq \exp [\operatorname{tr}(A^{-1}B)]$$

where A and $A + B$ are positive definite, with equality if and only if $B = 0$.

4. For any positive semidefinite $n \times n$ matrix A and $n \times 1$ vector b,

$$0 \leq b'(A + bb')^+ b \leq 1$$

with equality if and only if $b = 0$ or $b \notin \mathcal{M}(A)$.

5. Let A be positive definite and B symmetric, both of order n. Then

$$\min_{1 \leq i \leq n} \lambda_i(A^{-1}B) \leq \frac{x'Bx}{x'Ax} \leq \max_{1 \leq i \leq n} \lambda_i(A^{-1}B)$$

for every $x \neq 0$.

6. Let A be a symmetric $m \times m$ matrix and let B be an $m \times n$ matrix of rank n. Let $C = (B'B)^{-1}B'AB$. Then
$$\min_{1 \leqslant t \leqslant n} \lambda_t(C) \leqslant \frac{x'Ax}{x'x} \leqslant \max_{1 \leqslant t \leqslant n} \lambda_t(C)$$
for every $x \in \mathcal{M}(B)$.
7. Let A be an $m \times n$ matrix with full column rank, and let $\imath$ be the $m \times 1$ vector consisting of ones only. Assume that $\imath$ is the first column of A. Then
$$x'(A'A)^{-1}x \geqslant 1/m$$
for every x satisfying $x_1 = 1$, with equality if and only if $x = (1/m)A'\imath$.
8. Let A be an $m \times n$ matrix of rank r. Let $\delta_1, \ldots, \delta_r$ be the *singular values* of A (that is, the positive square roots of the non-zero eigenvalues of AA'), and let $\delta = \delta_1 + \delta_2 + \ldots + \delta_r$. Then
$$-\delta \leqslant \operatorname{tr} AX \leqslant \delta$$
for every $n \times m$ matrix X satisfying $X'X = I_m$.
9. Let A be a positive definite $n \times n$ matrix and B an $m \times n$ matrix of rank m. Then
$$x'Ax \geqslant b'(BA^{-1}B')^{-1}b$$
for every x satisfying $Bx = b$, with equality if and only if $x = A^{-1}B'(BA^{-1}B')^{-1}b$.
10. Let A be a positive definite $n \times n$ matrix and B an $m \times n$ matrix. Then
$$\operatorname{tr} X'AX \geqslant \operatorname{tr}(BA^{-1}B')^{-1}$$
for every $n \times m$ matrix X satisfying $BX = I_m$, with equality if and only if $X = A^{-1}B'(BA^{-1}B')^{-1}$.
11. Let A and B be matrices of the same order, and assume that A has full row rank. Define $C = A'(AA')^{-1}B$. Then
$$\operatorname{tr} X^2 \geqslant 2 \operatorname{tr} C(I - A^+A)C' + 2 \operatorname{tr} C^2$$
for every *symmetric* matrix X satisfying $AX = B$, with equality if and only if $X = C + C' - CA'(AA')^{-1}A$.
12. For any positive semidefinite matrix M, let $\mu(M)$ denote the largest of its eigenvalues. Then, for any positive semidefinite $n \times n$ matrix V and $m \times n$ matrix A we have

    (i) $\mu(AVA') \leqslant \mu(V)\mu(AA')$
    (ii) $\mu(V)AA' - AVA'$ is positive semidefinite
    (iii) $\operatorname{tr} AVA' \leqslant \mu(V) \operatorname{tr} AA' \leqslant (\operatorname{tr} V)(\operatorname{tr} AA')$
    (iv) $\operatorname{tr} V^2 \leqslant \mu(V) \operatorname{tr} V$
    (v) $\operatorname{tr} (AVA')^2 \leqslant \mu^2(V)\mu(AA') \operatorname{tr} AA'$.

13. Let A and B be positive semidefinite matrices of the same order. Show that

$$\sqrt{\operatorname{tr} AB} \leqslant \tfrac{1}{2}(\operatorname{tr} A + \operatorname{tr} B)$$

with equality if and only if $A = B$ and $r(A) \leqslant 1$ (Yang 1988, Neudecker 1992)

14. For any two matrices A and B of the same order,
    (i) $2(AA'+BB')-(A+B)(A+B)'$ is positive semidefinite
    (ii) $\mu[(A+B)(A+B)'] \leqslant 2[\mu(AA')+\mu(BB')]$
    (iii) $\operatorname{tr}(A+B)(A+B)' \leqslant 2(\operatorname{tr} AA' + \operatorname{tr} BB')$.
15. Let A, B and C be matrices of the same order. Show that

$$\mu(ABC+C'B'A') \leqslant 2[\mu(AA')\,\mu(BB')\,\mu(CC')]^{1/2}$$

where $\mu(.)$ now denotes the largest eigenvalue in absolute value. In particular, if A, B and C are symmetric,

$$\mu(ABC+CBA) \leqslant 2\mu(A)\mu(B)\mu(C).$$

16. Let A be a positive definite $n \times n$ matrix with eigenvalues $0 < \lambda_1 \leqslant \lambda_2 \leqslant \ldots \leqslant \lambda_n$. Show that the matrix

$$(\lambda_1 + \lambda_n)I_n - A - (\lambda_1\lambda_n)A^{-1}$$

is positive semidefinite with rank $\leqslant n-2$. [*Hint*: Use the fact that $x^2 - (a+b)x + ab \leqslant 0$ for all $x \in [a, b]$.]

17. (*Kantorovich inequality*) Let A be a positive definite $n \times n$ matrix with eigenvalues $0 < \lambda_1 \leqslant \lambda_2 \leqslant \ldots \leqslant \lambda_n$. Use the previous exercise to prove that

$$1 \leqslant (x'Ax)(x'A^{-1}x) \leqslant \frac{(\lambda_1 + \lambda_n)^2}{4\lambda_1\lambda_n}$$

for every $x \in \mathbb{R}^n$ satisfying $x'x = 1$ (Kantorovich 1948, Greub and Rheinboldt 1959).

18. For any two matrices A and B satisfying $A'B = I$, we have

$$B'B \geqslant (A'A)^{-1}, A'A \geqslant (B'B)^{-1}$$

[*Hint*: Use the fact that $I - A(A'A)^{-1}A' \geqslant 0$.]

19. Hence, for any positive definite $n \times n$ matrix A and $n \times k$ matrix X with $r(X) = k$, we have

$$(X'X)^{-1}X'AX(X'X)^{-1} \geqslant (X'A^{-1}X)^{-1}.$$

20. (*Kantorovich inequality*, matrix version). Let A be a positive definite matrix with eigenvalues $0 < \lambda_1 \leqslant \lambda_2 \leqslant \cdots \leqslant \lambda_n$. Show that

$$(X'A^{-1}X)^{-1} \leqslant X'AX \leqslant \frac{(\lambda_1 + \lambda_n)^2}{4\lambda_1\lambda_n}(X'A^{-1}X)^{-1}$$

for every X satisfying $X'X = I$.

21. (*Bergstrom's inequality*, matrix version). Let A and B be positive definite and X of full column rank. Then

$$(X'(A+B)^{-1}X)^{-1} \geqslant (X'A^{-1}X)^{-1} + (X'B^{-1}X)^{-1}$$

(Marshall and Olkin 1979, 469–473 and Neudecker and Liu 1995).

22. Let A and B be positive definite matrices of the same order. Show that

$$2(A^{-1}+B^{-1})^{-1} \leqslant A^{1/2}(A^{-1/2}BA^{-1/2})^{1/2}A^{1/2} \leqslant \tfrac{1}{2}(A+B).$$

This provides a matrix version of the harmonic–geometric–arithmetic mean inequality (Ando 1979, 1983).

23. Let A be positive definite and B symmetric such that $|A+B| \neq 0$. Prove that

$$A^{-1}-(A+B)^{-1}-(A+B)^{-1}B(A+B)^{-1}$$

is positive semidefinite. Prove further that the matrix is positive definite if and only if B is non-singular (see Olkin 1983).

24. Let A be positive definite and $V_1, V_2, \ldots, V_m$ positive semidefinite, all of the same order. Then

$$A^{-1}-\sum_{i=1}^{m}(A+V_1+\ldots+V_i)^{-1}V_i(A+V_1+\ldots+V_i)^{-1}$$

is positive definite (Olkin 1983).

25. Let A be a positive definite $n \times n$ matrix and let $B_1, \ldots, B_m$ be $n \times r$ matrices. Then

$$\sum_{i=1}^{m} \operatorname{tr} B_i'(A+B_1B_1'+\ldots+B_iB_i')^{-2}B_i < \operatorname{tr} A^{-1}$$

(Olkin 1983).

26. Let the $n \times n$ matrix A have real eigenvalues $\lambda_1 \leqslant \lambda_2 \leqslant \ldots \leqslant \lambda_n$. Show that

(a) $m - s(n-1)^{1/2} \leqslant \lambda_1 \leqslant m - \dfrac{s}{(n-1)^{1/2}}$

(b) $m + \dfrac{s}{(n-1)^{1/2}} \leqslant \lambda_n \leqslant m + s(n-1)^{1/2}$

where

$$m = (1/n)\operatorname{tr} A, \qquad s^2 = (1/n)\operatorname{tr} A^2 - m^2.$$

Equality holds on the left (right) of (a) if and only if equality holds on the left (right) of (b) if and only if the $n-1$ largest (smallest) eigenvalues are equal (Wolkowicz and Styan 1980).

BIBLIOGRAPHICAL NOTES

§1. The classic work on inequalities is Hardy *et al.* (1952), which was first published in 1934. Useful sources are also Beckenbach and Bellman (1961), Marcus and Minc (1964), Bellman (1970, chaps 7 and 8), and Wang and Chow (1994).
§5. The idea of expressing a nonlinear function as an envelope of linear functions goes back to Minkowski and was used extensively by Bellman and others.
§8. See Fischer (1905). An important extension by Courant can be found in Courant and Hilbert (1931).
§10. See Poincaré (1890).
§16. The inequality (5) was proved in 1932 by Karamata. Proofs and historical details can be found in Hardy *et al.* (1952, th. 108, p. 89) or Beckenbach and Bellman (1961, pp. 30–2). Hardy *et al.* also prove that if $\phi''(x)$ exists for all x, and is positive, then there is strict inequality in (5) unless $x = y$. Theorem 19 gives a weaker condition. See also Marshall and Olkin (1979, props 3.C.1 and 3.C.1.a).
§17–§20. See Magnus (1987). An alternative proof of Theorem 23 was given by Neudecker (1989a).
§21. Hölder's inequality is a very famous one and is discussed extensively by Hardy *et al.* (1952, pp. 22–4). Theorem 24 is due to Magnus (1987).
§22. See Fan (1949, 1950).
§23. A detailed discussion of Minkowski's inequality can be found in Hardy *et al.* (1952, pp. 30–1). Theorem 26 is due to Magnus (1987).
§24. The second proof of Theorem 27 is based on Neudecker (1974).
§26–§28. Weighted means are thoroughly treated in Hardy *et al.* (1952).

Part Five—

The linear model

CHAPTER 12

Statistical preliminaries

1. INTRODUCTION

The purpose of this chapter is to review briefly those statistical concepts and properties that we shall use in the remainder of this book. No attempt is made to be either exhaustive or rigorous.

It is assumed that the reader is familiar (however vaguely) with the concepts of probability and random variables and has a rudimentary knowledge of Riemann integration. Integrals are necessary in this chapter, but they will appear in no other chapter of the book.

2. THE CUMULATIVE DISTRIBUTION FUNCTION

If x is a real-valued random variable, we define the cumulative distribution function F of x by

$$F(\xi)=\Pr(x\leqslant\xi). \tag{1}$$

Thus, $F(\xi)$ specifies the probability that the random variable x is at most equal to a given number ξ.

It is clear that F is non-decreasing and that

$$\lim_{\xi\to-\infty} F(\xi)=0, \qquad \lim_{\xi\to\infty} F(\xi)=1. \tag{2}$$

Similarly, if $(x_1, \ldots, x_n)'$ is an $n\times 1$ vector of real random variables, we define the cumulative distribution function F by

$$F(\xi_1,\xi_2,\ldots,\xi_n)=\Pr(x_1\leqslant\xi_1,x_2\leqslant\xi_2,\ldots,x_n\leqslant\xi_n) \tag{3}$$

which specifies the probability of the joint occurrence $x_i\leqslant\xi_i$ for all i.

3. THE JOINT DENSITY FUNCTION

Let F be the cumulative distribution function of a real-valued random variable x. If there exists a non-negative real-valued (in fact, Lebesgue-measurable) function f such that

$$F(\xi)=\int_{-\infty}^{\xi} f(y)\,\mathrm{d}y \tag{1}$$

for all $y \in \mathbb{R}$, then we say that x is a *continuous random variable* and f is called its *density function.* In this case the derivative of F exists and we have

$$\mathrm{d}F(\xi)/\mathrm{d}\xi=f(\xi). \tag{2}$$

(Strictly speaking, (2) is true except for a set of values of ξ of probability zero.) The density function satisfies

$$f(\xi)\geqslant 0, \qquad \int_{-\infty}^{\infty} f(\xi)\,\mathrm{d}\xi=1. \tag{3}$$

In the case of a *continuous* $n\times 1$ *random vector* $(x_1, \ldots, x_n)'$, there exists a non-negative real-valued function f such that

$$F(\xi_1,\xi_2,\ldots,\xi_n)=\int_{-\infty}^{\xi_1}\int_{-\infty}^{\xi_2}\cdots\int_{-\infty}^{\xi_n} f(y_1,y_2,\ldots,y_n)\,\mathrm{d}y_1\,\mathrm{d}y_2\ldots\mathrm{d}y_n \tag{4}$$

for all $(y_1,y_2,\ldots,y_n)\in\mathbb{R}^n$, in which case

$$\frac{\partial^n F(\xi_1,\xi_2,\ldots,\xi_n)}{\partial\xi_1\,\partial\xi_2\ldots\partial\xi_n}=f(\xi_1,\xi_2,\ldots,\xi_n) \tag{5}$$

at all points in $\mathbb{R}^n$ (except possibly for a set of probability 0).

The function f defined by (4) is called the *joint density function* of $(x_1, \ldots, x_n)$.

In this and subsequent chapters we shall only be concerned with continuous random variables.

4. EXPECTATIONS

The expectation (or expected value) of any function g of a random variable x is defined as

$$\mathscr{E}g(x)=\int_{-\infty}^{\infty} g(\xi)f(\xi)\,\mathrm{d}\xi, \tag{1}$$

if the integral exists. More generally, let $x=(x_1, \ldots, x_n)'$ be a random $n\times 1$ vector with joint density function f. Then the expectation of any function g of x is defined as

$$\mathscr{E}g(x)=\int_{-\infty}^{\infty}\cdots\int_{-\infty}^{\infty} g(\xi_1,\ldots,\xi_n)f(\xi_1,\ldots,\xi_n)\,\mathrm{d}\xi_1\ldots\mathrm{d}\xi_n \tag{2}$$

if the n-fold integral exists.

If $G=(g_{ij})$ is an $m\times p$ matrix function, then we define the expectation of the matrix G as the $m\times p$ matrix of the expectations:

$$\mathscr{E}G(x)=(\mathscr{E}g_{ij}(x)). \tag{3}$$

Below we list some useful elementary facts about expectations when they exist. The first of these is

$$\mathscr{E}A=A \tag{4}$$

where A is a matrix of constants. Next

$$\mathscr{E}AG(x)B=A(\mathscr{E}G(x))B \tag{5}$$

where A and B are matrices of constants and G is a matrix function. Finally,

$$\mathscr{E}\sum_i \alpha_i G_i(x)=\sum_i \alpha_i \mathscr{E}G_i(x) \tag{6}$$

where the α_i are constants and the G_i are matrix functions. This last property characterizes expectation as a linear operator.

5. VARIANCE AND COVARIANCE

If x is a random variable, we define its *variance* as

$$\mathscr{V}(x)=\mathscr{E}(x-\mathscr{E}x)^2. \tag{1}$$

If x and y are two random variables with a joint density function, we define their *covariance* as

$$\mathscr{C}(x,y)=\mathscr{E}(x-\mathscr{E}x)(y-\mathscr{E}y). \tag{2}$$

If $\mathscr{C}(x,y)=0$ we say that x and y are *uncorrelated.*

We note the following facts about two random variables x and y and two constants α and β:

$$\mathscr{V}(x+\alpha)=\mathscr{V}(x) \tag{3}$$

$$\mathscr{V}(\alpha x)=\alpha^2\mathscr{V}(x) \tag{4}$$

$$\mathscr{V}(x+y)=\mathscr{V}(x)+\mathscr{V}(y)+2\mathscr{C}(x,y) \tag{5}$$

$$\mathscr{C}(\alpha x,\beta y)=\alpha\beta\mathscr{C}(x,y). \tag{6}$$

If x and y are uncorrelated, we obtain as a special case of (5):

$$\mathscr{V}(x+y)=\mathscr{V}(x)+\mathscr{V}(y). \tag{7}$$

Let us now consider the multivariate case. We define the *variance (matrix)* of an $n \times 1$ random vector x as the $n \times n$ matrix

$$\mathscr{V}(x) = \mathscr{E}(x - \mathscr{E}x)(x - \mathscr{E}x)'. \tag{8}$$

It is clear that the ij-th ($i \neq j$) element of $\mathscr{V}(x)$ is just the covariance between x_i and x_j, and that the i-th diagonal element of $\mathscr{V}(x)$ is just the variance of x_i.

Theorem 1

Each variance matrix is symmetric and positive semidefinite.

Proof. Symmetry is obvious. To prove that $\mathscr{V}(x)$ is positive semidefinite, define a real-valued random variable $y = a'(x - \mathscr{E}x)$, where a is an arbitrary $n \times 1$ vector. Then

$$a'\mathscr{V}(x)a = a'\mathscr{E}(x - \mathscr{E}x)(x - \mathscr{E}x)'a = \mathscr{E}a'(x - \mathscr{E}x)(x - \mathscr{E}x)'a = \mathscr{E}y^2 \geqslant 0, \tag{9}$$

and hence $\mathscr{V}(x)$ is positive semidefinite. □

The determinant $|\mathscr{V}(x)|$ is sometimes called the *generalized variance* of x. The variance matrix of an $m \times n$ random matrix X is defined as the $mn \times mn$ variance matrix of $\operatorname{vec} X$.

If x is a random $n \times 1$ vector and y a random $m \times 1$ vector, then we define the *covariance (matrix) between x and y* as the $n \times m$ matrix

$$\mathscr{C}(x, y) = \mathscr{E}(x - \mathscr{E}x)(y - \mathscr{E}y)'. \tag{10}$$

If $\mathscr{C}(x, y) = 0$ we say that the two vectors x and y are uncorrelated.

The next two results generalize properties (3)–(7) to the multivariate case.

Theorem 2

Let x be a random $n \times 1$ vector and define $y = Ax + b$, where A is a constant $m \times n$ matrix and b a constant $m \times 1$ vector. Then

$$\mathscr{E}y = A\mathscr{E}x + b, \qquad \mathscr{V}(y) = A\mathscr{V}(x)A'. \tag{11}$$

Proof. The proof is left as an exercise for the reader. □

Theorem 3

Let x and y be random $n \times 1$ vectors and let z be a random $m \times 1$ vector. Let $A(p \times n)$ and $B(q \times m)$ be matrices of constants. Then

$$\mathscr{V}(x + y) = \mathscr{V}(x) + \mathscr{V}(y) + \mathscr{C}(x, y) + \mathscr{C}(y, x), \tag{12}$$

$$\mathscr{C}(Ax, Bz) = A\mathscr{C}(x, z)B', \tag{13}$$

and, if x and y are uncorrelated,

$$\mathscr{V}(x + y) = \mathscr{V}(x) + \mathscr{V}(y). \tag{14}$$

Proof. The proof is easy and again left as an exercise. □

Finally, we present the following useful result regarding the expected value of a quadratic form.

Theorem 4

Let x be a random $n \times 1$ vector with $\mathscr{E}x = \mu$ and $\mathscr{V}(x) = \Omega$. Let A be an $n \times n$ matrix. Then

$$\mathscr{E}x'Ax = \operatorname{tr} A\Omega + \mu'A\mu. \tag{15}$$

Proof. We have

$$\begin{aligned} \mathscr{E}x'Ax &= \mathscr{E} \operatorname{tr} x'Ax = \mathscr{E} \operatorname{tr} Axx' \\ &= \operatorname{tr} \mathscr{E} Axx' = \operatorname{tr} A(\mathscr{E}xx') \\ &= \operatorname{tr} A(\Omega + \mu\mu') = \operatorname{tr} A\Omega + \mu'A\mu. \end{aligned} \tag{16}$$

□

Exercises

1. Show that x has a degenerate distribution if and only if $\mathscr{V}(x) = 0$. (A random vector x is said to have a degenerate distribution if $\Pr(x = \xi) = 1$. If x has a degenerate distribution we also say that $x = \xi$ *almost surely* (a.s.) or *with probability one.*)
2. Show that $\mathscr{V}(x)$ is positive definite if and only if the distribution of $a'x$ is non-degenerate for all $a \neq 0$.

6. INDEPENDENCE OF TWO RANDOM VARIABLES (VECTORS)

Let $f(x, y)$ be the joint density function of two random variables x and y. Suppose we wish to calculate a probability that concerns only x, say the probability of the event

$$a < x < b \tag{1}$$

where $a < b$. We then have

$$\begin{aligned} \Pr(a < x < b) &= \Pr(a < x < b, -\infty < y < \infty) \\ &= \int_a^b \int_{-\infty}^{\infty} f(x, y)\, \mathrm{d}y\, \mathrm{d}x = \int_a^b f_x(x)\, \mathrm{d}x \end{aligned} \tag{2}$$

where we have defined

$$f_x(x) = \int_{-\infty}^{\infty} f(x, y)\, \mathrm{d}y. \tag{3}$$

This function $f_x(x)$ defined in (3) is called the *marginal density function of x*. Similarly we define

$$f_y(y) = \int_{-\infty}^{\infty} f(x, y)\,\mathrm{d}x \tag{4}$$

as the *marginal density function of y*. We proceed to define the important concept of independence.

Definition 1

Let $f(x, y)$ be the joint density function of two random variables x and y and let $f_x(x)$ and $f_y(y)$ denote the marginal density functions of x and y respectively. Then we say that x and y are (stochastically) independent if

$$f(x, y) = f_x(x) f_y(y). \tag{5}$$

The following result states that functions of independent random variables are uncorrelated.

Theorem 5

Let x and y be two independent random variables. Then, for any functions g and h,

$$\mathscr{E}g(x)h(y) = (\mathscr{E}g(x))(\mathscr{E}h(y)) \tag{6}$$

if the expectations exist.

Proof. We have

$$\begin{aligned} \mathscr{E}g(x)h(y) &= \int_{-\infty}^{\infty}\int_{-\infty}^{\infty} g(x)h(y)f_x(x)f_y(y)\,\mathrm{d}x\,\mathrm{d}y \\ &= \left(\int_{-\infty}^{\infty} g(x)f_x(x)\,\mathrm{d}x\right)\left(\int_{-\infty}^{\infty} h(y)f_y(y)\,\mathrm{d}y\right) = \mathscr{E}g(x)\mathscr{E}h(y). \end{aligned} \tag{7}$$

□

As an immediate consequence of Theorem 5 we obtain Theorem 6.

Theorem 6

If two random variables are independent, they are uncorrelated.

The converse of Theorem 6 is not, in general, true (see exercise 1). A partial converse is given in Theorem 8.

If x and y are random vectors rather than random variables, straightforward extensions of Definition 1 and Theorems 5 and 6 hold.

Exercise

1. Let x be a random variable with $\mathscr{E}x=\mathscr{E}x^3=0$. Show that x and x^2 are uncorrelated, but not in general independent.

7. INDEPENDENCE OF *n* RANDOM VARIABLES (VECTORS)

The notion of independence can be extended in an obvious manner to the case of three or more random variables (vectors).

Definition 2

Let the random variables $x_1, \ldots, x_n$ have the joint density function $f(x_1, \ldots, x_n)$ and marginal density functions $f_1(x_1), \ldots, f_n(x_n)$, respectively. Then we say that $x_1, \ldots, x_n$ are (mutually) independent if

$$f(x_1, \ldots, x_n)=f_1(x_1) \ldots f_n(x_n). \tag{1}$$

We note that, if $x_1, \ldots, x_n$ are independent in the sense of Definition 2, they are pairwise independent (that is, x_i and x_j are independent for all $i \neq j$), but that the converse is not true. Thus pairwise independence does not necessarily imply mutual independence.

Again the extension to random vectors is straightforward.

8. SAMPLING

Definition 3

Let $x_1, \ldots, x_n$ be independent random variables (vectors), each with the same density function $f(x)$. Then we say that $x_1, \ldots, x_n$ are *independent and identically distributed* (i.i.d.) or, equivalently, that they constitute a (random) *sample* (of size n) from a distribution with density function $f(x)$.

Thus, if we have a sample $x_1, \ldots, x_n$ from a distribution with density $f(x)$, the joint density function of the sample is

$$f(x_1)f(x_2) \ldots f(x_n).$$

9. THE ONE-DIMENSIONAL NORMAL DISTRIBUTION

The most important of all distributions—and the only one that will play a role in the subsequent chapters of this book—is the *normal distribution*. Its density function is defined as

$$f(x)=\frac{1}{(2\pi\sigma^2)^{1/2}}\exp\left(-\frac{(x-\mu)^2}{2\sigma^2}\right) \tag{1}$$

for $-\infty < x < \infty$, where μ and σ^2 are the parameters of the distribution. If x is distributed as in (1) we write

$$x \sim \mathcal{N}(\mu, \sigma^2). \tag{2}$$

If $\mu = 0$ and $\sigma^2 = 1$ we say that x is *standard-normally distributed.*

Without proof we present the following theorem.

Theorem 7

If $x \sim \mathcal{N}(\mu, \sigma^2)$, then

$$\mathscr{E}x = \mu, \qquad \mathscr{E}x^2 = \mu^2 + \sigma^2, \tag{3}$$

$$\mathscr{E}x^3 = \mu(\mu^2 + 3\sigma^2), \quad \mathscr{E}x^4 = \mu^4 + 6\mu^2\sigma^2 + 3\sigma^4, \tag{4}$$

and hence

$$\mathscr{V}(x) = \sigma^2, \qquad \mathscr{V}(x^2) = 2\sigma^4 + 4\mu^2\sigma^2. \tag{5}$$

10. THE MULTIVARIATE NORMAL DISTRIBUTION

A random $n \times 1$ vector x is said to be normally distributed if its density function is given by

$$f(x) = (2\pi)^{-n/2}|\Omega|^{-1/2}\exp\left[-\tfrac{1}{2}(x-\mu)'\Omega^{-1}(x-\mu)\right] \tag{1}$$

for $x \in \mathbb{R}^n$, where μ is an $n \times 1$ vector and Ω a non-singular symmetric $n \times n$ matrix.

It is easily verified that (1) reduces to the one-dimensional normal density (9.1) in the case $n = 1$.

If x is distributed as in (1) we write

$$x \sim \mathcal{N}(\mu, \Omega) \tag{2}$$

or, occasionally, if we wish to emphasize the dimension of x,

$$x \sim \mathcal{N}_n(\mu, \Omega). \tag{3}$$

The parameters μ and Ω are just the expectation and variance matrix of x:

$$\mathscr{E}x = \mu, \qquad \mathscr{V}(x) = \Omega. \tag{4}$$

We shall present (without proof) five theorems concerning the multivariate normal distribution which we shall need in the following chapters. The first of these provides a partial converse of Theorem 6.

Theorem 8

If x and y are normally distributed with $\mathscr{C}(x, y) = 0$, then they are independent.

Next, let us consider the marginal distributions associated with the multivariate normal distribution.

Theorem 9

The marginal distributions associated with a normally distributed vector are also normal. That is, if $x \sim \mathcal{N}(\mu, \Omega)$ is partitioned as

$$\begin{pmatrix} x_1 \\ x_2 \end{pmatrix} \sim \mathcal{N}\left[\begin{pmatrix} \mu_1 \\ \mu_2 \end{pmatrix}, \begin{pmatrix} \Omega_{11} & \Omega_{12} \\ \Omega_{21} & \Omega_{22} \end{pmatrix}\right], \tag{5}$$

then the marginal distribution of x_1 is $\mathcal{N}(\mu_1, \Omega_{11})$ and the marginal distribution of x_2 is $\mathcal{N}(\mu_2, \Omega_{22})$.

A crucial property of the normal distribution is given in Theorem 10.

Theorem 10

An affine transformation of a normal vector is again normal. That is, if $x \sim \mathcal{N}(\mu, \Omega)$ and $y = Ax + b$ where A has full row rank, then

$$y \sim \mathcal{N}(A\mu + b, A\Omega A'). \tag{6}$$

If $\mu = 0$ and $\Omega = I_n$ we say that x is standard-normally distributed and we write

$$x \sim \mathcal{N}(0, I_n). \tag{7}$$

Theorem 11

If $x \sim \mathcal{N}(0, I_n)$, then x and $x \otimes x$ are uncorrelated.

Proof. This follows from the fact that

$$\mathcal{E} x_i x_j x_k = 0 \qquad \text{for all } i, j, k. \tag{8}$$

□

Let us conclude this section with two results on quadratic forms in normal variables, the first of which is a special case of Theorem 4.

Theorem 12

If $x \sim \mathcal{N}_n(\mu, \Omega)$ and A is a symmetric $n \times n$ matrix, then

$$\mathcal{E} x'Ax = \operatorname{tr} A\Omega + \mu' A \mu \tag{9}$$

and

$$\mathcal{V}(x'Ax) = 2 \operatorname{tr} (A\Omega)^2 + 4\mu' A\Omega A \mu. \tag{10}$$

Exercise

1. (Proof of (10)) Let $x \sim \mathcal{N}(\mu, \Omega)$ and $A = A'$. Let T be an orthogonal matrix and Λ a diagonal matrix such that

$$T'\Omega^{1/2} A \Omega^{1/2} T = \Lambda \tag{11}$$

and define

$$y = T'\Omega^{-1/2}(x-\mu), \qquad \omega = T'\Omega^{1/2}A\mu. \tag{12}$$

Prove that:

(a) $y \sim \mathcal{N}(0, I_n)$
(b) $x'Ax = y'\Lambda y + 2\omega' y + \mu' A\mu$
(c) $y'\Lambda y$ and $\omega' y$ are uncorrelated
(d) $\mathcal{V}(y'\Lambda y) = 2\operatorname{tr}\Lambda^2 = 2\operatorname{tr}(A\Omega)^2$
(e) $\mathcal{V}(\omega' y) = \omega'\omega = \mu' A\Omega A\mu$
(f) $\mathcal{V}(x'Ax) = \mathcal{V}(y'\Lambda y) + \mathcal{V}(2\omega' y) = 2\operatorname{tr}(A\Omega)^2 + 4\mu' A\Omega A\mu.$

11. ESTIMATION

Statistical inference asks the question: Given a sample, what can be inferred about the population from which it was drawn? Most textbooks distinguish between point estimation, interval estimation and hypothesis testing. In the following we shall only be concerned with point estimation.

In the theory of point estimation we seek to select a function of the observations that will approximate a parameter of the population in some well defined sense. A function of the *hypothetical* observations used to approximate a parameter (vector) is called an *estimator*. An estimator is thus a random variable. The realized value of the estimator, i.e. the value taken when a specific set of sample observations is inserted in the function, is called an *estimate*.

Let θ be the parameter (vector) in question and let $\hat\theta$ be an estimator of θ. The *sampling error* of an estimator $\hat\theta$ is defined as

$$\hat\theta - \theta \tag{1}$$

and, of course, we seek estimators whose sampling errors are small. The expectation of the sampling error,

$$\mathcal{E}(\hat\theta - \theta), \tag{2}$$

is called the *bias* of $\hat\theta$. An unbiased estimator is one whose bias is zero. The expectation of the square of the sampling error,

$$\mathcal{E}(\hat\theta - \theta)(\hat\theta - \theta)', \tag{3}$$

is called the *mean squared error* of $\hat\theta$, and denoted MSE ($\hat\theta$). We always have

$$\text{MSE}(\hat\theta) \geqslant \mathcal{V}(\hat\theta) \tag{4}$$

with equality if and only if $\hat\theta$ is an unbiased estimator of θ.

Two constructive methods of obtaining estimators with desirable properties are the method of best linear (affine, quadratic) unbiased estimation (introduced and employed in Chapters 13 and 14) and the method of maximum likelihood (Chapters 15–17).

MISCELLANEOUS EXERCISES

1. Let ϕ be a density function depending on a vector parameter θ and define

$$f = \partial \log \phi / \partial \theta, \qquad F = \frac{\partial^2 \log \phi}{\partial \theta \, \partial \theta'}, \qquad G = \frac{\partial \operatorname{vec} F}{\partial \theta'}.$$

Show that

$$-\mathscr{E}G = \mathscr{E}((\operatorname{vec} F + f \otimes f)f') + \mathscr{E}(f \otimes F + F \otimes f)$$

if differentiating under the integral sign is permitted (Lancaster 1984).

2. Let $x_1, \ldots, x_n$ be a sample from the $\mathscr{N}_p(\mu, V)$ distribution, and let X be the $n \times p$ matrix $X = (x_1, \ldots, x_n)'$. Let $A = (a_{ij})$ be a symmetric $n \times n$ matrix, and let

$$\alpha = \imath' A \imath, \qquad \beta = \imath' A^2 \imath,$$

where $\imath$ denotes the $n \times 1$ sum-vector $(1, 1, \ldots, 1)'$. Prove that

$$\mathscr{E}(X'AX) = (\operatorname{tr} A) V + \alpha \mu \mu'$$

$$\mathscr{V}(\operatorname{vec} X'AX) = (I + K_p)[(\operatorname{tr} A^2)(V \otimes V) + \beta(V \otimes \mu\mu' + \mu\mu' \otimes V)]$$

(Neudecker 1985a).

3. Let the $p \times 1$ random vectors x_i $(i = 1, \ldots, n)$ be independently distributed as $\mathscr{N}_p(\mu_i, V)$. Let $X = (x_1, \ldots, x_n)'$ and $M = (\mu_1, \ldots, \mu_n)'$. Let A be an arbitrary $n \times n$ matrix, not necessarily symmetric. Prove that

$$\mathscr{E}(X'AX) = M'AM + (\operatorname{tr} A)V,$$

$$\begin{aligned}\mathscr{V}(\operatorname{vec} X'AX) = {} & (\operatorname{tr} A'A)(V \otimes V) + (\operatorname{tr} A^2) K_{pp}(V \otimes V) \\ & + M'A'AM \otimes V + V \otimes M'AA'M \\ & + K_{pp}(M'A^2M \otimes V + (V \otimes M'A^2M)')\end{aligned}$$

(Neudecker 1985b).

BIBLIOGRAPHICAL NOTES

Two good texts at the intermediate level are Mood *et al.* (1974) and Hogg and Craig (1970). More advanced treatments can be found in Anderson (1984), Wilks (1962), or Rao (1973).

CHAPTER 13

The linear regression model

1. INTRODUCTION

In this chapter we consider the general linear regression model

$$y = X\beta + \varepsilon, \qquad \beta \in \mathscr{B}, \tag{1}$$

where y is an $n \times 1$ vector of observable random variables, X is a non-stochastic $n \times k$ matrix ($n \geqslant k$) of observations on the regressors and ε is an $n \times 1$ vector of (non-observable) random disturbances with

$$\mathscr{E}\varepsilon = 0, \qquad \mathscr{E}\varepsilon\varepsilon' = \sigma^2 V, \tag{2}$$

where V is a known positive semidefinite $n \times n$ matrix and σ^2 is unknown. The $k \times 1$ vector β of regression coefficients is supposed to be a fixed but unknown point in the parameter space $\mathscr{B}$. The problem is that of estimating (linear combinations of) β on the basis of the vector of observations y.

To save space we shall denote the linear regression model by the triplet

$$(y, X\beta, \sigma^2 V). \tag{3}$$

We shall make varying assumptions about the rank of X and the rank of V.

We assume that the parameter space $\mathscr{B}$ is either the k-dimensional Euclidean space,

$$\mathscr{B} = \mathbb{R}^k, \tag{4}$$

or a non-empty affine subspace of $\mathbb{R}^k$, having the representation

$$\mathscr{B} = \{\beta : R\beta = r, \beta \in \mathbb{R}^k\}, \tag{5}$$

where the matrix R and the vector r are non-stochastic. Of course, by putting $R = 0$ and $r = 0$, we obtain (4) as a special case of (5); nevertheless, distinguishing between the two cases is useful.

The purpose of this chapter is to *derive* the 'best' affine unbiased estimator of (linear combinations of) β. The emphasis is on 'derive'. We are not satisfied with simply presenting an estimator and then showing its optimality; rather we wish to describe a method by which estimators can be *constructed*. The constructive device that we seek is the method of affine minimum-trace unbiased estimation.

2. AFFINE MINIMUM-TRACE UNBIASED ESTIMATION

Let $(y, X\beta, \sigma^2 V)$ be the linear regression model and consider, for a given matrix W, the parametric function $W\beta$. An estimator of $W\beta$ is said to be *affine* if it is of the form

$$Ay + c, \tag{1}$$

where the matrix A and the vector c are fixed and non-stochastic. An *unbiased* estimator of $W\beta$ is an estimator, say $\widehat{W\beta}$, such that

$$\mathscr{E}(\widehat{W\beta}) = W\beta \qquad \text{for all } \beta \in \mathscr{B}. \tag{2}$$

If there exists at least one affine unbiased estimator of $W\beta$ (that is, if the class of affine unbiased estimators is not empty), then we say that $W\beta$ is *estimable*. A complete characterization of the class of estimable functions is given in section 7. If $W\beta$ is estimable, we are interested in the 'best' estimator among its affine unbiased estimators. The following definition makes this concept precise.

Definition 1

The *best affine unbiased estimator* of an estimable parametric function $W\beta$ is an affine unbiased estimator of $W\beta$, say $\widehat{W\beta}$, such that

$$\mathscr{V}(\hat{\theta}) - \mathscr{V}(\widehat{W\beta}) \tag{3}$$

is positive semidefinite for all affine unbiased estimators $\hat{\theta}$ of $W\beta$.

As yet there is no guarantee that there exists a best affine unbiased estimator, nor that, if it exists, it is unique. In what follows we shall see that in all cases considered such an estimator exists and is unique.

We shall find that when the parameter space $\mathscr{B}$ is the whole of $\mathbb{R}^k$, then the best affine unbiased estimator turns out to be *linear* (that is, of the form Ay); hence the more common name 'best linear unbiased estimator' or BLUE. However, when $\mathscr{B}$ is restricted, then the best affine unbiased estimator is in general affine.

An obvious drawback of the optimality criterion (3) is that it is not operational—we cannot minimize a matrix. We can, however, minimize a scalar function of a matrix: its trace, its determinant, or its largest eigenvalue. The trace criterion appears to be the most practicable.

Definition 2

The *affine minimum-trace unbiased estimator* of an estimable parametric function $W\beta$ is an affine unbiased estimator of $W\beta$, say $\widehat{W\beta}$, such that

$$\operatorname{tr} \mathscr{V}(\hat{\theta}) \geq \operatorname{tr} \mathscr{V}(\widehat{W\beta}) \tag{4}$$

for all affine unbiased estimators $\hat{\theta}$ of $W\beta$.

Now, for any two square matrices B and C, if $B-C$ is positive semidefinite, then $\operatorname{tr} B \geq \operatorname{tr} C$. Hence the best affine unbiased estimator is also an affine minimum-trace unbiased estimator, but not vice versa. If, therefore, the affine minimum-trace unbiased estimator is *unique* (which is always the case in this chapter), then the affine minimum-trace unbiased estimator is the best affine unbiased estimator, unless the latter does not exist.

Thus the method of affine minimum-trace unbiased estimation is both practicable and powerful.

3. THE GAUSS–MARKOV THEOREM

Let us consider the simplest case, that of the linear regression model

$$y = X\beta + \varepsilon, \tag{1}$$

where X has full column rank k and the disturbances $\varepsilon_1, \varepsilon_2, \ldots, \varepsilon_n$ are uncorrelated, i.e.

$$\mathscr{E}\varepsilon = 0, \qquad \mathscr{E}\varepsilon\varepsilon' = \sigma^2 I_n. \tag{2}$$

We shall first demonstrate the following proposition.

Proposition 1

Consider the linear regression model $(y, X\beta, \sigma^2 I)$. The affine minimum-trace unbiased estimator $\hat{\beta}$ of β exists if, and only if, $r(X)=k$, in which case

$$\hat{\beta} = (X'X)^{-1}X'y \tag{3}$$

with variance matrix

$$\mathscr{V}(\hat{\beta}) = \sigma^2(X'X)^{-1}. \tag{4}$$

Proof. We seek an affine estimator $\hat{\beta}$ of β, that is an estimator of the form

$$\hat{\beta} = Ay + c, \tag{5}$$

where A is a constant $k \times n$ matrix and c is a constant $k \times 1$ vector. The unbiasedness requirement is

$$\beta = \mathscr{E}\hat{\beta} = AX\beta + c \qquad \text{for all } \beta \text{ in } \mathbb{R}^k, \tag{6}$$

which yields

$$AX = I_k, \qquad c = 0. \tag{7}$$

The constraint $AX = I$ can only be imposed if $r(X) = k$. Necessary, therefore, for the existence of an affine unbiased estimator of β is that $r(X) = k$. It is sufficient, too, as we shall see.

The variance matrix of $\hat{\beta}$ is

$$\mathscr{V}(\hat{\beta}) = \mathscr{V}(Ay) = \sigma^2 AA'. \tag{8}$$

Hence the affine minimum-trace unbiased estimator (that is, the estimator whose sampling variance has minimum trace within the class of affine unbiased estimators) is obtained by solving the deterministic problem

$$\text{minimize} \quad \tfrac{1}{2}\operatorname{tr} AA' \tag{9}$$

$$\text{subject to} \quad AX = I. \tag{10}$$

To solve this problem we define the Lagrangian function ψ by

$$\psi(A) = \tfrac{1}{2}\operatorname{tr} AA' - \operatorname{tr} L'(AX - I), \tag{11}$$

where L is a $k \times k$ matrix of Lagrange multipliers. Differentiating ψ with respect to A yields

$$\begin{aligned} \mathrm{d}\psi &= \tfrac{1}{2}\operatorname{tr}(\mathrm{d}A)A' + \tfrac{1}{2}\operatorname{tr} A(\mathrm{d}A)' - \operatorname{tr} L'(\mathrm{d}A)X \\ &= \operatorname{tr} A'\mathrm{d}A - \operatorname{tr} XL'\mathrm{d}A = \operatorname{tr}(A' - XL')\mathrm{d}A. \end{aligned} \tag{12}$$

The first-order conditions are therefore

$$A' = XL' \tag{13}$$

$$AX = I_k. \tag{14}$$

These equations are easily solved. From

$$I_k = X'A' = X'XL' \tag{15}$$

we find $L' = (X'X)^{-1}$, so that

$$A' = XL' = X(X'X)^{-1}. \tag{16}$$

Since ψ is strictly convex (Why?), $\frac{1}{2}\operatorname{tr} AA'$ has a strict absolute minimum at $A = (X'X)^{-1}X'$ under the constraint $AX = I$ (see Theorem 7.13). Hence

$$\hat{\beta} = Ay = (X'X)^{-1}X'y \tag{17}$$

is the affine minimum-trace unbiased estimator. Its variance matrix is

$$\mathcal{V}(\hat{\beta}) = (X'X)^{-1}X'(\mathcal{V}(y))X(X'X)^{-1} = \sigma^2(X'X)^{-1}. \tag{18}$$

□

Proposition 1 shows that there exists a *unique* affine minimum-trace unbiased estimator $\hat{\beta}$ of β. Hence, *if* there exists a best affine unbiased estimator of β, it can only be $\hat{\beta}$.

Theorem 1 (Gauss–Markov)

Consider the linear regression model $(y, X\beta, \sigma^2 I)$. The best affine unbiased estimator $\hat{\beta}$ of β exists if, and only if, $r(X) = k$, in which case

$$\hat{\beta} = (X'X)^{-1}X'y \tag{19}$$

with variance matrix

$$\mathcal{V}(\hat{\beta}) = \sigma^2(X'X)^{-1}. \tag{20}$$

Proof. The only candidate for the best affine unbiased estimator of β is the affine minimum-trace unbiased estimator $\hat{\beta}=(X'X)^{-1}X'y$. Consider an arbitrary affine estimator $\tilde{\beta}$ of β which we write as

$$\tilde{\beta}=\hat{\beta}+Cy+d. \tag{21}$$

The estimator $\tilde{\beta}$ is unbiased if, and only if,

$$CX=0, \qquad d=0. \tag{22}$$

Imposing unbiasedness the variance matrix of $\tilde{\beta}$ is

$$\begin{aligned}\mathscr{V}(\tilde{\beta})=\mathscr{V}(\hat{\beta}+Cy)&=\sigma^2[(X'X)^{-1}X'+C][X(X'X)^{-1}+C']\\ &=\sigma^2(X'X)^{-1}+\sigma^2CC',\end{aligned} \tag{23}$$

which exceeds the variance matrix of $\hat{\beta}$ by σ^2CC', a positive semidefinite matrix. □

Exercises

1. Show that the function ψ defined in (11) is strictly convex.
2. Show that the constrained minimization problem

$$\begin{array}{lll}\text{minimize} & \tfrac{1}{2}x'x & \\ \text{subject to} & Cx=b & \text{(consistent)}\end{array}$$

has a unique solution $x^*=C^+b$.
3. The problem (9) subject to (10) is equivalent to k separate minimization problems. The i-th subproblem is

$$\begin{array}{ll}\text{minimize} & \tfrac{1}{2}a_i'a_i \\ \text{subject to} & X'a_i=e_i,\end{array}$$

where a_i' is the i-th row of A and e_i' is the i-th row of I_k. Show that

$$a_i=X(X'X)^{-1}e_i$$

is the unique solution, and compare this result with (16).
4. Consider the model $(y, X\beta, \sigma^2 I)$. The estimator $\hat{\beta}$ of β which, in the class of affine unbiased estimators, minimizes the *determinant* of $\mathscr{V}(\hat{\beta})$ (rather than its trace) is also $\hat{\beta}=(X'X)^{-1}X'y$. There are, however, certain disadvantages in using the minimum-determinant criterion instead of the minimum-trace criterion. Discuss these possible disadvantages.

4. THE METHOD OF LEAST SQUARES

Suppose we are given an $n\times 1$ vector y and an $n\times k$ matrix X with linearly independent columns. The vector y and the matrix X are assumed to be known (and non-stochastic). The problem is to determine the $k\times 1$ vector b that satisfies the equation

$$y = Xb. \tag{1}$$

If $X(X'X)^{-1}X'y = y$, then equation (1) is consistent and has a unique solution $b^* = (X'X)^{-1}X'y$. If $X(X'X)^{-1}X'y \neq y$, then equation (1) has no solution. In that case we may seek a vector b^* which, in a sense, minimizes the 'error' vector

$$e = y - Xb. \tag{2}$$

A convenient scalar measure of the 'error' would be

$$e'e = (y - Xb)'(y - Xb). \tag{3}$$

It follows from Theorem 11.34 that

$$b^* = (X'X)^{-1}X'y \tag{4}$$

minimizes $e'e$ over all b in $\mathbb{R}^k$. The vector b^* is called the *least squares solution* and Xb^* is called the *least squares approximation* to y. Thus b^* is the 'best' choice for b whether the equation $y = Xb$ is consistent or not. If $y = Xb$ is consistent, then b^* is the solution; if $y = Xb$ is not consistent, then b^* is the least squares solution.

The surprising fact that the least squares solution and the Gauss–Markov estimator are identical expressions has led to the unfortunate usage of the term '(ordinary) least squares *estimator*' meaning the Gauss–Markov estimator. The method of least squares, however, is a purely deterministic method which has to do with approximation, *not* with estimation.

Exercise

1. Show that the least squares approximation to y is y itself if, and only if, the equation $y = Xb$ is consistent.

5. AITKEN'S THEOREM

In Theorem 1 we considered the regression model $(y, X\beta, \sigma^2 I)$, where the random components $y_1, y_2, \ldots, y_n$ of the vector y are uncorrelated (but not identically distributed, since their expectations differ). A slightly more general set-up, first considered by Aitken (1935), is the regression model $(y, X\beta, \sigma^2 V)$, where V is a known positive definite matrix. In Aitken's model the observations $y_1, \ldots, y_n$ are neither independent nor identically distributed.

Theorem 2 (Aitken)

Consider the linear regression model $(y, X\beta, \sigma^2 V)$, and assume that $|V| \neq 0$. The best affine unbiased estimator $\widehat{W\beta}$ of $W\beta$ exists for *every* matrix W (with k columns) if, and only if, $r(X) = k$ in which case

$$\widehat{W\beta} = W(X'V^{-1}X)^{-1}X'V^{-1}y \tag{1}$$

with variance matrix

$$\mathcal{V}(\widehat{W\beta})=\sigma^2 W(X'V^{-1}X)^{-1}W'. \tag{2}$$

Note. In fact, Theorem 2 generalizes Theorem 1 in two ways. First, it is assumed that the variance matrix of y is $\sigma^2 V$ rather than $\sigma^2 I$. This then leads to the best affine unbiased estimator $\hat{\beta}=(X'V^{-1}X)^{-1}X'V^{-1}y$ of β, if $r(X)=k$. The estimator $\hat{\beta}$ is usually called Aitken's estimator (or the generalized least squares estimator). Secondly, we prove that the best affine unbiased estimator of an arbitrary linear combination of β, say $W\beta$, is $W\hat{\beta}$.

Proof. Let $\widehat{W\beta}=Ay+c$ be an affine estimator of $W\beta$. The estimator is unbiased if, and only if,

$$W\beta=AX\beta+c \qquad \text{for all } \beta \text{ in } \mathbb{R}^k, \tag{3}$$

that is, if and only if

$$AX=W, \qquad c=0. \tag{4}$$

The constraint $AX=W$ implies $r(W)\leqslant r(X)$. Since this must hold for every matrix W, X must have full column rank k.

The variance matrix of $\widehat{W\beta}$ is $\sigma^2 AVA'$. Hence the constrained minimization problem is

$$\text{minimize} \qquad \tfrac{1}{2}\operatorname{tr} AVA' \tag{5}$$

$$\text{subject to} \qquad AX=W. \tag{6}$$

Differentiating the appropriate Lagrangian function

$$\psi(A)=\tfrac{1}{2}\operatorname{tr} AVA'-\operatorname{tr} L'(AX-W), \tag{7}$$

yields the first-order conditions

$$VA'=XL' \tag{8}$$

$$AX=W. \tag{9}$$

Solving these two matrix equations we obtain

$$L=W(X'V^{-1}X)^{-1} \tag{10}$$

and

$$A=W(X'V^{-1}X)^{-1}X'V^{-1}. \tag{11}$$

Since the Lagrangian function is strictly convex, it follows that

$$\widehat{W\beta}=Ay=W(X'V^{-1}X)^{-1}X'V^{-1}y \tag{12}$$

is the affine minimum-trace unbiased estimator of $W\beta$. Its variance matrix is

$$\mathcal{V}(\widehat{W\beta})=\sigma^2 AVA'=W(X'V^{-1}X)^{-1}W'. \tag{13}$$

Let us now show that $\widehat{W\beta}$ is not merely the affine minimum-trace unbiased estimator of $W\beta$, but the *best* affine unbiased estimator. Let c be an arbitrary column vector (such that $W'c$ is defined), and let $\beta^*=(X'V^{-1}X)^{-1}X'V^{-1}y$. Then $c'W\beta^*$ is the affine minimum-trace unbiased estimator of $c'W\beta$. Let $\hat{\theta}$ be an alternative affine unbiased estimator of $W\beta$. Then $c'\hat{\theta}$ is an affine unbiased estimator of $c'W\beta$, and so

$$\operatorname{tr} \mathscr{V}(c'\hat{\theta}) \geqslant \operatorname{tr} \mathscr{V}(c'W\beta^*), \tag{14}$$

that is,

$$c'(\mathscr{V}(\hat{\theta}))c \geqslant c'(\mathscr{V}(W\beta^*))c. \tag{15}$$

Since c is arbitrary, it follows that $\mathscr{V}(\hat{\theta}) - \mathscr{V}(W\beta^*)$ is positive semidefinite.

□

The proof that $W(X'V^{-1}X)^{-1}X'V^{-1}y$ is the affine minimum-trace unbiased estimator of $W\beta$ is similar to the proof of Proposition 1. But the proof that this estimator is indeed the best affine unbiased estimator of $W\beta$ is essentially different from the corresponding proof of Theorem 1, and much more useful as a general device.

Exercise

1. Show that the model $(y, X\beta, \sigma^2 V)$, $|V| \neq 0$, is equivalent to the model $(V^{-1/2}y, V^{-1/2}X\beta, \sigma^2 I)$, and obtain Aitken's estimator $\hat{\beta} = (X'V^{-1}X)^{-1}X'V^{-1}y$ as a special case of Theorem 1.

6. MULTICOLLINEARITY

It is easy to see that Theorem 2 does not cover the topic completely. In fact, complications of three types may occur, and we shall discuss each of these in detail. The first complication is that the k columns of X may be linearly dependent; the second complication arises if we have *a priori* knowledge that the parameters satisfy a linear constraint of the form $R\beta = r$; and the third complication is that the $n \times n$ variance matrix $\sigma^2 V$ may be singular.

We shall take each of these complications in turn. Thus we assume in this and the next section that V is non-singular and that no *a priori* knowledge as to constraints of the form $R\beta = r$ is available, but that X fails to have full column rank. This problem (that the columns of X are linearly dependent) is called *multicollinearity*.

If $r(X) < k$, then no affine unbiased estimator of β can be found, let alone a best affine unbiased estimator. This is easy to see. Let the affine estimator be

$$\hat{\beta} = Ay + c. \tag{1}$$

Then unbiasedness requires

$$AX = I_k, \qquad c = 0, \tag{2}$$

which is impossible if $r(X) < k$. Not all hope is lost, however. We shall show that an affine unbiased estimator of $X\beta$ always exists, and derive the best estimator of $X\beta$ in the class of affine unbiased estimators.

Theorem 3

Consider the linear regression model $(y, X\beta, \sigma^2 V)$, and assume that $|V| \neq 0$. Then the estimator

$$\widehat{X\beta} = X(X'V^{-1}X)^{+}X'V^{-1}y \tag{3}$$

is the best affine unbiased estimator of $X\beta$, and its variance matrix is

$$\mathscr{V}(\widehat{X\beta}) = \sigma^2 X(X'V^{-1}X)^{+}X'. \tag{4}$$

Proof. Let the estimator be $\widehat{X\beta} = Ay + c$. The estimator is unbiased if, and only if,

$$X\beta = AX\beta + c \qquad \text{for all } \beta \text{ in } \mathbb{R}^k, \tag{5}$$

which implies

$$AX = X, \qquad c = 0. \tag{6}$$

Notice that the equation $AX = X$ always has a solution for A, whatever the rank of X. The variance matrix of $\widehat{X\beta}$ is

$$\mathscr{V}(\widehat{X\beta}) = \sigma^2 AVA'. \tag{7}$$

Hence we consider the following minimization problem:

$$\text{minimize} \qquad \tfrac{1}{2}\operatorname{tr} AVA' \tag{8}$$

$$\text{subject to} \qquad AX = X, \tag{9}$$

the solution of which will yield the affine minimum-trace unbiased estimator of $X\beta$. The appropriate Lagrangian function is

$$\psi(A) = \tfrac{1}{2}\operatorname{tr} AVA' - \operatorname{tr} L'(AX - X). \tag{10}$$

Differentiating (10) with respect to A yields the following first-order conditions:

$$VA' = XL' \tag{11}$$

$$AX = X. \tag{12}$$

From (11) we have $A' = V^{-1}XL'$. Hence

$$X = AX = LX'V^{-1}X. \tag{13}$$

Equation (13) always has a solution for L (Why?), but this solution is not unique unless X has full rank. However LX' *does* have a unique solution, namely

$$LX' = X(X'V^{-1}X)^{+}X' \tag{14}$$

(see exercise 2). Hence A also has a unique solution:

$$A = LX'V^{-1} = X(X'V^{-1}X)^{+}X'V^{-1}. \tag{15}$$

It follows that $X(X'V^{-1}X)^{+}X'V^{-1}y$ is the affine minimum-trace unbiased estimator of $X\beta$. Hence, if there is a best affine unbiased estimator of $X\beta$, this is it.

Now consider an arbitrary affine estimator

$$[X(X'V^{-1}X)^{+}X'V^{-1}+C]y+d \tag{16}$$

of $X\beta$. This estimator is unbiased if, and only if, $CX=0$ and $d=0$. Imposing unbiasedness, the variance matrix is

$$\sigma^2 X(X'V^{-1}X)^{+}X'+\sigma^2 CVC', \tag{17}$$

which exceeds the variance matrix of $X(X'V^{-1}X)^{+}X'V^{-1}y$ by $\sigma^2 CVC'$, a positive semidefinite matrix. □

Exercises

1. Show that the solution A in (15) satisfies $AX=X$.
2. Prove that (13) implies (14). [*Hint*: Post-multiply both sides of (13) by $(X'V^{-1}X)^{+}X'V^{-1/2}$.]

7. ESTIMABLE FUNCTIONS

Recall from section 2 that, in the framework of the linear regression model $(y, X\beta, \sigma^2 V)$, a parametric function $W\beta$ is said to be *estimable* if there exists an affine unbiased estimator of $W\beta$. In the previous section we have seen that $X\beta$ is always estimable. We shall now show that any linear combination of $X\beta$ is also estimable and, in fact, that *only* linear combinations of $X\beta$ are estimable. Thus we obtain a complete characterization of the class of estimable functions.

Proposition 2

In the linear regression model $(y, X\beta, \sigma^2 V)$, the parametric function $W\beta$ is estimable if, and only if, $\mathcal{M}(W')\subset\mathcal{M}(X')$.

Note. Proposition 2 holds true whatever the rank of V. If X has full column rank k, then $\mathcal{M}(W')\subset\mathcal{M}(X')$ is true for every W, in particular for $W=I$; if $r(X)<k$, then $\mathcal{M}(W')\subset\mathcal{M}(X')$ is not true for every W, and in particular not for $W=I$.

Proof. Let $Ay+c$ be an affine estimator of $W\beta$. Unbiasedness requires that

$$W\beta=\mathcal{E}(Ay+c)=AX\beta+c \qquad \text{for all } \beta \text{ in } \mathbb{R}^k, \tag{1}$$

which leads to

$$AX=W, \qquad c=0. \tag{2}$$

Hence the matrix A exists if, and only if, the rows of W are linear combinations of the rows of X, that is, iff $\mathcal{M}(W')\subset\mathcal{M}(X')$. □

Let us now demonstrate Theorem 4.

Theorem 4

Consider the linear regression model $(y, X\beta, \sigma^2 V)$, and assume that $|V| \neq 0$. Then the best affine unbiased estimator $\widehat{W\beta}$ of $W\beta$ exists if, and only if, $\mathcal{M}(W') \subset \mathcal{M}(X')$, in which case

$$\widehat{W\beta} = W(X'V^{-1}X)^{+}X'V^{-1}y \tag{3}$$

with variance matrix

$$\mathscr{V}(\widehat{W\beta}) = \sigma^2 W(X'V^{-1}X)^{+}W'. \tag{4}$$

Proof. To prove that $\widehat{W\beta}$ is the affine minimum-trace unbiased estimator of $W\beta$ we proceed along the same lines as in the proof of Theorem 3. To prove that this is the *best* affine unbiased estimator, we use the same argument as in the corresponding part of the proof of Theorem 2. □

Exercises

1. Let $r(X) = r < k$. Then there exists a $k \times (k-r)$ matrix C of full column rank such that $XC = 0$. Show that $W\beta$ is estimable if, and only if, $WC = 0$.
2. (Season dummies) Let X' be given by

$$X' = \begin{bmatrix} 1 & 1 & 1 & 1 & 1 & 1 & 1 & 1 & 1 & 1 & 1 & 1 \\ 1 & 1 & 1 & & & & & & & & & \\ & & & 1 & 1 & 1 & & & & & & \\ & & & & & & 1 & 1 & 1 & & & \\ & & & & & & & & & 1 & 1 & 1 \end{bmatrix}$$

   where all undesignated elements are zero. Show that $W\beta$ is estimable if, and only if, $(1, -1, -1, -1)W' = 0$.
3. Let $\tilde{\beta}$ be any solution of the equation $X'V^{-1}X\beta = X'V^{-1}y$. Then the following three statements are equivalent:
   (i) $W\beta$ is estimable,
   (ii) $W\tilde{\beta}$ is an unbiased estimator of $W\beta$,
   (iii) $W\tilde{\beta}$ is unique.

8. LINEAR CONSTRAINTS: THE CASE $\mathcal{M}(R') \subset \mathcal{M}(X')$

Suppose now that we have *a priori* information consisting of exact linear constraints on the coefficients:

$$R\beta = r, \tag{1}$$

where the matrix R and the vector r are known. Some authors require that the constraints are linearly independent, that is, that R has full row rank, but this is not assumed here. Of course, we must assume that (1) is a consistent equation, that is, $r \in \mathcal{M}(R)$ or equivalently

$$RR^+r = r. \tag{2}$$

To incorporate this extraneous information is clearly desirable, since the resulting estimator will become more efficient.

In this section we discuss the special case where $\mathcal{M}(R') \subset \mathcal{M}(X')$; the complete solution is given in section 9. This means, in effect, that we impose linear constraints not on β but on $X\beta$. Of course, the condition is automatically fulfilled when X has full column rank.

Theorem 5

Consider the linear regression model $(y, X\beta, \sigma^2 V)$, where β satisfies the consistent linear constraints $R\beta = r$. Assume that $|V| \neq 0$, and that $\mathcal{M}(R') \subset \mathcal{M}(X')$. Then the best affine unbiased estimator $\widehat{W\beta}$ of $W\beta$ exists if, and only if, $\mathcal{M}(W') \subset \mathcal{M}(X')$, in which case

$$\widehat{W\beta} = W\beta^* + W(X'V^{-1}X)^+R'[R(X'V^{-1}X)^+R']^+(r - R\beta^*) \tag{3}$$

where

$$\beta^* = (X'V^{-1}X)^+X'V^{-1}y. \tag{4}$$

Its variance matrix is

$$\begin{aligned}\mathscr{V}(\widehat{W\beta}) &= \sigma^2 W(X'V^{-1}X)^+W' \\ &\quad - \sigma^2 W(X'V^{-1}X)^+R'[R(X'V^{-1}X)^+R']^+R(X'V^{-1}X)^+W'.\end{aligned} \tag{5}$$

Note. If X has full column rank, we have $\mathcal{M}(R') \subset \mathcal{M}(X')$ for every R, $\mathcal{M}(W') \subset \mathcal{M}(X')$ for every W (in particular for $W = I$) and $X'V^{-1}X$ is non-singular. If, in addition, R has full row rank, then $R\beta = r$ is always consistent and $R(X'V^{-1}X)^{-1}R'$ non-singular.

Proof. We write the affine estimator of $W\beta$ again as

$$\widehat{W\beta} = Ay + c. \tag{6}$$

Unbiasedness requires

$$W\beta = AX\beta + c \qquad \text{for all } \beta \text{ satisfying } R\beta = r. \tag{7}$$

The general solution of $R\beta = r$ is

$$\beta = R^+r + (I - R^+R)q, \tag{8}$$

where q is an arbitrary $k \times 1$ vector. Replacing β in (7) by its 'solution' (8), we obtain

$$(W - AX)[R^+r + (I - R^+R)q] = c \qquad \text{for all } q, \tag{9}$$

which implies

$$(W - AX)R^+r = c \tag{10}$$

and

$$(W-AX)(I-R^+R)=0. \tag{11}$$

We now solve $W-AX$ from (11). This gives

$$W-AX=BR \tag{12}$$

where B is an arbitrary $k\times m$ matrix. Inserting (12) in (10) yields

$$c=BRR^+r=Br, \tag{13}$$

using (2). If follows that the estimator (6) can be written as

$$\widehat{W\beta}=Ay+Br, \tag{14}$$

while the unbiasedness condition boils down to

$$AX+BR=W. \tag{15}$$

Equation (15) can only be satisfied if $\mathcal{M}(W')\subset\mathcal{M}(X':R')$. Since $\mathcal{M}(R')\subset\mathcal{M}(X')$ by assumption, it follows that $\mathcal{M}(W')\subset\mathcal{M}(X')$ is a necessary condition for the existence of an affine unbiased estimator of $W\beta$.

The variance matrix of $\widehat{W\beta}$ is σ^2AVA'. Hence the relevant minimization problem to find the affine minimum-trace unbiased estimator of $W\beta$ is

$$\begin{array}{ll}\text{minimize} & \tfrac{1}{2}\operatorname{tr}AVA' \\ \text{subject to} & AX+BR=W.\end{array} \tag{16}$$

Let us define the Lagrangian function ψ by

$$\psi(A,\,B)=\tfrac{1}{2}\operatorname{tr}AVA'-\operatorname{tr}L'(AX+BR-W), \tag{17}$$

where L is a matrix of Lagrange multipliers. Differentiating ψ with respect to A and B yields

$$\begin{aligned}\mathrm{d}\psi&=\operatorname{tr}AV(\mathrm{d}A)'-\operatorname{tr}L'(\mathrm{d}A)X-\operatorname{tr}L'(\mathrm{d}B)R\\&=\operatorname{tr}(VA'-XL')(\mathrm{d}A)-\operatorname{tr}RL'(\mathrm{d}B).\end{aligned} \tag{18}$$

Hence we obtain the first-order conditions

$$VA'=XL' \tag{19}$$

$$RL'=0 \tag{20}$$

$$AX+BR=W. \tag{21}$$

From (19) we obtain

$$L(X'V^{-1}X)=AX. \tag{22}$$

Regarding (22) as an equation in L, given A, we notice that it has a solution for every A, because

$$X(X'V^{-1}X)^+(X'V^{-1}X)=X. \tag{23}$$

As in the passage from (6.13) to (6.14), this solution is not, in general, unique. LX'

however *does* have a unique solution:

$$LX' = AX(X'V^{-1}X)^{+}X'. \tag{24}$$

Since $\mathcal{M}(R') \subset \mathcal{M}(X')$ and using (23) we obtain

$$\begin{aligned} 0 = LR' &= AX(X'V^{-1}X)^{+}R' \\ &= (W - BR)(X'V^{-1}X)^{+}R' \end{aligned} \tag{25}$$

from (20) and (21). This leads to the equation in B

$$BR(X'V^{-1}X)^{+}R' = W(X'V^{-1}X)^{+}R'. \tag{26}$$

Post-multiplying both sides of (26) by $[R(X'V^{-1}X)^{+}R']^{+}R$, and using the fact that

$$R(X'V^{-1}X)^{+}R'[R(X'V^{-1}X)^{+}R']^{+}R = R \tag{27}$$

(see exercise 2), we obtain

$$BR = W(X'V^{-1}X)^{+}R'[R(X'V^{-1}X)^{+}R']^{+}R. \tag{28}$$

Equation (28) provides the solution for BR and, in view of (21), AX. From these we could obtain (non-unique) solutions for A and B. But these explicit solutions are not needed since we can write the estimator $\widehat{W\beta}$ of $W\beta$ as

$$\begin{aligned} \widehat{W\beta} &= Ay + Br \\ &= LX'V^{-1}y + BRR^{+}r \\ &= AX(X'V^{-1}X)^{+}X'V^{-1}y + BRR^{+}r, \end{aligned} \tag{29}$$

using (19) and (24). Inserting the solutions for AX and BR in (29) we find

$$\widehat{W\beta} = W\beta^{*} + W(X'V^{-1}X)^{+}R'[R(X'V^{-1}X)^{+}R']^{+}(r - R\beta^{*}). \tag{30}$$

It is easy to derive the variance matrix of $\widehat{W\beta}$. Finally, to prove that $\widehat{W\beta}$ is not only the minimum-trace estimator but also the best estimator among the affine unbiased estimators of $W\beta$, we use the same argument as in the proof of Theorem 2. □

Exercises

1. Prove that

$$R(X'V^{-1}X)^{+}R'[R(X'V^{-1}X)^{+}R']^{+}R(X'V^{-1}X)^{+} = R(X'V^{-1}X)^{+}.$$

2. Show that $\mathcal{M}(R') \subset \mathcal{M}(X')$ implies $R(X'V^{-1}X)^{+}X'V^{-1}X = R$, and use this and exercise 1 to prove (27).

9. LINEAR CONSTRAINTS: THE GENERAL CASE

Recall from section 7 that a parametric function $W\beta$ is called *estimable* if there exists an affine unbiased estimator of $W\beta$. In Proposition 2 we established the class of estimable functions $W\beta$ for the linear regression model $(y, X\beta, \sigma^2 V)$ without constraints on β. Let us now characterize the estimable functions $W\beta$ for the linear regression model, assuming that β satisfies certain linear constraints.

Proposition 3

In the linear regression model $(y, X\beta, \sigma^2 V)$ where β satisfies the consistent linear constraints $R\beta = r$, the parametric function $W\beta$ is estimable if, and only if, $\mathcal{M}(W') \subset \mathcal{M}(X' : R')$.

Proof. We can write the linear regression model with exact linear constraints as

$$\begin{pmatrix} y \\ r \end{pmatrix} = \begin{pmatrix} X \\ R \end{pmatrix} \beta + u, \tag{1}$$

with

$$\mathscr{E}u = 0, \qquad \mathscr{E}uu' = \sigma^2 \begin{pmatrix} V & 0 \\ 0 & 0 \end{pmatrix}. \tag{2}$$

Proposition 3 then follows from Proposition 2. □

Not surprisingly, there are more estimable functions in the constrained case than there are in the unconstrained case.

Having established which functions are estimable, we now want to find the 'best' estimator for such functions.

Theorem 6

Consider the linear regression model $(y, X\beta, \sigma^2 V)$, where β satisfies the consistent linear constraints $R\beta = r$. Assume that $|V| \neq 0$. Then the best affine unbiased estimator $\widehat{W\beta}$ of $W\beta$ exists if, and only if, $\mathcal{M}(W') \subset \mathcal{M}(X' : R')$, in which case

$$\widehat{W\beta} = W\beta^* + WG^+R'(RG^+R')^+(r - R\beta^*), \tag{3}$$

where

$$G = X'V^{-1}X + R'R \qquad \text{and} \qquad \beta^* = G^+X'V^{-1}y. \tag{4}$$

Its variance matrix is

$$\mathscr{V}(\widehat{W\beta}) = \sigma^2 WG^+W' - \sigma^2 WG^+R'(RG^+R')^+RG^+W'. \tag{5}$$

Proof. The proof is similar to the proof of Theorem 5. As there, the estimator can be written as

$$\widehat{W\beta} = Ay + Br, \tag{6}$$

and we obtain the following first-order conditions:

$$VA' = XL' \tag{7}$$

$$RL' = 0 \tag{8}$$

$$AX + BR = W. \tag{9}$$

From (7) and (8) we obtain

$$LG=AX, \tag{10}$$

where G is the positive semidefinite matrix defined in (4). It is easy to prove that

$$GG^+X'=X', \qquad GG^+R'=R'. \tag{11}$$

Post-multiplying both sides of (10) by G^+X' and G^+R', respectively, we thus obtain

$$LX'=AXG^+X' \tag{12}$$

and

$$0=LR'=AXG^+R', \tag{13}$$

in view of (8). Using (9) we obtain from (13) the following equation in B:

$$BRG^+R'=WG^+R'. \tag{14}$$

Post-multiplying both sides of (14) by $(RG^+R')^+R$, we obtain, using (11),

$$BR=WG^+R'(RG^+R')^+R. \tag{15}$$

We can now solve for A as follows:

$$\begin{aligned} A&=LX'V^{-1}=AXG^+X'V^{-1}=(W-BR)G^+X'V^{-1}\\ &=WG^+X'V^{-1}-BRG^+X'V^{-1}\\ &=WG^+X'V^{-1}-WG^+R'(RG^+R')^+RG^+X'V^{-1}, \end{aligned} \tag{16}$$

using (7), (12), (9) and (16). The estimator $\widehat{W\beta}$ of $W\beta$ then becomes

$$\begin{aligned} \widehat{W\beta}&=Ay+Br=Ay+BRR^+r\\ &=WG^+X'V^{-1}y+WG^+R'(RG^+R')^+(r-RG^+X'V^{-1}y). \end{aligned} \tag{17}$$

The variance matrix of $\widehat{W\beta}$ is easily derived. Finally, to prove that $\widehat{W\beta}$ is the best affine unbiased estimator of $W\beta$ (and not merely the affine minimum-trace unbiased estimator) we use the same argument that concludes the proof of Theorem 2. □

Exercises

1. Prove that Theorem 6 remains valid when we replace G by $\bar{G}=X'V^{-1}X+R'ER$, where E is a positive semidefinite matrix such that $\mathcal{M}(R')\subset\mathcal{M}(\bar{G})$. Obtain Theorems 5 and 6 as special cases by letting $E=0$ and $E=I$, respectively.
2. We shall say that a parametric function $W\beta$ is *strictly estimable* if there exists a *linear* (rather than affine) unbiased estimator of $W\beta$. Show that in the linear regression model without constraints the parametric function $W\beta$ is estimable if, and only if, it is strictly estimable.
3. In the linear regression model $(y, X\beta, \sigma^2V)$ where β satisfies the consistent linear constraints $R\beta=r$, the parametric function $W\beta$ is strictly estimable if, and only if, $\mathcal{M}(W')\subset\mathcal{M}(X':R'N)$, where $N=I-rr^+$.

4. Consider the linear regression model $(y, X\beta, \sigma^2 V)$, where β satisfies the consistent linear constraints $R\beta = r$. Assume that $|V| \neq 0$. Then the best *linear* unbiased estimator of a strictly estimable parametric function $W\beta$ is $W\hat{\beta}$, where
$$\hat{\beta} = [G^+ - G^+R'N(NRG^+R'N)^+NRG^+]X'V^{-1}y$$
with
$$G = X'V^{-1}X + R'NR, \qquad N = I - rr^+.$$

10. LINEAR CONSTRAINTS: THE CASE $\mathcal{M}(R') \cap \mathcal{M}(X') = \{0\}$

We have seen that if X fails to have full column rank, not all components of β are estimable; only the components of $X\beta$ (and linear combinations thereof) are estimable. Proposition 3 tells us that we can improve this situation by adding linear constraints. More precisely, Proposition 3 shows that every parametric function of the form
$$(AX + BR)\beta \tag{1}$$
is estimable when β satisfies consistent linear constraints $R\beta = r$. Thus if we add linear constraints in such a way that the rank of $(X' : R')$ increases, then more and more linear combinations of β will become estimable, until—when $(X' : R')$ has full rank k—all linear combinations of β are estimable.

In Theorem 5 we considered the case where every row of R is a linear combination of the rows of X, in which case $r(X' : R') = r(X')$, so that the class of estimable functions remains the same. In this section we shall consider the opposite situation where the rows of R are linearly independent of the rows of X, i.e. $\mathcal{M}(R') \cap \mathcal{M}(X') = \{0\}$. We shall see that the best affine unbiased estimator takes a particularly simple form.

Theorem 7

Consider the linear regression model $(y, X\beta, \sigma^2 V)$, where β satisfies the consistent linear constraints $R\beta = r$. Assume that $|V| \neq 0$ and that $\mathcal{M}(R') \cap \mathcal{M}(X') = \{0\}$. Then the best affine unbiased estimator $\widehat{W\beta}$ of $W\beta$ exists if, and only if, $\mathcal{M}(W') \subset \mathcal{M}(X' : R')$, in which case
$$\widehat{W\beta} = WG^+(X'V^{-1}y + R'r), \tag{2}$$
where
$$G = X'V^{-1}X + R'R. \tag{3}$$
Its variance matrix is
$$\mathcal{V}(\widehat{W\beta}) = \sigma^2 WG^+W' - \sigma^2 WG^+R'RG^+W' \tag{4}$$

Note. The requirement $\mathcal{M}(R') \cap \mathcal{M}(X') = \{0\}$ is equivalent to $r(X' : R') = r(X) + r(R)$, see Theorem 3.19.

Proof. Since $\mathscr{M}(R') \cap \mathscr{M}(X') = \{0\}$, it follows from Theorem 3.19 that

$$RG^+R' = RR^+, \tag{5}$$

$$XG^+X' = X(X'V^{-1}X)^+X', \tag{6}$$

and

$$RG^+X' = 0. \tag{7}$$

(In order to apply Theorem 3.19 let $A = X'V^{-1/2}$, $B = R'$.) Now define $\beta^* = G^+X'V^{-1}y$. Then $R\beta^* = 0$, and applying Theorem 6,

$$\begin{aligned} \widehat{W\beta} &= W\beta^* + WG^+R'(RG^+R')^+(r - R\beta^*) \\ &= W\beta^* + WG^+R'RR^+r = WG^+(X'V^{-1}y + R'r). \end{aligned} \tag{8}$$

The variance matrix of $\widehat{W\beta}$ is easily derived. □

Exercises

1. Suppose that the conditions of Theorem 7 are satisfied and, in addition, that $r(X' : R') = k$. Then the best affine unbiased estimator of β is
$$\hat{\beta} = (X'V^{-1}X + R'R)^{-1}(X'V^{-1}y + R'r).$$
2. Under the same conditions, show that an alternative expression for $\hat{\beta}$ is
$$\hat{\beta} = [(X'V^{-1}X)^2 + R'R]^{-1}(X'V^{-1}XX'V^{-1}y + R'r).$$
[*Hint*: Choose $W = I = [(X'V^{-1}X)^2 + R'R]^{-1}[(X'V^{-1}X)^2 + R'R]$.]
3. (Generalization) Under the same conditions, show that
$$\hat{\beta} = (X'EX + R'R)^{-1}[X'EX(X'V^{-1}X)^+X'V^{-1}y + R'r],$$
where E is a positive semidefinite matrix such that $r(X'EX) = r(X)$.
4. Obtain Theorem 4 as a special case of Theorem 7.

11. A SINGULAR VARIANCE MATRIX: THE CASE $\mathscr{M}(X) \subset \mathscr{M}(V)$

So far we have assumed that the variance matrix $\sigma^2 V$ of the disturbances is non-singular. Let us now relax this assumption. Thus we consider the linear regression model

$$y = X\beta + \varepsilon, \tag{1}$$

with

$$\mathscr{E}\varepsilon = 0, \qquad \mathscr{E}\varepsilon\varepsilon' = \sigma^2 V, \tag{2}$$

and V possibly singular. Pre-multiplication of the disturbance vector ε by $I - VV^+$ leads to

$$(I - VV^+)\varepsilon = 0 \quad \text{a.s.}, \tag{3}$$

because the expectation and variance matrix of $(I - VV^+)\varepsilon$ both vanish. Hence we can rewrite (1) as

$$y = X\beta + VV^{+}\varepsilon, \tag{4}$$

from which follows our next proposition.

Proposition 4 (consistency of the linear model)

In order for the linear regression model $(y, X\beta, \sigma^2 V)$ to be a consistent model, it is necessary and sufficient that $y \in \mathcal{M}(X:V)$ a.s.

Hence, in general, there are certain implicit restrictions on the dependent variable y; these are automatically satisfied when V is non-singular.

Since V is symmetric and positive semidefinite, there exists an orthogonal matrix $(S:T)$ and a diagonal matrix Λ with positive diagonal elements such that

$$VS = S\Lambda, \qquad VT = 0. \tag{5}$$

(If n' denotes the rank of V, then the orders of the matrices S, T and Λ are $n \times n'$, $n \times (n-n')$ and $n' \times n'$, respectively.) The orthogonality of $(S:T)$ implies that

$$S'S = I, \qquad T'T = I, \qquad S'T = 0, \tag{6}$$

and also

$$SS' + TT' = I. \tag{7}$$

Hence we can express V and V^{+} as

$$V = S\Lambda S', \qquad V^{+} = S\Lambda^{-1}S'. \tag{8}$$

After these preliminaries let us transform the regression model $y = X\beta + \varepsilon$ by means of the orthogonal matrix $(S:T)'$. This yields

$$S'y = S'X\beta + u, \qquad \mathcal{E}u = 0, \qquad \mathcal{E}uu' = \sigma^2\Lambda, \tag{9}$$

$$T'y = T'X\beta. \tag{10}$$

The vector $T'y$ is degenerate (has zero variance matrix), so that the equation $T'X\beta = T'y$ may be interpreted as a set of linear constraints on β.

We conclude that the model $(y, X\beta, \sigma^2 V)$, where V is singular, is equivalent to the model $(S'y, S'X\beta, \sigma^2\Lambda)$ where β satisfies the consistent (Why?) linear constraint $T'X\beta = T'y$.

Thus singularity of V implies some restrictions on the unknown parameter β, *unless* $T'X = 0$, or, equivalently, $\mathcal{M}(X) \subset \mathcal{M}(V)$. If we assume that $\mathcal{M}(X) \subset \mathcal{M}(V)$, then the model $(y, X\beta, \sigma^2 V)$, where V is singular, is equivalent to the *unconstrained* model $(S'y, S'X\beta, \sigma^2\Lambda)$, where Λ is non-singular, so that Theorem 4 applies. This leads to Theorem 8.

Theorem 8

Consider the linear regression model $(y, X\beta, \sigma^2 V)$, where $y \in \mathcal{M}(V)$ a.s. Assume that $\mathcal{M}(X) \subset \mathcal{M}(V)$. Then the best affine unbiased estimator $\widehat{W\beta}$ of $W\beta$ exists if, and only if, $\mathscr{M}(W') \subset \mathscr{M}(X')$, in which case

$$\widehat{W\beta} = W(X'V^+X)^+X'V^+y \tag{11}$$

with variance matrix

$$\mathscr{V}(\widehat{W\beta}) = \sigma^2 W(X'V^+X)^+W'. \tag{12}$$

Exercises

1. Show that the equation $T'X\beta = T'y$ in β has a solution if, and only if, the linear model is consistent.
2. Show that $T'X = 0$ if, and only if, $\mathscr{M}(X) \subset \mathscr{M}(V)$.
3. Show that $\mathscr{M}(X) \subset \mathscr{M}(V)$ implies $r(X'V^+X) = r(X)$.
4. Obtain Theorems 1–4 as special cases of Theorem 8.

12. A SINGULAR VARIANCE MATRIX: THE CASE $r(X'V^+X) = r(X)$

Somewhat weaker than the assumption $\mathscr{M}(X) \subset \mathscr{M}(V)$ made in the previous section is the condition

$$r(X'V^+X) = r(X). \tag{1}$$

With S and T as before, we shall show that (1) is equivalent to

$$\mathscr{M}(X'T) \subset \mathscr{M}(X'S). \tag{2}$$

(If $\mathscr{M}(X) \subset \mathscr{M}(V)$, then $X'T = 0$, so that (2) is automatically satisfied.) From $V^+ = S\Lambda^{-1}S'$ we obtain $X'V^+X = X'S\Lambda^{-1}S'X$ and hence

$$r(X'V^+X) = r(X'S). \tag{3}$$

Also, since $(S:T)$ is non-singular,

$$r(X) = r(X'S:X'T). \tag{4}$$

It follows that (1) and (2) are equivalent conditions.

Writing the model $(y, X\beta, \sigma^2 V)$ in its equivalent form

$$S'y = S'X\beta + u, \qquad \mathscr{E}u = 0, \qquad \mathscr{E}uu' = \sigma^2\Lambda, \tag{5}$$

$$T'y = T'X\beta, \tag{6}$$

and assuming that either (1) or (2) holds, we see that all conditions of Theorem 5 are satisfied. Thus we obtain Theorem 9.

Theorem 9

Consider the linear regression model $(y, X\beta, \sigma^2 V)$, where $y \in \mathscr{M}(X:V)$ a.s. Assume that $r(X'V^+X) = r(X)$. Then the best affine unbiased estimator $\widehat{W\beta}$ of $W\beta$

exists if, and only if, $\mathcal{M}(W') \subset \mathcal{M}(X')$, in which case

$$\widehat{W\beta} = W\beta^* + W(X'V^+X)^+ R_0'[R_0(X'V^+X)^+R_0']^+(r_0 - R_0\beta^*), \tag{7}$$

where

$$R_0 = T'X, \qquad r_0 = T'y, \qquad \beta^* = (X'V^+X)^+X'V^+y, \tag{8}$$

and T is a matrix of maximum rank such that $VT=0$. The variance matrix of $\widehat{W\beta}$ is

$$\begin{aligned}\mathcal{V}(\widehat{W\beta}) &= \sigma^2 W(X'V^+X)^+W' \\ &\quad - \sigma^2 W(X'V^+X)^+R_0'[R_0(X'V^+X)^+R_0']^+R_0(X'V^+X)^+W'.\end{aligned} \tag{9}$$

Exercises

1. $\mathcal{M}(X'V^+X) = \mathcal{M}(X')$ if, and only if, $r(X'V^+X) = r(X)$.
2. A necessary condition for $r(X'V^+X) = r(X)$ is that the rank of X does not exceed the rank of V. Show by means of a counter-example that this condition is not sufficient.
3. Show that $\mathcal{M}(X') = \mathcal{M}(X'S)$.

13. A SINGULAR VARIANCE MATRIX: THE GENERAL CASE, I

Let us now consider the general case of the linear regression model $(y, X\beta, \sigma^2 V)$, where X may not have full column rank and V may be singular.

Theorem 10

Consider the linear regression model $(y, X\beta, \sigma^2 V)$, where $y \in \mathcal{M}(X:V)$ a.s. The best affine unbiased estimator $\widehat{W\beta}$ of $W\beta$ exists if, and only if, $\mathcal{M}(W') \subset \mathcal{M}(X')$, in which case

$$\widehat{W\beta} = W\beta^* + WG^+R_0'(R_0G^+R_0')^+(r_0 - R_0\beta^*), \tag{1}$$

where

$$R_0 = T'X, \qquad r_0 = T'y, \qquad G = X'V^+X + R_0'R_0, \qquad \beta^* = G^+X'V^+y, \tag{2}$$

and T is a matrix of maximum rank such that $VT=0$. The variance matrix of $\widehat{W\beta}$ is

$$\mathcal{V}(\widehat{W\beta}) = \sigma^2 WG^+W' - \sigma^2 WG^+R_0'(R_0G^+R_0')^+R_0G^+W'. \tag{3}$$

Note. We give alternative expressions for (1) and (3) in Theorem 13.

Proof. Transform the model $(y, X\beta, \sigma^2 V)$ into the model $(S'y, S'X\beta, \sigma^2\Lambda)$, where β satisfies the consistent linear constraint $T'X\beta = T'y$, and S and T are defined in section 11. Then $|\Lambda| \neq 0$, and the result follows from Theorem 6. □

Exercises

1. Suppose that $\mathcal{M}(X'S)\cap\mathcal{M}(X'T)=\{0\}$ in the model $(y, X\beta, \sigma^2 V)$. Show that the best affine unbiased estimator of $AX\beta$ (which always exists) is ACy, where
$$C=SS'X(X'V^+X)^+X'V^+ + TT'XX'T(T'XX'T)^+T'.$$
[*Hint*: Use Theorem 7.]
2. Show that the variance matrix of this estimator is
$$\mathcal{V}(ACy)=\sigma^2 ASS'X(X'V^+X)^+X'SS'A'.$$

14. EXPLICIT AND IMPLICIT LINEAR CONSTRAINTS

Linear constraints on the parameter vector β arise in two ways. First, we may possess *a priori* knowledge that the parameters satisfy certain linear constraints
$$R\beta=r, \tag{1}$$
where the matrix R and vector r are known and non-stochastic. These are the *explicit* constraints.

Secondly, if the variance matrix $\sigma^2 V$ is singular, then β satisfies the linear constraints
$$T'X\beta=T'y \quad \text{a.s.}, \tag{2}$$
where T is a matrix of maximum column rank such that $VT=0$. These are the *implicit* constraints, due to the stochastic structure of the model. Implicit constraints exist whenever $T'X\neq 0$, that is, whenever $\mathcal{M}(X)\not\subset\mathcal{M}(V)$.

Let us combine the two sets of constraints (1) and (2) as follows:
$$R_0\beta=r_0 \quad \text{a.s.}, \qquad R_0=\begin{pmatrix}T'X\\ R\end{pmatrix}, \qquad r_0=\begin{pmatrix}T'y\\ r\end{pmatrix}. \tag{3}$$
We do not require that the matrix R_0 has full row rank; the constraints may thus be linearly dependent. We must require, however, that the model is consistent.

Proposition 5 (consistency of the linear model with constraints)

In order for the linear regression model $(y, X\beta, \sigma^2 V)$, where β satisfies the constraints $R\beta=r$, to be a consistent model it is necessary and sufficient that
$$\begin{pmatrix}y\\ r\end{pmatrix}\in\mathcal{M}\begin{pmatrix}X & V\\ R & 0\end{pmatrix} \quad \text{a.s.} \tag{4}$$

Proof. We write the model $(y, X\beta, \sigma^2 V)$ together with the constraints $R\beta=r$ as
$$\begin{pmatrix}y\\ r\end{pmatrix}=\begin{pmatrix}X\\ R\end{pmatrix}\beta+u, \tag{5}$$

where

$$\mathscr{E}u=0, \qquad \mathscr{E}uu'=\sigma^2\begin{pmatrix} V & 0 \\ 0 & 0 \end{pmatrix}. \tag{6}$$

Proposition 5 then follows from Proposition 4. □

The consistency condition (4) is equivalent (as, of course, it should be) to the requirement that (3) is a consistent equation, i.e.

$$r_0 \in \mathscr{M}(R_0). \tag{7}$$

Let us see why. If (7) holds, then there exists a vector c such that

$$T'y=T'Xc, \qquad r=Rc. \tag{8}$$

This implies that $T'(y-Xc)=0$ from which we solve

$$y-Xc=(I-TT')q, \tag{9}$$

where q is arbitrary. Further, since

$$I-TT'=SS'=S\Lambda S'S\Lambda^{-1}S'=VV^+, \tag{10}$$

we obtain

$$y=Xc+VV^+q, \qquad r=Rc, \tag{11}$$

and hence (4). It is easy to see that the converse is also true, that is, (4) implies (7).

The necessary consistency condition being established, let us now seek to find the best affine unbiased estimator of a parametric function $W\beta$ in the model $(y, X\beta, \sigma^2 V)$, where X may fail to have full column rank, V may be singular and explicit constraints $R\beta=r$ may be present.

We first prove a special case; the general result is discussed in the next section.

Theorem 11

Consider the linear regression model $(y, X\beta, \sigma^2 V)$, where β satisfies the consistent linear constraints $R\beta=r$, and

$$\begin{pmatrix} y \\ r \end{pmatrix} \in \mathscr{M}\begin{pmatrix} X & V \\ R & 0 \end{pmatrix} \quad \text{a.s.} \tag{12}$$

Assume that $r(X'V^+X)=r(X)$ and $\mathscr{M}(R')\subset\mathscr{M}(X')$. Then the best affine unbiased estimator $\widehat{W\beta}$ of $W\beta$ exists if, and only if, $\mathscr{M}(W')\subset\mathscr{M}(X')$, in which case

$$\widehat{W\beta}=W\beta^*+W(X'V^+X)^+R_0'[R_0(X'V^+X)^+R_0']^+(r_0-R_0\beta^*), \tag{13}$$

where

$$R_0=\begin{pmatrix} T'X \\ R \end{pmatrix}, \qquad r_0=\begin{pmatrix} T'y \\ r \end{pmatrix}, \qquad \beta^*=(X'V^+X)^+X'V^+y, \tag{14}$$

and T is a matrix of maximum rank such that $VT=0$. The variance matrix of $\widehat{W\beta}$ is

$$\mathscr{V}(\widehat{W\beta})=\sigma^2 W(X'V^+X)^+W' \\ -\sigma^2 W(X'V^+X)^+R_0'[R_0(X'V^+X)^+R_0']^+R_0(X'V^+X)^+W'. \tag{15}$$

Proof. We write the constrained model in its equivalent form

$$S'y=S'X\beta+u, \qquad \mathscr{E}u=0, \qquad \mathscr{E}uu'=\sigma^2\Lambda, \tag{16}$$

where β satisfies the combined implicit and explicit constraints

$$R_0\beta=r_0. \tag{17}$$

From section 12 we know that the three conditions

$$r(X'V^+X)=r(X), \tag{18}$$

$$\mathscr{M}(X'T)\subset\mathscr{M}(X'S) \tag{19}$$

and

$$\mathscr{M}(X'S)=\mathscr{M}(X') \tag{20}$$

are equivalent. Hence the two conditions $r(X'V^+X)=r(X)$ and $\mathscr{M}(R')\subset\mathscr{M}(X')$ are both satisfied if, and only if,

$$\mathscr{M}(R_0')\subset\mathscr{M}(X'S). \tag{21}$$

The result then follows from Theorem 5. □

15. THE GENERAL LINEAR MODEL, I

Now we consider the general linear model

$$(y, X\beta, \sigma^2 V), \tag{1}$$

where V is possibly singular, X may fail to have full column rank, and β satisfies certain *a priori* (explicit) constraints $R\beta=r$. As before, we transform the model into

$$(S'y, S'X\beta, \sigma^2\Lambda), \tag{2}$$

where Λ is a diagonal matrix with positive diagonal elements, and the parameter vector β satisfies

$$T'X\beta=T'y \qquad \text{(implicit constraints)} \tag{3}$$

and

$$R\beta=r \qquad \text{(explicit constraints)}, \tag{4}$$

which we combine as

$$R_0\beta=r_0, \qquad R_0=\begin{pmatrix}T'X\\R\end{pmatrix}, \qquad r_0=\begin{pmatrix}T'y\\r\end{pmatrix}. \tag{5}$$

The model is consistent (that is, the implicit and explicit linear constraints are consistent equations) if, and only if,

$$\begin{pmatrix} y \\ r \end{pmatrix} \in \mathcal{M}\begin{pmatrix} X & V \\ R & 0 \end{pmatrix} \quad \text{a.s.,} \tag{6}$$

according to Proposition 5.

We want to find the best affine unbiased estimator of a parametric function $W\beta$. According to Proposition 3, the class of affine unbiased estimators of $W\beta$ is not empty (that is, $W\beta$ is estimable) if, and only if,

$$\mathcal{M}(W') \subset \mathcal{M}(X':R'). \tag{7}$$

Notice that we can apply Proposition 3 to model (1) subject to the explicit constraints, or to model (2) subject to the explicit *and* implicit constraints; in either case we find (7).

A direct application of Theorem 6 now yields the following theorem.

Theorem 12

Consider the linear regression model $(y, X\beta, \sigma^2 V)$, where β satisfies the consistent linear constraints $R\beta = r$, and

$$\begin{pmatrix} y \\ r \end{pmatrix} \in \mathcal{M}\begin{pmatrix} X & V \\ R & 0 \end{pmatrix} \quad \text{a.s.} \tag{8}$$

The best affine unbiased estimator $\widehat{W\beta}$ of $W\beta$ exists if, and only if, $\mathcal{M}(W') \subset \mathcal{M}(X':R')$, in which case

$$\widehat{W\beta} = W\beta^* + WG^+R_0'(R_0G^+R_0')^+(r_0 - R_0\beta^*), \tag{9}$$

where

$$R_0 = \begin{pmatrix} T'X \\ R \end{pmatrix}, \qquad r_0 = \begin{pmatrix} T'y \\ r \end{pmatrix}, \tag{10}$$

$$G = X'V^+X + R_0'R_0, \qquad \beta^* = G^+X'V^+y, \tag{11}$$

and T is a matrix of maximum rank such that $VT = 0$. The variance matrix of $\widehat{W\beta}$ is

$$\mathcal{V}(\widehat{W\beta}) = \sigma^2 WG^+W' - \sigma^2 WG^+R_0'(R_0G^+R_0')^+R_0G^+W'. \tag{12}$$

Note. We give alternative expressions for (9) and (12) in Theorem 14.

16. A SINGULAR VARIANCE MATRIX: THE GENERAL CASE, II

We have now discussed every single case and combination of cases. Hence we could stop here. There is, however, an alternative route that is of interest, and leads to different (although equivalent) expressions for the estimators.

The route we have followed is this: first we considered the estimation of a parametric function $W\beta$ with explicit restrictions $R\beta=r$, assuming that V is non-singular; then we transformed the model with singular V into a model with non-singular variance matrix and explicit restrictions, thereby making the implicit restrictions (due to the singularity of V) explicit. Thus we have treated the singular model as a special case of the constrained model.

An alternative procedure is to reverse this route, and to look first at the model

$$(y, X\beta, \sigma^2 V) \tag{1}$$

where V is possibly singular (and X may not have full column rank). In the case of *a priori* constraints $R\beta=r$ we then consider

$$y_e=\begin{pmatrix} y \\ r \end{pmatrix}, \qquad X_e=\begin{pmatrix} X \\ R \end{pmatrix}. \tag{2}$$

in which case

$$\sigma^2 V_e=\mathscr{V}(y_e)=\sigma^2\begin{pmatrix} V & 0 \\ 0 & 0 \end{pmatrix} \tag{3}$$

so that the extended model can be written as

$$(y_e, X_e\beta, \sigma^2 V_e), \tag{4}$$

which is in the same form as (1). In this set-up the constrained model is a special case of the singular model.

Thus we consider the model $(y, X\beta, \sigma^2 V)$, where V is possibly singular, X may have linearly dependent columns, but no explicit constraints are given. We know, however, that the singularity of V implies certain constraints on β, which we have called implicit constraints,

$$T'X\beta=T'y, \tag{5}$$

where T is a matrix of maximum column rank such that $VT=0$. In the present approach, *the implicit constraints need not be taken into account* (they are automatically satisfied, see exercise 5), because we consider the whole V matrix and the constraints are embodied in V.

According to Proposition 2, the parametric function $W\beta$ is estimable if, and only if,

$$\mathscr{M}(W')\subset\mathscr{M}(X'). \tag{6}$$

According to Proposition 4, the model is consistent if, and only if,

$$y\in\mathscr{M}(X:V) \qquad \text{a.s.} \tag{7}$$

(Recall that (7) is equivalent to the requirement that the implicit constraint (5) is a consistent equation in β.) Let $Ay+c$ be the affine estimator of $W\beta$. The estimator is unbiased if, and only if,

$$AX\beta+c=W\beta \qquad \text{for all } \beta \text{ in } \mathbb{R}^k, \tag{8}$$

which implies

$$AX = W, \qquad c = 0. \tag{9}$$

Since the variance matrix of Ay is $\sigma^2 AVA'$, the affine minimum-trace unbiased estimator of $W\beta$ is found by solving the problem

$$\text{minimize} \qquad \operatorname{tr} AVA' \tag{10}$$

$$\text{subject to} \qquad AX = W. \tag{11}$$

Theorem 11.37 provides the solution

$$A^* = W(X'V_0^+X)^+X'V_0^+ + Q(I - V_0V_0^+), \tag{12}$$

where $V_0 = V + XX'$ and Q is arbitrary. Since $y \in \mathcal{M}(V_0)$ a.s., because of (7), it follows that

$$A^*y = W(X'V_0^+X)^+X'V_0^+y \tag{13}$$

is the unique affine minimum-trace unbiased estimator of $W\beta$. If, in addition, $\mathcal{M}(X) \subset \mathcal{M}(V)$, then A^*y simplifies to

$$A^*y = W(X'V^+X)^+X'V^+y. \tag{14}$$

Summarizing we have proved our next theorem.

Theorem 13

Consider the linear regression model $(y, X\beta, \sigma^2 V)$, where $y \in \mathcal{M}(X:V)$ a.s. The best affine unbiased estimator $\widehat{W\beta}$ of $W\beta$ exists if, and only if, $\mathcal{M}(W') \subset \mathcal{M}(X')$, in which case

$$\widehat{W\beta} = W(X'V_0^+X)^+X'V_0^+y, \tag{15}$$

where $V_0 = V + XX'$. Its variance matrix is

$$\mathcal{V}(\widehat{W\beta}) = \sigma^2 W[(X'V_0^+X)^+ - I]W'. \tag{16}$$

Moreover, if $\mathcal{M}(X) \subset \mathcal{M}(V)$, then the estimator simplifies to

$$\widehat{W\beta} = W(X'V^+X)^+X'V^+y \tag{17}$$

with variance matrix

$$\mathcal{V}(\widehat{W\beta}) = \sigma^2 W(X'V^+X)^+W'. \tag{18}$$

Note. Theorem 13 gives another (but equivalent) expression for the estimator of Theorem 10. The special case $\mathcal{M}(X) \subset \mathcal{M}(V)$ is identical to Theorem 8.

Exercises

1. Show that $V_0V_0^+X = X$.
2. Show that $X(X'V_0^+X)(X'V_0^+X)^+ = X$.

3. Let T be any matrix such that $VT=0$. Then

$$T'X(X'V_0^+X)=T'X=T'X(X'V_0^+X)^+.$$

4. Suppose that we replace the unbiasedness condition (8) by

$$AX\beta+c=W\beta \qquad \text{for all } \beta \text{ satisfying } T'X\beta=T'y.$$

Show that this yields the same constrained minimization problem (10) and (11) and hence the same estimator for $W\beta$.

5. Show that the best affine unbiased estimator of $T'X\beta$ is $T'y$ with $\mathcal{V}(T'y)=0$. Conclude that the implicit constraints $T'X\beta=T'y$ are automatically satisfied and need not be imposed.

17. THE GENERAL LINEAR MODEL, II

Let us look at the general linear model

$$(y, X\beta, \sigma^2 V), \tag{1}$$

where V is possibly singular, X may fail to have full column rank and β satisfies explicit *a priori* constraints $R\beta=r$. As discussed in the previous section, we write the constrained model as

$$(y_e, X_e\beta, \sigma^2 V_e), \tag{2}$$

where

$$y_e=\begin{pmatrix} y \\ r \end{pmatrix}, \qquad X_e=\begin{pmatrix} X \\ R \end{pmatrix}, \qquad V_e=\begin{pmatrix} V & 0 \\ 0 & 0 \end{pmatrix}. \tag{3}$$

Applying Theorem 13 to model (2) we obtain Theorem 14, which provides a different (though equivalent) expression for the estimator of Theorem 12.

Theorem 14

Consider the linear regression model $(y, X\beta, \sigma^2 V)$, where β satisfies the consistent linear constraints $R\beta=r$, and

$$\begin{pmatrix} y \\ r \end{pmatrix} \in \mathcal{M}\begin{pmatrix} X & V \\ R & 0 \end{pmatrix} \qquad \text{a.s.} \tag{4}$$

The best affine unbiased estimator $\widehat{W\beta}$ of $W\beta$ exists if, and only if, $\mathcal{M}(W') \subset \mathcal{M}(X':R')$, in which case

$$\widehat{W\beta}=W(X_e'V_0^+X_e)^+X_e'V_0^+y_e, \tag{5}$$

where y_e, X_e and V_e are defined in (3), and $V_0=V_e+X_eX_e'$. Its variance matrix is

$$\mathcal{V}(\widehat{W\beta})=\sigma^2 W[(X_e'V_0^+X_e)^+-I]W'. \tag{6}$$

18. GENERALIZED LEAST SQUARES

Consider the Gauss–Markov set-up $(y, X\beta, \sigma^2 I)$ where $r(X)=k$. In section 3 we obtained the best affine unbiased estimator of β, $\hat{\beta}=(X'X)^{-1}X'y$ (the Gauss–Markov estimator), by minimizing a quadratic form (the trace of the estimator's variance matrix) subject to a linear constraint (unbiasedness). In section 4 we showed that the Gauss–Markov estimator can also be obtained by minimizing $(y-X\beta)'(y-X\beta)$ over all β in $\mathbb{R}^k$. The fact that the principle of least squares (which is not a method of *estimation* but a method of *approximation*) produces best affine unbiased estimators is rather surprising and by no means trivial.

We now ask the question whether this relationship stands up against the introduction of more general assumptions such as $|V|=0$, or $r(X)<k$. The answer to this question is in the affirmative.

To see why, we recall from Theorem 11.35 that for a given positive semidefinite matrix M the problem

$$\text{minimize } (y-X\beta)'M(y-X\beta) \tag{1}$$

has a unique solution for $W\beta$ if, and only if,

$$\mathcal{M}(W') \subset \mathcal{M}(X'M^{1/2}), \tag{2}$$

in which case

$$W\beta^* = W(X'MX)^+X'My. \tag{3}$$

Choosing $M=(V+XX')^+$ and comparing with Theorem 13 yields the following.

Theorem 15

Consider the linear regression model $(y, X\beta, \sigma^2 V)$, where $y \in \mathcal{M}(X:V)$ a.s. Let W be a matrix such that $\mathcal{M}(W') \subset \mathcal{M}(X')$. Then the best affine unbiased estimator of $W\beta$ is $W\hat{\beta}$, where $\hat{\beta}$ minimizes

$$(y-X\beta)'(V+XX')^+(y-X\beta). \tag{4}$$

In fact we may, instead of (4), minimize the quadratic form

$$(y-X\beta)'(V+XEX')^+(y-X\beta), \tag{5}$$

where E is a positive semidefinite matrix such that $\mathcal{M}(X) \subset \mathcal{M}(V+XEX')$. The estimator $W\hat{\beta}$ will be independent of the actual value of E. For $E=I$ the requirement $\mathcal{M}(X) \subset \mathcal{M}(V+XX')$ is obviously satisfied; this leads to Theorem 15. If $\mathcal{M}(X) \subset \mathcal{M}(V)$, which includes the case of non-singular V, we can choose $E=0$ and minimize, instead of (4),

$$(y-X\beta)'V^+(y-X\beta). \tag{6}$$

In the case of *a priori* linear constraints, the following corollary applies.

Corollary 1

Consider the linear regression model $(y, X\beta, \sigma^2 V)$, where β satisfies the consistent linear constraints $R\beta = r$, and

$$\begin{pmatrix} y \\ r \end{pmatrix} \in \mathcal{M}\begin{pmatrix} X & V \\ R & 0 \end{pmatrix} \quad \text{a.s.} \tag{7}$$

Let W be a matrix such that $\mathcal{M}(W') \subset \mathcal{M}(X' : R')$. Then the best affine unbiased estimator of $W\beta$ is $W\hat{\beta}$, where $\hat{\beta}$ minimizes

$$\begin{pmatrix} y - X\beta \\ r - R\beta \end{pmatrix}' \begin{pmatrix} V + XX' & XR' \\ RX' & RR' \end{pmatrix}^{+} \begin{pmatrix} y - X\beta \\ r - R\beta \end{pmatrix}. \tag{8}$$

Proof. Define

$$y_e = \begin{pmatrix} y \\ r \end{pmatrix}, \qquad X_e = \begin{pmatrix} X \\ R \end{pmatrix}, \qquad V_e = \begin{pmatrix} V & 0 \\ 0 & 0 \end{pmatrix}, \tag{9}$$

and apply Theorem 15 to the extended model $(y_e, X_e\beta, \sigma^2 V_e)$. □

19. RESTRICTED LEAST SQUARES

Alternatively, we can use the method of restricted least squares.

Theorem 16

Consider the linear regression model $(y, X\beta, \sigma^2 V)$, where $|V| \neq 0$ and β satisfies the consistent linear constraints $R\beta = r$. Let W be a matrix such that $\mathcal{M}(W') \subset \mathcal{M}(X' : R')$. Then the best affine unbiased estimator of $W\beta$ is $W\hat{\beta}$, where $\hat{\beta}$ is a solution of the constrained minimization problem

$$\text{minimize} \qquad (y - X\beta)' V^{-1} (y - X\beta) \tag{1}$$

$$\text{subject to} \qquad R\beta = r. \tag{2}$$

Proof. From Theorem 11.36 we know that $(y - X\beta)' V^{-1} (y - X\beta)$ is minimized over all β satisfying $R\beta = r$, where β takes the value

$$\hat{\beta} = \beta^* + G^+ R'(RG^+R')^+(r - R\beta^*) + (I - G^+G)q, \tag{3}$$

where

$$G = X'V^{-1}X + R'R, \qquad \beta^* = G^+X'V^{-1}y \tag{4}$$

and q is arbitrary. Since $\mathcal{M}(W') \subset \mathcal{M}(X' : R') = \mathcal{M}(G)$, we obtain the unique expression

$$W\hat{\beta} = W\beta^* + WG^+R'(RG^+R')^+(r - R\beta^*) \tag{5}$$

which is identical to the best affine unbiased estimator of $W\beta$, see Theorem 6. □

The model where V is singular can be treated as a special case of the non-singular model with constraints.

Corollary 2

Consider the linear regression model $(y, X\beta, \sigma^2 V)$ where β satisfies the consistent linear constraints $R\beta = r$, and

$$\begin{pmatrix} y \\ r \end{pmatrix} \in \mathcal{M}\begin{pmatrix} X & V \\ R & 0 \end{pmatrix} \quad \text{a.s.} \tag{6}$$

Let W be a matrix such that $\mathcal{M}(W') \subset \mathcal{M}(X' : R')$. Then the best affine unbiased estimator of $W\beta$ is $W\hat{\beta}$, where $\hat{\beta}$ is a solution of the constrained minimization problem

$$\text{minimize} \quad (y - X\beta)' V^+ (y - X\beta) \tag{7}$$

$$\text{subject to} \quad (I - VV^+)X\beta = (I - VV^+)y \tag{8}$$

$$\text{and} \quad R\beta = r. \tag{9}$$

Proof. As in section 11 we introduce the orthogonal matrix $(S:T)$ which diagonalizes V:

$$(S:T)'V(S:T) = \begin{pmatrix} \Lambda & 0 \\ 0 & 0 \end{pmatrix}, \tag{10}$$

where Λ is a diagonal matrix containing the positive eigenvalues of V. Transforming the model $(y, X\beta, \sigma^2 V)$ by means of the matrix $(S:T)'$ yields the equivalent model $(S'y, S'X\beta, \sigma^2\Lambda)$, where β now satisfies the (implicit) constraints $T'X\beta = T'y$ in addition to the (explicit) constraints $R\beta = r$. Condition (6) shows that the combined constraints are consistent (section 14, Proposition 5). Applying Theorem 16 to the transformed model shows that the best affine unbiased estimator of $W\beta$ is $W\hat{\beta}$ where $\hat{\beta}$ is a solution of the constrained minimization problem

$$\text{minimize} \quad (S'y - S'X\beta)'\Lambda^{-1}(S'y - S'X\beta) \tag{11}$$

$$\text{subject to} \quad T'X\beta = T'y \tag{12}$$

$$\text{and} \quad R\beta = r. \tag{13}$$

It is easy to see that this constrained minimization problem is equivalent to the constrained minimization problem (7)–(9). □

Theorems 15 and 16 and their corollaries prove the striking and by no means trivial fact that the principle of (restricted) least squares provides best affine unbiased estimators.

Exercises

1. Show that the unconstrained problem

$$\text{minimize} \quad (y - X\beta)'(V + XX')^+(y - X\beta)$$

and the constrained problem

$$\begin{array}{ll} \text{minimize} & (y-X\beta)'V^{+}(y-X\beta) \\ \text{subject to} & (I-VV^{+})X\beta=(I-VV^{+})y \end{array}$$

have the same solution for β.

2. Show further that if $\mathcal{M}(X) \subset \mathcal{M}(V)$, both problems reduce to the unconstrained problem

$$\text{minimize} \quad (y-X\beta)'V^{+}(y-X\beta).$$

MISCELLANEOUS EXERCISES

Consider the model $(y, X\beta, \sigma^2 V)$. Define the mean squared error (MSE) matrix of an estimator $\hat{\beta}$ of β as $\text{MSE}(\hat{\beta})=\mathscr{E}(\hat{\beta}-\beta)(\hat{\beta}-\beta)'$.

1. If $\hat{\beta}$ is a linear estimator, say $\hat{\beta}=Ay$, show that

$$\text{MSE}(\hat{\beta})=(AX-I)\beta\beta'(AX-I)'+\sigma^2 AVA'.$$

2. Let $\phi(A)=\operatorname{tr}\text{MSE}(\hat{\beta})$ and consider the problem of minimizing ϕ with respect to A. Show that

$$\mathrm{d}\phi=2\operatorname{tr}(\mathrm{d}A)X\beta\beta'(AX-I)'+2\sigma^2\operatorname{tr}(\mathrm{d}A)VA',$$

and obtain the first-order condition

$$(\sigma^2 V+X\beta\beta'X')A'=X\beta\beta'.$$

3. Conclude that the matrix A which minimizes $\phi(A)$ is a function of the unknown parameter vector β, unless $\hat{\beta}$ is unbiased.
4. Show that

$$(\sigma^2 V+X\beta\beta'X')(\sigma^2 V+X\beta\beta'X')^{+}X\beta=X\beta$$

and conclude that the first-order condition is a consistent equation in A.

5. The matrices A which minimize $\phi(A)$ are then given by

$$A=\beta\beta'X'C^{+}+Q(I-CC^{+}),$$

where $C=\sigma^2 V+X\beta\beta'X'$ and Q is an arbitrary matrix.

6. Show that $CC^{+}V=V$, and hence that $(I-CC^{+})\varepsilon=0$ a.s.
7. Conclude from exercises 4 and 6 above that $(I-CC^{+})y=0$ a.s.
8. The 'estimator' which, in the class of linear estimators, minimizes the trace of the MSE matrix is therefore

$$\hat{\beta}=\lambda\beta$$

where

$$\lambda=\beta'X'(\sigma^2 V+X\beta\beta'X')^{+}y.$$

9. Let $\mu=\beta'X'(\sigma^2 V+X\beta\beta'X')^{+}X\beta$. Show that

$$\mathscr{E}\lambda=\mu, \qquad \mathscr{V}(\lambda)=\mu(1-\mu).$$

10. Show that $0 \leq \mu \leq 1$, so that $\hat{\beta}$ will in general 'underestimate' β.
11. Discuss the usefulness of the 'estimator' $\hat{\beta} = \lambda\beta$ in an iterative procedure.

BIBLIOGRAPHICAL NOTES

§1. The linear model is treated in every econometrics and most statistics books. See, e.g., Theil (1971) and Rao (1973). The theory originated with Gauss (1809) and Markov (1900).
§2. See Schönfeld (1971) for an alternative optimality criterion.
§3–§4. See Gauss (1809) and Markov (1900).
§5. See Aitken (1935).
§8. See also Rao (1945).
§11. See Zyskind and Martin (1969) and Albert (1973).
§14. The special case of Theorem 11 where X has full column rank and R has full row rank was considered by Kreijger and Neudecker (1977).
§16–§17. Theorems 13 and 14 are due to Rao (1971a, 1973) in the context of a unified theory of linear estimation.

CHAPTER 14

Further topics in the linear model

1. INTRODUCTION

In the preceding chapter we derived the 'best' affine unbiased estimator of β in the linear regression model $(y, X\beta, \sigma^2 V)$ under various assumptions about the ranks of X and V. In this chapter we discuss some other topics relating to the linear model.

Sections 2–7 are devoted to constructing the 'best' quadratic estimator of σ^2. The multivariate analogue is discussed in section 8. The estimator

$$\hat{\sigma}^2 = \frac{1}{n-k} y'(I - XX^+)y, \tag{1}$$

known as the least squares estimator of σ^2, is the best quadratic unbiased estimator in the model $(y, X\beta, \sigma^2 I)$. But if $\mathscr{V}(y) \neq \sigma^2 I_n$, then $\hat{\sigma}^2$ in (1) will, in general, be biased. Bounds for this bias which do not depend on X are obtained in sections 9 and 10.

The statistical analysis of the disturbances $\varepsilon = y - X\beta$ is taken up in sections 11–14, where predictors that are best linear unbiased with scalar variance matrix (BLUS) and with fixed variance matrix (BLUF) are derived.

Finally, we show how matrix differential calculus can be useful in sensitivity analysis. In particular, we study the sensitivities of the posterior moments of β in a Bayesian framework.

2. BEST QUADRATIC UNBIASED ESTIMATION OF σ^2

Let $(y, X\beta, \sigma^2 V)$ be the linear regression model. In the previous chapter we considered the estimation of β as a *linear* function of the observation vector y. Since the variance σ^2 is a *quadratic* concept, we now consider the estimation of σ^2

as a quadratic function of y, that is, a function of the form

$$y'Ay \tag{1}$$

where A is non-stochastic and symmetric. Any estimator satisfying (1) is called a *quadratic estimator.*

If, in addition, the matrix A is positive (semi)definite and $AV \neq 0$, and if y is a *continuous* random vector, then

$$\Pr(y'Ay > 0) = 1, \tag{2}$$

and we say that the estimator is *quadratic and positive* (almost surely).

An unbiased estimator of σ^2 is an estimator, say $\hat{\sigma}^2$, such that

$$\mathscr{E}\hat{\sigma}^2 = \sigma^2 \qquad \text{for all } \beta \in \mathbb{R}^k \text{ and } \sigma^2 > 0. \tag{3}$$

In (3) it is implicitly assumed that β and σ^2 are not restricted (for example, by $R\beta = r$) apart from the requirement that σ^2 is positive.

We now propose the following definition.

Definition 1

The *best quadratic (and positive) unbiased estimator* of σ^2 in the linear regression model $(y, X\beta, \sigma^2 I_n)$ is a quadratic (and positive) unbiased estimator of σ^2, say $\hat{\sigma}^2$, such that

$$\mathscr{V}(\hat{t}^2) \geq \mathscr{V}(\hat{\sigma}^2) \tag{4}$$

for all quadratic (and positive) unbiased estimators $\hat{t}^2$ of σ^2.

In the following two sections we shall derive the best quadratic unbiased estimator of σ^2 for the *normal* linear regression model where

$$y \sim \mathscr{N}(X\beta, \sigma^2 V), \tag{5}$$

first requiring that the estimator is positive, then dropping this requirement.

3. THE BEST QUADRATIC AND POSITIVE UNBIASED ESTIMATOR OF σ^2

Our first result is the following well known theorem.

Theorem 1

The best quadratic and positive unbiased estimator of σ^2 in the normal linear regression model $(y, X\beta, \sigma^2 I_n)$ is

$$\hat{\sigma}^2 = \frac{1}{n-r} y'(I - XX^+)y \tag{1}$$

where r denotes the rank of X.

Proof. We consider a quadratic estimator $y'Ay$. To ensure that the estimator is positive we write $A=C'C$. The problem is to determine an $n\times n$ matrix C such that $y'C'Cy$ is unbiased and has the smallest variance in the class of unbiased estimators.

Unbiasedness requires

$$\mathscr{E}y'C'Cy=\sigma^2 \qquad \text{for all } \beta \text{ and } \sigma^2, \tag{2}$$

that is,

$$\beta'X'C'CX\beta+\sigma^2 \operatorname{tr} C'C=\sigma^2 \qquad \text{for all } \beta \text{ and } \sigma^2. \tag{3}$$

This leads to the conditions

$$CX=0, \qquad \operatorname{tr} C'C=1. \tag{4}$$

Given the condition $CX=0$ we can write

$$y'C'Cy=\varepsilon'C'C\varepsilon \tag{5}$$

where $\varepsilon\sim\mathscr{N}(0,\sigma^2 I_n)$, and hence, by Theorem 12.12,

$$\mathscr{V}(y'C'Cy)=2\sigma^4 \operatorname{tr}(C'C)^2. \tag{6}$$

Our optimization problem thus becomes

$$\text{minimize} \qquad \operatorname{tr}(C'C)^2 \tag{7}$$

$$\text{subject to} \qquad CX=0 \quad \text{and} \quad \operatorname{tr} C'C=1. \tag{8}$$

To solve (7) and (8) we form the Lagrangian function

$$\psi(C)=\tfrac{1}{4}\operatorname{tr}(C'C)^2-\tfrac{1}{2}\lambda(\operatorname{tr} C'C-1)-\operatorname{tr} L'CX \tag{9}$$

where λ is a Lagrange multiplier and L is a matrix of Lagrange multipliers. Differentiating ψ gives

$$\begin{aligned} \mathrm{d}\psi &= \tfrac{1}{2}\operatorname{tr} CC'C(\mathrm{d}C)'+\tfrac{1}{2}\operatorname{tr} C'CC'(\mathrm{d}C)-\tfrac{1}{2}\lambda(\operatorname{tr}(\mathrm{d}C)'C+\operatorname{tr} C'\mathrm{d}C) \\ &\quad -\operatorname{tr} L'(\mathrm{d}C)X \\ &= \operatorname{tr} C'CC'\mathrm{d}C-\lambda\operatorname{tr} C'\mathrm{d}C-\operatorname{tr} XL'\mathrm{d}C, \end{aligned} \tag{10}$$

so that we obtain as our first-order conditions

$$C'CC'=\lambda C'+XL' \tag{11}$$

$$\operatorname{tr} C'C=1 \tag{12}$$

$$CX=0. \tag{13}$$

Pre-multiplying (11) with XX^+ and using (13) gives

$$XL'=0. \tag{14}$$

Inserting (14) in (11) gives

$$C'CC'=\lambda C'. \tag{15}$$

Condition (15) implies that $\lambda>0$. Also, defining

$$B=(1/\lambda)C'C \tag{16}$$

we obtain from (12), (13) and (15)

$$B^2=B \tag{17}$$

$$\operatorname{tr} B=1/\lambda \tag{18}$$

$$BX=0. \tag{19}$$

Hence B is an idempotent symmetric matrix. Now, since by (12) and (15)

$$\operatorname{tr}(C'C)^2=\lambda,$$

it appears that we must choose λ as small as possible, that is, we must choose the rank of B as large as possible. The only constraint on the rank of B is (19), which implies

$$r(B)\leqslant n-r \tag{20}$$

where r is the rank of X. Since we want to maximize $r(B)$ we take

$$1/\lambda=r(B)=n-r. \tag{21}$$

From (17), (19) and (21) we find, using Theorem 2.9,

$$B=I_n-XX^+ \tag{22}$$

and hence

$$A=C'C=\lambda B=\frac{1}{n-r}(I_n-XX^+). \tag{23}$$

□

4. THE BEST QUADRATIC UNBIASED ESTIMATOR OF σ^2

The estimator obtained in the preceding section is, in fact, the best in a wider class of estimators: the class of quadratic unbiased estimators. In other words, the constraint that $\hat{\sigma}^2$ be positive is not binding. Thus we obtain the following generalization of Theorem 1.

Theorem 2

The best quadratic unbiased estimator of σ^2 in the normal linear regression model $(y, X\beta, \sigma^2 I_n)$ is

$$\hat{\sigma}^2=\frac{1}{n-r}y'(I-XX^+)y \tag{1}$$

where r denotes the rank of X.

Proof. Let $\hat{\sigma}^2=y'Ay$ be the quadratic estimator of σ^2, and let $\varepsilon\sim\mathcal{N}(0,\sigma^2 I_n)$.

Then

$$\hat{\sigma}^2 = \beta' X' A X \beta + 2\beta' X' A \varepsilon + \varepsilon' A \varepsilon \tag{2}$$

so that $\hat{\sigma}^2$ is an unbiased estimator of σ^2 for all β and σ^2 if and only if

$$X'AX = 0 \qquad \text{and} \qquad \operatorname{tr} A = 1. \tag{3}$$

The variance of $\hat{\sigma}^2$ is

$$\mathscr{V}(\hat{\sigma}^2) = 2\sigma^4(\operatorname{tr} A^2 + 2\gamma' X' A^2 X \gamma) \tag{4}$$

where $\gamma = \beta/\sigma$. Hence the optimization problem is

$$\begin{aligned} &\text{minimize} && \operatorname{tr} A^2 + 2\gamma' X' A^2 X \gamma && (5)\\ &\text{subject to} && X'AX = 0 \quad \text{and} \quad \operatorname{tr} A = 1. && (6) \end{aligned}$$

We notice that the function in (5) to be minimized depends on γ so that we would expect the optimal value of A to depend on γ as well. This, however, turns out not to be the case. We form the Lagrangian (taking into account the symmetry of A, see section 3.8)

$$\psi(\mathrm{v}(A)) = \tfrac{1}{2} \operatorname{tr} A^2 + \gamma' X' A^2 X \gamma - \lambda(\operatorname{tr} A - 1) - \operatorname{tr} L' X' A X, \tag{7}$$

where λ is a Lagrange multiplier and L is a matrix of Lagrange multipliers. Since the constraint function $X'AX$ is symmetric, we may take L to be symmetric too.

Differentiating ψ gives

$$\begin{aligned} \mathrm{d}\psi &= \operatorname{tr} A\mathrm{d}A + 2\gamma' X' A(\mathrm{d}A) X \gamma - \lambda \operatorname{tr} \mathrm{d}A - \operatorname{tr} LX'(\mathrm{d}A)X \\ &= \operatorname{tr}(A + X\gamma\gamma' X' A + AX\gamma\gamma' X' - \lambda I - XLX')\mathrm{d}A, \end{aligned} \tag{8}$$

so that the first-order conditions are

$$A - \lambda I_n + AX\gamma\gamma' X' + X\gamma\gamma' X' A = XLX' \tag{9}$$

$$X'AX = 0 \tag{10}$$

$$\operatorname{tr} A = 1. \tag{11}$$

Pre- and post-multiplying (9) with XX^+ gives, in view of (10),

$$-\lambda XX^+ = XLX'. \tag{12}$$

Inserting (12) in (9) we obtain

$$A = \lambda M - P \tag{13}$$

where

$$M = I_n - XX^+ \qquad \text{and} \qquad P = AX\gamma\gamma' X' + X\gamma\gamma' X' A. \tag{14}$$

Since $\operatorname{tr} P = 0$, because of (10), we have

$$\operatorname{tr} A = \lambda \operatorname{tr} M \tag{15}$$

and hence

$$\lambda = 1/(n - r). \tag{16}$$

Also, since

$$MP + PM = P, \tag{17}$$

we obtain

$$A^2 = \lambda^2 M + P^2 - \lambda P \tag{18}$$

so that

$$\begin{aligned}\operatorname{tr} A^2 &= \lambda^2 \operatorname{tr} M + \operatorname{tr} P^2 \\ &= 1/(n-r) + 2(\gamma' X' X\gamma)(\gamma' X' A^2 X\gamma).\end{aligned} \tag{19}$$

The objective function (5) can now be written as

$$\operatorname{tr} A^2 + 2\gamma' X' A^2 X\gamma = 1/(n-r) + 2(\gamma' X' A^2 X\gamma)(1 + \gamma' X' X\gamma) \tag{20}$$

which is minimized for $AX\gamma = 0$, that is, for $P = 0$. Inserting $P = 0$ in (13) and using (16) gives

$$A = \frac{1}{n-r} M. \tag{21}$$

□

5. BEST QUADRATIC INVARIANT ESTIMATION OF σ^2

Unbiasedness, though a useful property for linear estimators in linear models, is somewhat suspect for nonlinear estimators. Another, perhaps more useful, criterion is *invariance*. In the context of the linear regression model

$$y = X\beta + \varepsilon, \tag{1}$$

let us consider, instead of β, a translation $\beta - \beta_0$. Then (1) is equivalent to

$$y - X\beta_0 = X(\beta - \beta_0) + \varepsilon, \tag{2}$$

and we say that a quadratic estimator $y'Ay$ is *invariant under translation of* β if

$$(y - X\beta_0)' A (y - X\beta_0) = y'Ay \qquad \text{for all } \beta_0. \tag{3}$$

This, clearly, is the case if and only if

$$AX = 0. \tag{4}$$

We can obtain (4) in another, though closely related, way if we assume that the disturbance vector ε is normally distributed, $\varepsilon \sim \mathcal{N}(0, \sigma^2 V)$, V positive definite. Then, by Theorem 12.12,

$$\mathscr{E}(y'Ay) = \beta' X' AX\beta + \sigma^2 \operatorname{tr} AV \tag{5}$$

and

$$\mathscr{V}(y'Ay) = 4\sigma^2 \beta' X' AVAX\beta + 2\sigma^4 \operatorname{tr}(AV)^2, \tag{6}$$

so that, under normality, the distribution of $y'Ay$ is independent of β if and only if $AX = 0$.

If the estimator is biased we replace the minimum variance criterion by the minimum mean squared error criterion. Thus we obtain Definition 2.

Definition 2

The *best quadratic (and positive) invariant estimator* of σ^2 in the linear regression model $(y, X\beta, \sigma^2 I_n)$ is a quadratic (and positive) estimator of σ^2, say $\hat{\sigma}^2$, which is invariant under translation of β, such that

$$\mathscr{E}(\hat{\tau}^2-\sigma^2)^2 \geqslant \mathscr{E}(\hat{\sigma}^2-\sigma^2)^2 \tag{7}$$

for all quadratic (and positive) invariant estimators $\hat{\tau}^2$ of σ^2.

In sections 6 and 7 we shall derive the best quadratic invariant estimator of σ^2, assuming normality, first requiring that $\hat{\sigma}^2$ is positive, then that $\hat{\sigma}^2$ is merely quadratic.

6. THE BEST QUADRATIC AND POSITIVE INVARIANT ESTIMATOR OF σ^2

Given invariance instead of unbiasedness we obtain Theorem 3 instead of Theorem 1.

Theorem 3

The best quadratic and positive invariant estimator of σ^2 in the normal linear regression model $(y, X\beta, \sigma^2 I_n)$ is

$$\hat{\sigma}^2 = \frac{1}{n-r+2} y'(I-XX^+)y \tag{1}$$

where r denotes the rank of X.

Proof. Again, let $\hat{\sigma}^2 = y'Ay$ be the quadratic estimator of σ^2 and write $A = C'C$. Invariance requires $C'CX = 0$, that is,

$$CX = 0. \tag{2}$$

Letting $\varepsilon \sim \mathscr{N}(0, \sigma^2 I_n)$, the estimator for σ^2 can then be written as

$$\hat{\sigma}^2 = \varepsilon' C' C \varepsilon \tag{3}$$

so that the mean squared error becomes

$$\mathscr{E}(\hat{\sigma}^2-\sigma^2)^2 = \sigma^4(1-\operatorname{tr} C'C)^2 + 2\sigma^4 \operatorname{tr}(C'C)^2. \tag{4}$$

The minimization problem is thus

$$\text{minimize} \quad (1-\operatorname{tr} C'C)^2 + 2\operatorname{tr}(C'C)^2 \tag{5}$$

$$\text{subject to} \quad CX = 0. \tag{6}$$

The Lagrangian is

$$\psi(C) = \tfrac{1}{4}(1-\operatorname{tr} C'C)^2 + \tfrac{1}{2}\operatorname{tr}(C'C)^2 - \operatorname{tr} L'CX, \tag{7}$$

where L is a matrix of Lagrange multipliers, leading to the first-order conditions

$$2C'CC' - (1 - \operatorname{tr} C'C)C' = XL' \tag{8}$$

$$CX = 0. \tag{9}$$

Pre-multiplying both sides of (8) with XX^+ gives, in view of (9),

$$XL' = 0. \tag{10}$$

Inserting (10) in (8) gives

$$2C'CC' = (1 - \operatorname{tr} C'C)C'. \tag{11}$$

Now define

$$B = \left(\frac{2}{1 - \operatorname{tr} C'C}\right)C'C = \left(\frac{2}{1 - \operatorname{tr} A}\right)A. \tag{12}$$

(Notice that $\operatorname{tr} C'C \neq 1$. (Why?)) Then, from (9) and (11),

$$B^2 = B \tag{13}$$

$$BX = 0. \tag{14}$$

We also obtain from (12),

$$\operatorname{tr} A = \frac{\operatorname{tr} B}{2 + \operatorname{tr} B}, \qquad \operatorname{tr} A^2 = \frac{\operatorname{tr} B^2}{(2 + \operatorname{tr} B)^2}. \tag{15}$$

Let ρ denote the rank of B. Then $\operatorname{tr} B = \operatorname{tr} B^2 = \rho$ and hence

$$\tfrac{1}{4}(1 - \operatorname{tr} A)^2 + \tfrac{1}{2} \operatorname{tr} A^2 = \frac{1}{2(2 + \rho)}. \tag{16}$$

The left-hand side of (16) is the function we wish to minimize; thus we must choose ρ as large as possible. In view of (14) therefore

$$\rho = n - r. \tag{17}$$

From (13), (14) and (17) we find, using Theorem 2.9,

$$B = I_n - XX^+ \tag{18}$$

and hence

$$A = \left(\frac{1}{2 + \operatorname{tr} B}\right)B = \frac{1}{n - r + 2}(I_n - XX^+). \tag{19}$$

□

7. THE BEST QUADRATIC INVARIANT ESTIMATOR OF σ^2

A generalization of Theorem 2 is obtained by dropping the requirement that the quadratic estimator of σ^2 be positive. In this wider class of estimators we find that the estimator of Theorem 3 is again the best (smallest mean squared error), thus showing that the requirement of positiveness is not binding.

Comparing Theorems 2 and 4 we see that the best quadratic invariant estimator has a larger bias (it underestimates σ^2) but a smaller variance than the best quadratic unbiased estimator, and altogether a smaller mean squared error.

Theorem 4

The best quadratic invariant estimator of σ^2 in the normal linear regression model $(y, X\beta, \sigma^2 I_n)$ is

$$\hat{\sigma}^2 = \frac{1}{n-r+2} y'(I - XX^+)y \tag{1}$$

where r denotes the rank of X.

Proof. Here we must solve the following problem:

$$\text{minimize} \quad (1 - \operatorname{tr} A)^2 + 2 \operatorname{tr} A^2 \tag{2}$$

$$\text{subject to} \quad AX = 0. \tag{3}$$

This is the same as in the proof of Theorem 3, except that A is now symmetric and not necessarily positive definite. The Lagrangian is

$$\psi(\mathrm{v}(A)) = \tfrac{1}{2}(1 - \operatorname{tr} A)^2 + \operatorname{tr} A^2 - \operatorname{tr} L'AX \tag{4}$$

and the first-order conditions are

$$2A - (1 - \operatorname{tr} A)I_n = (XL' + LX')/2 \tag{5}$$

$$AX = 0. \tag{6}$$

Pre-multiplying (5) with A gives, in view of (6),

$$2A^2 - (1 - \operatorname{tr} A)A = \tfrac{1}{2} ALX'. \tag{7}$$

Post-multiplying (7) with XX^+ gives, again using (6), $ALX' = 0$. Inserting $ALX' = 0$ in (7) then shows that the matrix

$$B = \left[\frac{2}{1 - \operatorname{tr} A}\right] A \tag{8}$$

is symmetric idempotent. Furthermore, by (6), $BX = 0$. The remainder of the proof follows in the same way as in the proof of Theorem 3 (from (15) onwards). □

8. BEST QUADRATIC UNBIASED ESTIMATION IN THE MULTIVARIATE NORMAL CASE

Extending Definition 1 to the multivariate case we obtain Definition 3.

Definition 3

Let $y_1, y_2, \ldots, y_n$ be a random sample from an m-dimensional distribution with positive definite variance matrix Ω. Let $Y = (y_1, y_2, \ldots, y_n)'$. The *best quadratic unbiased estimator* of Ω, say $\hat{\Omega}$, is a quadratic estimator (that is, an estimator of the form $Y'AY$ where A is symmetric) such that $\hat{\Omega}$ is unbiased and

$$\mathscr{V}(\operatorname{vec}\hat{\Psi})-\mathscr{V}(\operatorname{vec}\hat{\Omega}) \tag{1}$$

is positive semidefinite for all quadratic unbiased estimators $\hat{\Psi}$ of Ω.

We can now generalize Theorem 2 to the multivariate case. We see again that the estimator is positive semidefinite, even though this was not required.

Theorem 5

Let $y_1, y_2, \ldots, y_n$ be a random sample from the m-dimensional normal distribution with mean μ and positive definite variance matrix Ω. Let $Y=(y_1, y_2, \ldots, y_n)'$. The best quadratic unbiased estimator of Ω is

$$\hat{\Omega}=\frac{1}{n-1}Y'\left(I_n-\frac{1}{n}\imath\imath'\right)Y \tag{2}$$

where $\imath=(1, 1, \ldots, 1)'$.

Proof. Consider a quadratic estimator $Y'AY$. From Chapter 12 (miscellaneous exercise 2) we know that

$$\mathscr{E}Y'AY=(\operatorname{tr}A)\Omega+(\imath'A\imath)\mu\mu' \tag{3}$$

and

$$\begin{aligned}\mathscr{V}(\operatorname{vec}Y'AY)&=(I+K_{mm})[(\operatorname{tr}A^2)(\Omega\otimes\Omega)+(\imath'A^2\imath)(\Omega\otimes\mu\mu'+\mu\mu'\otimes\Omega)]\\&=(I+K_{mm})[\tfrac{1}{2}(\operatorname{tr}A^2)(\Omega\otimes\Omega)+(\imath'A^2\imath)(\Omega\otimes\mu\mu')](I+K_{mm}).\end{aligned} \tag{4}$$

The estimator $Y'AY$ is unbiased if and only if

$$(\operatorname{tr}A)\Omega+(\imath'A\imath)\mu\mu'=\Omega \qquad \text{for all } \mu \text{ and } \Omega, \tag{5}$$

that is,

$$\operatorname{tr}A=1, \qquad \imath'A\imath=0. \tag{6}$$

Let $T\neq 0$ be an arbitrary $m\times m$ matrix and let $\tilde{T}=T+T'$. Then

$$(\operatorname{vec}T)'[\mathscr{V}(\operatorname{vec}Y'AY)]\operatorname{vec}T$$

As the objective function (9) is strictly convex, this solution provides the required minimum. □

9. BOUNDS FOR THE BIAS OF THE LEAST SQUARES ESTIMATOR OF σ^2, I

Let us again consider the linear regression model $(y, X\beta, \sigma^2V)$ where X has full column rank k and V is positive semidefinite.

where α and β are fixed numbers. If the optimal matrix A which minimizes (9) subject to (10) does not depend on α and β—and this will turn out to be the case—then this matrix A must be the best quadratic unbiased estimator according to Definition 3.

Define the Lagrangian function

$$\psi(A)=\alpha \operatorname{tr} A^2+\beta\imath'A^2\imath-\lambda_1(\operatorname{tr} A-1)-\lambda_2\imath'A\imath \tag{11}$$

where λ_1 and λ_2 are Lagrange multipliers. Differentiating ψ gives

$$\begin{aligned} \mathrm{d}\psi &= 2\alpha \operatorname{tr} A\mathrm{d}A+2\beta\imath'A(\mathrm{d}A)\imath-\lambda_1 \operatorname{tr} \mathrm{d}A-\lambda_2\imath'(\mathrm{d}A)\imath \\ &= \operatorname{tr}[2\alpha A+\beta(\imath\imath'A+A\imath\imath')-\lambda_1 I-\lambda_2\imath\imath']\mathrm{d}A. \end{aligned} \tag{12}$$

Since the matrix in square brackets in (12) is symmetric, we do not have to impose the symmetry condition on A. Thus we find as our first-order conditions:

$$2\alpha A+\beta(\imath\imath'A+A\imath\imath')-\lambda_1 I_n-\lambda_2\imath\imath'=0 \tag{13}$$

$$\operatorname{tr} A=1 \tag{14}$$

$$\imath'A\imath=0. \tag{15}$$

Taking the trace in (13), yields

$$2\alpha=n(\lambda_1+\lambda_2). \tag{16}$$

Pre- and post-multiplying (13) with $\imath$ gives

$$\lambda_1+n\lambda_2=0. \tag{17}$$

Hence

$$\lambda_1=\frac{2\alpha}{n-1}, \qquad \lambda_2=\frac{-2\alpha}{n(n-1)}. \tag{18}$$

Post-multiplying (13) with $\imath$ gives, in view of (17),

$$(2\alpha+n\beta)A\imath=0. \tag{19}$$

Since $\alpha>0$ (Why?) and $\beta\geqslant 0$ we obtain

$$A\imath=0 \tag{20}$$

and hence

$$A=\frac{1}{n-1}\left(I_n-\frac{1}{n}\imath\imath'\right). \tag{21}$$

As the objective function (9) is strictly convex, this solution provides the required minimum. □

9. BOUNDS FOR THE BIAS OF THE LEAST SQUARES ESTIMATOR OF σ^2, I

Let us again consider the linear regression model $(y, X\beta, \sigma^2 V)$ where X has full column rank k and V is positive semidefinite.

If $V=I_n$, then we know from Theorem 2 that

$$\hat{\sigma}^2=\frac{1}{n-k}y'[I_n-X(X'X)^{-1}X']y \tag{1}$$

is the best quadratic unbiased estimator of σ^2, also known as the least squares (LS) estimator of σ^2. If $V\neq I_n$, then (1) is no longer an unbiased estimator of σ^2, because, in general,

$$\mathscr{E}\hat{\sigma}^2=\frac{\sigma^2}{n-k}\operatorname{tr}[I_n-X(X'X)^{-1}X']V\neq\sigma^2. \tag{2}$$

If both V and X are known, we can calculate the relative bias

$$\frac{\mathscr{E}\hat{\sigma}^2-\sigma^2}{\sigma^2} \tag{3}$$

exactly. Here we are concerned with the case where V is known (at least in structure, say first-order autocorrelation) while X is not known. Of course we cannot calculate the exact relative bias in this case. We can, however, find a lower and an upper bound for the relative bias of $\hat{\sigma}^2$ over all possible values of X.

Theorem 6

Consider the linear regression model $(y, X\beta, \sigma^2V)$, where V is a positive semidefinite $n\times n$ matrix with eigenvalues $\lambda_1\leq\lambda_2\leq\ldots\leq\lambda_n$, and X is a non-stochastic $n\times k$ matrix of rank k. Let $\hat{\sigma}^2$ be the least squares estimator of σ^2, that is

$$\hat{\sigma}^2=\frac{1}{n-k}y'[I_n-X(X'X)^{-1}X']y. \tag{4}$$

Then

$$\sum_{i=1}^{n-k}\lambda_i\leq\frac{(n-k)\mathscr{E}\hat{\sigma}^2}{\sigma^2}\leq\sum_{i=k+1}^{n}\lambda_i. \tag{5}$$

Proof. Let $M=I-X(X'X)^{-1}X'$. Then

$$\mathscr{E}\hat{\sigma}^2=\frac{\sigma^2}{n-k}\operatorname{tr}MV=\frac{\sigma^2}{n-k}\operatorname{tr}MVM. \tag{6}$$

Now, M is an idempotent symmetric $n\times n$ matrix of rank $n-k$. Let us denote the eigenvalues of MVM, apart from k zeros, by

$$\mu_1\leq\mu_2\leq\ldots\leq\mu_{n-k}. \tag{7}$$

Then, by Theorem 11.11,

$$\lambda_i\leq\mu_i\leq\lambda_{k+i}\qquad(i=1,2,\ldots,n-k) \tag{8}$$

and hence

$$\sum_{i=1}^{n-k} \lambda_i \leqslant \sum_{i=1}^{n-k} \mu_i \leqslant \sum_{i=1}^{n-k} \lambda_{k+i} \tag{9}$$

and the result follows. □

10. BOUNDS FOR THE BIAS OF THE LEAST SQUARES ESTIMATOR OF σ^2, II

Suppose now that X is not completely unknown. In particular, suppose that the regression contains a constant term, so that X contains a column of ones. Surely this additional information must lead to a tighter interval for the relative bias of $\hat{\sigma}^2$. Theorem 7 shows that this is indeed the case. Somewhat surprisingly perhaps only the upper bound of the relative bias is affected, not the lower bound.

Theorem 7

Consider the linear regression model $(y, X\beta, \sigma^2 V)$, where V is a positive semidefinite $n \times n$ matrix with eigenvalues $\lambda_1 \leqslant \lambda_2 \leqslant \ldots \leqslant \lambda_n$, and X is a non-stochastic $n \times k$ matrix of rank k. Assume that X contains a column $\imath = (1, 1, \ldots, 1)'$. Let $A = I_n - (1/n)\imath\imath'$ and let $0 = \lambda_1^* \leqslant \lambda_2^* \leqslant \ldots \leqslant \lambda_n^*$ be the eigenvalues of AVA. Let $\hat{\sigma}^2$ be the least squares estimator σ^2, that is

$$\hat{\sigma}^2 = \frac{1}{n-k} y'[I_n - X(X'X)^{-1}X']y. \tag{1}$$

Then

$$\sum_{i=1}^{n-k} \lambda_i \leqslant \frac{(n-k)\mathscr{E}\hat{\sigma}^2}{\sigma^2} \leqslant \sum_{i=k+1}^{n} \lambda_i^*. \tag{2}$$

Proof. Let $M = I_n - X(X'X)^{-1}X'$. Since $MA = M$ we have $MVM = MAVAM$ and hence

$$\mathscr{E}\hat{\sigma}^2 = \frac{\sigma^2}{n-k} \operatorname{tr} MVM = \frac{\sigma^2}{n-k} \operatorname{tr} MAVAM. \tag{3}$$

We obtain, just as in the proof of Theorem 6,

$$\sum_{i=2}^{n-k} \lambda_i^* \leqslant \operatorname{tr} MAVAM \leqslant \sum_{i=k+1}^{n} \lambda_i^*. \tag{4}$$

We also have, by Theorem 6,

$$\sum_{i=1}^{n-k} \lambda_i \leqslant \operatorname{tr} MAVAM \leqslant \sum_{i=k+1}^{n} \lambda_i. \tag{5}$$

In order to select the smallest upper bound and largest lower bound we use the inequality

$$\lambda_i \leqslant \lambda^*_{i+1} \leqslant \lambda_{i+1} \qquad (i=1, \ldots, n-1), \tag{6}$$

which follows from Theorem 11.11. We then find

$$\sum_{i=2}^{n-k} \lambda^*_i \leqslant \sum_{i=2}^{n-k} \lambda_i \leqslant \sum_{i=1}^{n-k} \lambda_i \tag{7}$$

and

$$\sum_{i=k+1}^{n} \lambda^*_i \leqslant \sum_{i=k+1}^{n} \lambda_i, \tag{8}$$

so that

$$\sum_{i=1}^{n-k} \lambda_i \leqslant \operatorname{tr} MAVAM \leqslant \sum_{i=k+1}^{n} \lambda^*_i. \tag{9}$$

The result follows. □

11. THE PREDICTION OF DISTURBANCES

Let us write the linear regression model $(y, X\beta, \sigma^2 I_n)$ as

$$y = X\beta + \varepsilon, \qquad \mathscr{E}\varepsilon = 0, \qquad \mathscr{E}\varepsilon\varepsilon' = \sigma^2 I_n. \tag{1}$$

We have seen how the unknown parameters β and σ^2 can be optimally estimated by linear or quadratic functions of y.

We now turn our attention to the 'estimation' of the disturbance vector ε. Since ε (unlike β) is a random vector, it cannot, strictly speaking, be estimated. Furthermore, ε (unlike y) is unobservable.

If we try to find an *observable* random vector, say e, which approximates the unobservable ε as closely as possible in some sense, it is appealing to minimize

$$\mathscr{E}(e-\varepsilon)'(e-\varepsilon) \tag{2}$$

subject to the constraints

(i) (linearity)	$e = Ay$	for some square matrix A	(3)
(ii) (unbiasedness)	$\mathscr{E}(e-\varepsilon)=0$	for all β.	(4)

This leads to the *best linear unbiased predictor* of ε,

$$e = (I - XX^+)y, \tag{5}$$

which we recognize as the least squares residual vector (see exercises 1 and 2).

A major drawback of the best linear unbiased predictor given in (5) is that its variance matrix is non-scalar. In fact,

$$\mathscr{V}(e) = \sigma^2(I - XX^+), \tag{6}$$

whereas the variance matrix of ε, which e hopes to resemble, is $\sigma^2 I_n$. This drawback is especially serious if we wish to use e in testing the hypothesis $\mathscr{V}(\varepsilon) = \sigma^2 I_n$.

For this reason we wish to find a predictor of ε (or more generally, $S\varepsilon$) which, in addition to being linear and unbiased, has a scalar variance matrix.

Exercises

1. Show that the minimization problem (2) subject to (3) and (4) amounts to:

$$\begin{array}{ll} \text{minimize} & \operatorname{tr} A'A - 2\operatorname{tr} A \\ \text{subject to} & AX = 0. \end{array}$$

2. Solve this problem and show that the minimizer $\hat{A}$ satisfies

$$\hat{A} = I - XX^{+}.$$

3. Show that, while ε is unobservable, certain linear combinations of ε are observable. In fact, show that $c'\varepsilon$ is observable if and only if $X'c = 0$, in which case $c'\varepsilon = c'y$.

12. PREDICTORS THAT ARE BEST LINEAR UNBIASED WITH SCALAR VARIANCE MATRIX (BLUS)

Thus motivated, we propose the following definition of the predictor of $S\varepsilon$ that is best linear unbiased with scalar variance matrix (BLUS).

Definition 4

Consider the linear regression model $(y, X\beta, \sigma^2 I)$. Let S be a given $m \times n$ matrix. A random $m \times 1$ vector w will be called a *BLUS predictor* of $S\varepsilon$ if

$$\mathcal{E}(w - S\varepsilon)'(w - S\varepsilon) \tag{1}$$

is minimized subject to the constraints

(i) (linearity) $w = Ay$ for some $m \times n$ matrix A,
(ii) (unbiasedness) $\mathcal{E}(w - S\varepsilon) = 0$ for all β,
(iii) (scalar variance matrix) $\mathcal{V}(w) = \sigma^2 I_m$.

Our next task, of course, is to find the BLUS predictor of $S\varepsilon$.

Theorem 8

Consider the linear regression model $(y, X\beta, \sigma^2 I)$ and let $M = I - XX^{+}$. Let S be a given $m \times n$ matrix such that

$$r(SMS') = m. \tag{2}$$

Then the BLUS predictor of $S\varepsilon$ is

$$(SMS')^{-1/2} SMy, \tag{3}$$

where $(SMS')^{-1/2}$ is the *positive definite* square root of $(SMS')^{-1}$.

Proof. We seek a linear predictor w of $S\varepsilon$, that is a predictor of the form

$$w = Ay \tag{4}$$

where A is a constant $m \times n$ matrix. Unbiasedness of the prediction error requires

$$0 = \mathscr{E}(Ay - S\varepsilon) = AX\beta \qquad \text{for all } \beta \text{ in } \mathbb{R}^k, \tag{5}$$

which yields

$$AX = 0. \tag{6}$$

The variance matrix of w is

$$\mathscr{E}ww' = \sigma^2 AA'. \tag{7}$$

In order to satisfy condition (iii) of Definition 4, we thus require

$$AA' = I. \tag{8}$$

Under the constraints (6) and (8), the prediction error variance is

$$\mathscr{V}(Ay - S\varepsilon) = \sigma^2(I + SS' - AS' - SA'). \tag{9}$$

Hence the BLUS predictor of $S\varepsilon$ is obtained by minimizing the trace of (9) with respect to A subject to the constraints (6) and (8). This amounts to solving the following problem:

$$\text{maximize} \qquad \operatorname{tr}(AS') \tag{10}$$

$$\text{subject to} \qquad AX = 0 \quad \text{and} \quad AA' = I. \tag{11}$$

We define the Lagrangian function

$$\psi(A) = \operatorname{tr} AS' - \operatorname{tr} L_1' AX - \tfrac{1}{2}\operatorname{tr} L_2(AA' - I) \tag{12}$$

where L_1 and L_2 are matrices of Lagrange multipliers and L_2 is symmetric. Differentiating ψ with respect to A yields

$$\begin{aligned} \mathrm{d}\psi &= \operatorname{tr}(\mathrm{d}A)S' - \operatorname{tr} L_1'(\mathrm{d}A)X - \tfrac{1}{2}\operatorname{tr} L_2(\mathrm{d}A)A' - \tfrac{1}{2}\operatorname{tr} L_2 A(\mathrm{d}A)' \\ &= \operatorname{tr} S'\mathrm{d}A - \operatorname{tr} XL_1'\mathrm{d}A - \operatorname{tr} A'L_2\mathrm{d}A. \end{aligned} \tag{13}$$

The first-order conditions are

$$S' = XL_1' + A'L_2 \tag{14}$$

$$AX = 0 \tag{15}$$

$$AA' = I. \tag{16}$$

Pre-multiplying (14) with XX^+ yields

$$XL_1' = XX^+S' \tag{17}$$

because $X^+A' = 0$ in view of (15). Inserting (17) in (14) gives

$$MS' = A'L_2. \tag{18}$$

Also, pre-multiplying (14) with A gives

$$AS' = SA' = L_2 \tag{19}$$

in view of (15) and (16) and the symmetry of L_2. Pre-multiplying (18) with S and using (19) we find

$$SMS' = L_2^2 \tag{20}$$

and hence

$$L_2 = (SMS')^{1/2}. \tag{21}$$

It follows from (10) and (19) that our objective is to maximize the trace of L_2; therefore we must choose in (21) the *positive definite* square root of SMS'. Inserting (21) in (18) yields

$$A = (SMS')^{-1/2} SM. \tag{22}$$

□

13. PREDICTORS THAT ARE BEST LINEAR UNBIASED WITH FIXED VARIANCE MATRIX (BLUF), I

We can generalize the BLUS approach in two directions. First, we may assume that the variance matrix of the linear unbiased predictor is not scalar, but some *fixed* known positive semidefinite matrix, say Ω. This is useful, because for many purposes the requirement that the variance matrix of the predictor is scalar is unnecessary; it is sufficient that the variance matrix does not depend on X.

Secondly, we may wish to generalize the criterion function to

$$\mathscr{E}(w - S\varepsilon)' Q (w - S\varepsilon), \tag{1}$$

where Q is some given positive definite matrix.

Definition 5

Consider the linear regression model $(y, X\beta, \sigma^2 V)$ where V is a given positive definite $n \times n$ matrix. Let S be a given $m \times n$ matrix, Ω a given positive semidefinite $m \times m$ matrix and Q a given positive definite $m \times m$ matrix. A random $m \times 1$ vector w will be called a *BLUF*(Ω, Q) *predictor* of $S\varepsilon$ if

$$\mathscr{E}(w - S\varepsilon)' Q (w - S\varepsilon) \tag{2}$$

is minimized subject to the constraints

(i) (linearity) $\quad w = Ay \quad$ for some $m \times n$ matrix A

(ii) (unbiasedness) $\quad \mathscr{E}(w - S\varepsilon) = 0 \quad$ for all β

(iii) (fixed variance matrix) $\quad \mathscr{V}(w) = \sigma^2 \Omega$.

In Theorem 9 we consider the first generalization where the criterion function is unchanged, but where the variance matrix of the predictor is assumed to be some fixed known positive semidefinite matrix.

Theorem 9

Consider the linear regression model $(y, X\beta, \sigma^2 I)$ and let $M = I - XX^+$. Let S be a given $m \times n$ matrix and Ω a given positive semidefinite $m \times m$ matrix such that

$$r(SMS'\Omega) = r(\Omega). \tag{3}$$

Then the BLUF(Ω, I_m) predictor of $S\varepsilon$ is

$$P(P'SMS'P)^{-1/2}P'SMy, \tag{4}$$

where P is a matrix with full column rank satisfying $PP' = \Omega$ and $(P'SMS'P)^{-1/2}$ is the *positive definite* square root of $(P'SMS'P)^{-1}$.

Proof. Proceeding as in the proof of Theorem 8, we seek a linear predictor Ay of $S\varepsilon$ such that

$$\operatorname{tr} \mathscr{V}(Ay - S\varepsilon) \tag{5}$$

is minimized subject to the conditions

$$\mathscr{E}(Ay - S\varepsilon) = 0 \qquad \text{for all } \beta \text{ in } \mathbb{R}^k \tag{6}$$

and

$$\mathscr{V}(Ay) = \sigma^2\Omega. \tag{7}$$

This leads to the maximization problem

$$\text{maximize} \qquad \operatorname{tr} AS' \tag{8}$$

$$\text{subject to} \qquad AX = 0 \quad \text{and} \quad AA' = \Omega. \tag{9}$$

The first-order conditions are

$$S' = XL_1' + A'L_2 \tag{10}$$

$$AX = 0 \tag{11}$$

$$AA' = \Omega, \tag{12}$$

where L_1 and L_2 are matrices of Lagrange multipliers and L_2 is symmetric. Premultiplying (10) with XX^+ and A, respectively, yields

$$XL_1' = XX^+S' \tag{13}$$

and

$$AS' = \Omega L_2, \tag{14}$$

in view of (11) and (12). Inserting (13) in (10) gives

$$MS' = A'L_2. \tag{15}$$

Hence,

$$SMS' = SA'L_2 = L_2\Omega L_2 = L_2PP'L_2 \tag{16}$$

using (15), (14) and the fact that $\Omega = PP'$. This gives

$$P'SMS'P = (P'L_2P)^2 \tag{17}$$

and hence

$$P'L_2P=(P'SMS'P)^{1/2}. \tag{18}$$

By assumption, the matrix $P'SMS'P$ is positive definite. Also it follows from (8) and (14) that we must maximize the trace of $P'L_2P$, so that we must choose the *positive definite* square root of $P'SMS'P$.

So far the proof is very similar to the proof of Theorem 8. However, contrary to that proof we cannot now obtain A directly from (15) and (18). Instead, we proceed as follows. From (15), (12) and (18) we have

$$AMS'P=AA'L_2P=PP'L_2P=P(P'SMS'P)^{1/2}. \tag{19}$$

The general solution for A in (19) is

$$\begin{aligned} A&=P(P'SMS'P)^{1/2}(MS'P)^{+}+Q(I-MS'P(MS'P)^{+})\\ &=P(P'SMS'P)^{-1/2}P'SM+Q(I-MS'P(P'SMS'P)^{-1}P'SM), \end{aligned} \tag{20}$$

where Q is an arbitrary $m\times n$ matrix. From (20) we obtain

$$AA'=PP'+Q(I-MS'P(P'SMS'P)^{-1}P'SM)Q' \tag{21}$$

and hence, in view of (12),

$$Q(I-MS'P(P'SMS'P)^{-1}P'SM)Q'=0. \tag{22}$$

Since the matrix in the middle is idempotent, (22) implies

$$Q(I-MS'P(P'SMS'P)^{-1}P'SM)=0 \tag{23}$$

and hence, from (20),

$$A=P(P'SMS'P)^{-1/2}P'SM. \tag{24}$$

□

14. PREDICTORS THAT ARE BEST LINEAR UNBIASED WITH FIXED VARIANCE MATRIX (BLUF), II

Let us now present the full generalization of Theorem 8.

Theorem 10

Consider the linear regression model $(y, X\beta, \sigma^2 V)$, where V is positive definite, and let

$$R=V-X(X'V^{-1}X)^{+}X'. \tag{1}$$

Let S be a given $m\times n$ matrix and Ω a given positive semidefinite $m\times m$ matrix such that

$$r(SRS'\Omega)=r(\Omega). \tag{2}$$

Then, for any positive definite $m \times m$ matrix Q, the BLUF(Ω, Q) predictor of $S\varepsilon$ is

$$P(P'QSRS'QP)^{-1/2}P'QSRV^{-1}y, \tag{3}$$

where P is a matrix with full column rank satisfying $PP'=\Omega$, and $(P'QSRS'QP)^{-1/2}$ is the *positive definite* square root of $(P'QSRS'QP)^{-1}$.

Proof. The maximization problem amounts to

$$\text{maximize} \quad \operatorname{tr} QAVS' \tag{4}$$
$$\text{subject to} \quad AX=0 \quad \text{and} \quad AVA'=\Omega. \tag{5}$$

We define

$$A^*=QAV^{1/2}, \qquad S^*=SV^{1/2}, \qquad X^*=V^{-1/2}X, \tag{6}$$
$$\Omega^*=Q\Omega Q, \qquad P^*=QP, \qquad M^*=I-X^*X^{*+}. \tag{7}$$

Then we rewrite the maximization problem (4) subject to (5) as a maximization problem in A^*:

$$\text{maximize} \quad \operatorname{tr} A^*S^{*\prime} \tag{8}$$
$$\text{subject to} \quad A^*X^*=0 \quad \text{and} \quad A^*A^{*\prime}=\Omega^*. \tag{9}$$

We know from Theorem 9 that the solution is

$$A^*=P^*(P^{*\prime}S^*M^*S^{*\prime}P^*)^{-1/2}P^{*\prime}S^*M^*. \tag{10}$$

Hence, writing $M^*=V^{-1/2}RV^{-1/2}$, we obtain

$$QAV^{1/2}=QP(P'QSRS'QP)^{-1/2}P'QSRV^{-1/2} \tag{11}$$

and thus

$$A=P(P'QSRS'QP)^{-1/2}P'QSRV^{-1}. \tag{12}$$

□

15. LOCAL SENSITIVITY OF THE POSTERIOR MEAN

Let $(y, X\beta, V)$ be the normal linear regression model where V is positive definite. Suppose, however, that there is prior information concerning β:

$$\beta \sim \mathcal{N}(b^*, H^{*-1}). \tag{1}$$

Then, as Leamer (1978, p. 76) shows, the posterior distribution of β is

$$\beta \sim \mathcal{N}(b, H^{-1}) \tag{2}$$

with

$$b=H^{-1}(H^*b^*+X'V^{-1}y) \tag{3}$$

and

$$H=H^*+X'V^{-1}X. \tag{4}$$

We are interested in the effects of small changes in the precision matrix V^{-1}, the design matrix X and the prior moments b^* and H^{*-1} on the posterior mean b and the posterior precision H^{-1}.

We first study the effects on the posterior mean.

Theorem 11

Consider the normal linear regression model $(y, X\beta, V)$, V positive definite, with prior information $\beta \sim \mathcal{N}(b^*, H^{*-1})$. The local sensitivities of the posterior mean b given in (3) with respect to V^{-1}, X, and the prior moments b^* and H^{*-1} are

$$\partial b/\partial(\mathrm{v}(V^{-1}))' = [(y - Xb)' \otimes H^{-1}X']D_n \tag{5}$$

$$\partial b/\partial(\mathrm{vec}\, X)' = H^{-1} \otimes (y - Xb)'V^{-1} - b' \otimes H^{-1}X'V^{-1} \tag{6}$$

$$\partial b/\partial b^{*\prime} = H^{-1}H^* \tag{7}$$

$$\partial b/\partial(\mathrm{v}(H^{*-1}))' = [(b - b^*)'H^* \otimes H^{-1}H^*]D_k. \tag{8}$$

Note. The matrices D_n and D_k are 'duplication' matrices. See section 3.8.

Proof. We have

$$\begin{aligned}
\mathrm{d}b &= (\mathrm{d}H^{-1})(H^*b^* + X'V^{-1}y) + H^{-1}\mathrm{d}(H^*b^* + X'V^{-1}y)\\
&= -H^{-1}(\mathrm{d}H)b + H^{-1}\mathrm{d}(H^*b^* + X'V^{-1}y)\\
&= -H^{-1}[\mathrm{d}H^* + (\mathrm{d}X)'V^{-1}X + X'V^{-1}\mathrm{d}X + X'(\mathrm{d}V^{-1})X]b\\
&\quad + H^{-1}[(\mathrm{d}H^*)b^* + H^*\mathrm{d}b^* + (\mathrm{d}X)'V^{-1}y + X'(\mathrm{d}V^{-1})y]\\
&= H^{-1}[(\mathrm{d}H^*)(b^* - b) + H^*\mathrm{d}b^* + (\mathrm{d}X)'V^{-1}(y - Xb)\\
&\quad - X'V^{-1}(\mathrm{d}X)b + X'(\mathrm{d}V^{-1})(y - Xb)]\\
&= H^{-1}H^*(\mathrm{d}H^{*-1})H^*(b - b^*) + H^{-1}H^*\mathrm{d}b^*\\
&\quad + H^{-1}(\mathrm{d}X)'V^{-1}(y - Xb) - H^{-1}X'V^{-1}(\mathrm{d}X)b + H^{-1}X'(\mathrm{d}V^{-1})(y - Xb)\\
&= [(b - b^*)'H^* \otimes H^{-1}H^*]\mathrm{d}\,\mathrm{vec}\, H^{*-1} + H^{-1}H^*\mathrm{d}b^*\\
&\quad + \mathrm{vec}\,(y - Xb)'V^{-1}(\mathrm{d}X)H^{-1} - (b' \otimes H^{-1}X'V^{-1})\mathrm{d}\,\mathrm{vec}\, X\\
&\quad + [(y - Xb)' \otimes H^{-1}X']\mathrm{d}\,\mathrm{vec}\, V^{-1}\\
&= [(b - b^*)'H^* \otimes H^{-1}H^*]D_k\mathrm{d}\mathrm{v}(H^{*-1}) + H^{-1}H^*\mathrm{d}b^*\\
&\quad + [H^{-1} \otimes (y - Xb)'V^{-1} - b' \otimes H^{-1}X'V^{-1}]\mathrm{d}\,\mathrm{vec}\, X\\
&\quad + [(y - Xb)' \otimes H^{-1}X']D_n\mathrm{d}\mathrm{v}(V^{-1}).
\end{aligned}$$

The results follow. □

Exercise

1. Show that the local sensitivity of the least squares estimator $b = (X'X)^{-1}X'y$ with respect to X is given by

$$\frac{\partial b}{\partial(\mathrm{vec}\, X)'} = (X'X)^{-1} \otimes (y - Xb)' - b' \otimes (X'X)^{-1}X'.$$

16. LOCAL SENSITIVITY OF THE POSTERIOR PRECISION

In precisely the same manner we can obtain the local sensitivity of the posterior precision.

Theorem 12

Consider the normal linear regression model $(y, X\beta, V)$, V positive definite, with prior information $\beta \sim \mathcal{N}(b^*, H^{*-1})$. The local sensitivities of the posterior precision matrix H^{-1} given by

$$H^{-1} = (H^* + X'V^{-1}X)^{-1} \tag{1}$$

with respect to V^{-1}, X and the prior moments b^* and H^{*-1} are

$$\partial v(H^{-1})/\partial(v(V^{-1}))' = D_k^+(H^{-1}X' \otimes H^{-1}X')D_n \tag{2}$$

$$\partial v(H^{-1})/\partial(\text{vec}\,X)' = -2D_k^+(H^{-1} \otimes H^{-1}X'V^{-1}) \tag{3}$$

$$\partial v(H^{-1})/\partial b^{*\prime} = 0 \tag{4}$$

$$\partial v(H^{-1})/\partial(v(H^{*-1}))' = D_k^+(H^{-1}H^* \otimes H^{-1}H^*)D_k. \tag{5}$$

Proof. From $H = H^* + X'V^{-1}X$ we obtain

$$\begin{aligned} \mathrm{d}H^{-1} &= -H^{-1}(\mathrm{d}H)H^{-1} \\ &= -H^{-1}[\mathrm{d}H^* + (\mathrm{d}X)'V^{-1}X + X'V^{-1}\mathrm{d}X + X'(\mathrm{d}V^{-1})X]H^{-1} \\ &= H^{-1}H^*(\mathrm{d}H^{*-1})H^*H^{-1} - H^{-1}(\mathrm{d}X)'V^{-1}XH^{-1} \\ &\quad - H^{-1}X'V^{-1}(\mathrm{d}X)H^{-1} - H^{-1}X'(\mathrm{d}V^{-1})XH^{-1}. \end{aligned} \tag{6}$$

Hence

$$\begin{aligned} \mathrm{d}\,\text{vec}\,H^{-1} &= (H^{-1}H^* \otimes H^{-1}H^*)\mathrm{d}\,\text{vec}\,H^{*-1} - (H^{-1}X'V^{-1} \otimes H^{-1})\mathrm{d}\,\text{vec}\,X' \\ &\quad - (H^{-1} \otimes H^{-1}X'V^{-1})\mathrm{d}\,\text{vec}\,X - (H^{-1}X' \otimes H^{-1}X')\mathrm{d}\,\text{vec}\,V^{-1} \\ &= (H^{-1}H^* \otimes H^{-1}H^*)\mathrm{d}\,\text{vec}\,H^{*-1} \\ &\quad - [(H^{-1}X'V^{-1} \otimes H^{-1})K_{nk} + H^{-1} \otimes H^{-1}X'V^{-1}]\mathrm{d}\,\text{vec}\,X \\ &\quad - (H^{-1}X' \otimes H^{-1}X')\mathrm{d}\,\text{vec}\,V^{-1} \\ &= (H^{-1}H^* \otimes H^{-1}H^*)\mathrm{d}\,\text{vec}\,H^{*-1} \\ &\quad - (I_{k^2} + K_{kk})(H^{-1} \otimes H^{-1}X'V^{-1})\mathrm{d}\,\text{vec}\,X \\ &\quad - (H^{-1}X' \otimes H^{-1}X')\mathrm{d}\,\text{vec}\,V^{-1}, \end{aligned} \tag{7}$$

so that

$$\begin{aligned} \mathrm{d}v(H^{-1}) = D_k^+\,\mathrm{d}\,\text{vec}\,H^{-1} &= D_k^+(H^{-1}H^* \otimes H^{-1}H^*)D_k\mathrm{d}v(H^{*-1}) \\ &\quad - 2D_k^+(H^{-1} \otimes H^{-1}X'V^{-1})\mathrm{d}\,\text{vec}\,X \\ &\quad - D_k^+(H^{-1}X' \otimes H^{-1}X')D_n\mathrm{d}v(V^{-1}) \end{aligned} \tag{8}$$

and the results follow. □

BIBLIOGRAPHICAL NOTES

§2–§7. See Theil and Schweitzer (1961), Theil (1971), Rao (1971b) and Neudecker (1980a).
§8. See Balestra (1973), Neudecker (1980b, 1985), Neudecker and Liu (1993), and Rolle (1994).
§9–§10. See Neudecker (1977b, 1978). Theorem 7 corrects an error in Neudecker (1978).
§11–§14. See Abrahamse and Koerts (1971), Koerts and Abrahamse (1969), Theil (1965), Neudecker (1973, 1977a) and Dubbelman *et al.* (1972).
§15–§16. See Leamer (1978) and Polasek (1986).

Part Six—

Applications to maximum likelihood estimation

CHAPTER 15

Maximum likelihood estimation

1. INTRODUCTION

The method of maximum likelihood estimation has great intuitive appeal and generates estimators with desirable asymptotic properties. The estimators are obtained by maximization of the likelihood function, and the asymptotic precision of the estimators is measured by the inverse of the information matrix. Thus both the first and the second differential of the likelihood function need to be found and this provides an excellent example of the use of our techniques.

2. THE METHOD OF MAXIMUM LIKELIHOOD (ML)

Let $\{y_1, y_2, \ldots\}$ be a sequence of random variables, not necessarily independent or identically distributed. The joint density function of $y = (y_1, \cdots, y_n) \in \mathbb{R}^n$ is denoted $h_n(\cdot; \gamma_0)$ and is known except for γ_0, the true value of the parameter vector to be estimated. We assume that $\gamma_0 \in \Gamma$, where Γ (the parameter space) is a subset of a finite-dimensional Euclidean space. For every (fixed) $y \in \mathbb{R}^n$ the real-valued function

$$L_n(\gamma) = L_n(\gamma; y) = h_n(y; \gamma), \qquad \gamma \in \Gamma, \tag{1}$$

is called the *likelihood function*, and its logarithm

$$\Lambda_n(\gamma) = \log L_n(\gamma) \tag{2}$$

is called the *loglikelihood function.*

For fixed $y \in \mathbb{R}^n$ every value $\hat{\gamma}_n(y) \in \Gamma$ with

$$L_n(\hat{\gamma}_n(y); y) = \sup_{\gamma \in \Gamma} L_n(\gamma; y) \tag{3}$$

is called a *maximum likelihood* (ML) *estimate* of γ_0. In general there is no guarantee that an ML estimate of γ_0 exists for (almost) every $y \in \mathbb{R}^n$, but if it does and if the function $\hat{\gamma}_n : \mathbb{R}^n \to \Gamma$ so defined is measurable, then this function $\hat{\gamma}_n$ is called an *ML estimator* of γ_0.

When the supremum in (3) is attained at an interior point of Γ and $L_n(\gamma)$ is a differentiable function of γ, then the *score vector*

$$s_n(\gamma)=\partial\Lambda_n(\gamma)/\partial\gamma \tag{4}$$

vanishes at that point, so that $\hat{\gamma}_n$ is a solution of the vector equation $s_n(\gamma)=0$.

If $L_n(\gamma)$ is a twice differentiable function of γ, then the *Hessian matrix* is defined as

$$\mathsf{H}_n(\gamma)=\partial^2\Lambda_n(\gamma)/\partial\gamma\partial\gamma' \tag{5}$$

and the *information matrix* for γ_0 is

$$\mathcal{F}_n(\gamma_0)=-\mathcal{E}\,\mathsf{H}_n(\gamma_0). \tag{6}$$

Notice that the information matrix is evaluated at the true value γ_0. The *asymptotic information matrix* for γ_0 is defined as

$$\mathcal{F}(\gamma_0)=\lim_{n\to\infty}(1/n)\,\mathcal{F}_n(\gamma_0) \tag{7}$$

if the limit exists. If $\mathcal{F}(\gamma_0)$ is positive definite, its inverse $\mathcal{F}^{-1}(\gamma_0)$ is essentially a lower bound for the asymptotic variance matrix of any consistent estimator of γ_0 (asymptotic Cramér–Rao inequality). Under suitable regularity conditions the ML estimator attains this lower bound asymptotically. As a consequence we shall refer to $\mathcal{F}^{-1}(\gamma_0)$ as the *asymptotic variance matrix* of the ML estimator $\hat{\gamma}_n$. The precise meaning of this is that, under suitable conditions, the sequence of random variables

$$\sqrt{n}(\hat{\gamma}_n-\gamma_0) \tag{8}$$

converges in distribution to a normally distributed random vector with mean zero and variance matrix $\mathcal{F}^{-1}(\gamma_0)$. Thus, $\mathcal{F}^{-1}(\gamma_0)$ is the variance matrix of the asymptotic distribution, and an estimator of the variance matrix of $\hat{\gamma}_n$ is given by

$$(1/n)\,\mathcal{F}^{-1}(\hat{\gamma}_n) \qquad \text{or} \qquad \mathcal{F}_n^{-1}(\hat{\gamma}_n). \tag{9}$$

3. ML ESTIMATION OF THE MULTIVARIATE NORMAL DISTRIBUTION

Our first theorem is the following well known result concerning the multivariate normal distribution.

Theorem 1

Let the random $m\times 1$ vectors $y_1, y_2, \ldots, y_n$ be independently and identically distributed such that

$$y_i\sim\mathcal{N}_m(\mu_0,\Omega_0) \qquad (i=1,\ldots,n), \tag{1}$$

where Ω_0 is positive definite, and let $n \geq m+1$. The ML estimators of μ_0 and Ω_0 are

$$\hat{\mu} = (1/n) \sum_{i=1}^{n} y_i \equiv \bar{y} \tag{2}$$

$$\hat{\Omega} = (1/n) \sum_{i=1}^{n} (y_i - \bar{y})(y_i - \bar{y})'. \tag{3}$$

Let us give four proofs of this theorem. The first proof ignores the fact that Ω is symmetric.

First proof of Theorem 1. The loglikelihood function is

$$\Lambda_n(\mu, \Omega) = -\tfrac{1}{2}mn \log 2\pi - \tfrac{1}{2}n \log |\Omega| - \tfrac{1}{2} \operatorname{tr} \Omega^{-1} Z, \tag{4}$$

where

$$Z = \sum_{i=1}^{n} (y_i - \mu)(y_i - \mu)'. \tag{5}$$

The first differential of Λ_n is

$$\begin{aligned} \mathrm{d}\Lambda_n &= -\tfrac{1}{2}n \,\mathrm{d}\log|\Omega| - \tfrac{1}{2}\operatorname{tr}(\mathrm{d}\Omega^{-1})Z - \tfrac{1}{2}\operatorname{tr}\Omega^{-1}\mathrm{d}Z \\ &= -\tfrac{1}{2}n \operatorname{tr}\Omega^{-1}\mathrm{d}\Omega + \tfrac{1}{2}\operatorname{tr}\Omega^{-1}(\mathrm{d}\Omega)\Omega^{-1}Z \\ &\quad + \tfrac{1}{2}\operatorname{tr}\Omega^{-1}\left(\sum_i (y_i-\mu)(\mathrm{d}\mu)' + (\mathrm{d}\mu)\sum_i (y_i-\mu)'\right) \\ &= \tfrac{1}{2}\operatorname{tr}(\mathrm{d}\Omega)\Omega^{-1}(Z - n\Omega)\Omega^{-1} + (\mathrm{d}\mu)'\Omega^{-1}\sum_i (y_i - \mu) \\ &= \tfrac{1}{2}\operatorname{tr}(\mathrm{d}\Omega)\Omega^{-1}(Z - n\Omega)\Omega^{-1} + n(\mathrm{d}\mu)'\Omega^{-1}(\bar{y} - \mu). \end{aligned} \tag{6}$$

If we *ignore* the symmetry constraint on Ω, we obtain the first-order conditions

$$\Omega^{-1}(Z - n\Omega)\Omega^{-1} = 0, \qquad \Omega^{-1}(\bar{y} - \mu) = 0, \tag{7}$$

from which (2) and (3) follow immediately. To prove that we have in fact found the maximum of (4), we differentiate (6) again. This yields

$$\begin{aligned} \mathrm{d}^2\Lambda_n = {} & \tfrac{1}{2}\operatorname{tr}(\mathrm{d}\Omega)(\mathrm{d}\Omega^{-1})(Z - n\Omega)\Omega^{-1} + \tfrac{1}{2}\operatorname{tr}(\mathrm{d}\Omega)\Omega^{-1}(Z - n\Omega)\mathrm{d}\Omega^{-1} \\ & + \tfrac{1}{2}\operatorname{tr}(\mathrm{d}\Omega)\Omega^{-1}(\mathrm{d}Z - n\,\mathrm{d}\Omega)\Omega^{-1} + n(\mathrm{d}\mu)'(\mathrm{d}\Omega^{-1})(\bar{y} - \mu) \\ & - n(\mathrm{d}\mu)'\Omega^{-1}\mathrm{d}\mu. \end{aligned} \tag{8}$$

At the point $(\hat{\mu}, \hat{\Omega})$ we have $\hat{\mu} = \bar{y}$, $\hat{Z} - n\hat{\Omega} = 0$ and $\mathrm{d}\hat{Z} = 0$ (see exercise 1), and hence

$$\mathrm{d}^2\Lambda_n(\hat{\mu}, \hat{\Omega}) = -\frac{n}{2}\operatorname{tr}(\mathrm{d}\Omega)\hat{\Omega}^{-1}(\mathrm{d}\Omega)\hat{\Omega}^{-1} - n(\mathrm{d}\mu)'\hat{\Omega}^{-1}\mathrm{d}\mu < 0 \tag{9}$$

unless $\mathrm{d}\mu = 0$ and $\mathrm{d}\Omega = 0$. It follows that Λ_n has a strict local maximum at $(\hat{\mu}, \hat{\Omega})$.

□

Exercises

1. Show that $dZ = -n(d\mu)(\bar{y}-\mu)' - n(\bar{y}-\mu)(d\mu)'$, and conclude that $d\hat{Z}=0$.
2. Show that $\mathscr{E}\hat{\Omega} = [(n-1)/n]\Omega$.
3. Show that $\hat{\Omega} = (1/n)Y'[I-(1/n)\imath\imath']Y$, where $Y=(y_1, \ldots, y_n)'$ and $\imath$ is the $n\times 1$ vector $(1, 1, \ldots, 1)'$.
4. Hence show that $\hat{\Omega}$ is positive definite (almost surely) if and only if $n-1 \geqslant m$.

4. IMPLICIT VERSUS EXPLICIT TREATMENT OF SYMMETRY

The first proof of Theorem 1 shows that, even if we do not impose symmetry (or positive definiteness) on Ω, the solution $\hat{\Omega}$ is symmetric and positive semidefinite (in fact, positive definite with probability 1). Hence there is no need to impose symmetry at this stage. Nevertheless, we shall give two proofs of Theorem 1 where the symmetry is properly taken into account. We shall need these results in any case when we discuss the second-order conditions (Hessian matrix and information matrix).

Second proof of Theorem 1. Starting from (3.6) we have

$$\begin{aligned} d\Lambda_n &= \tfrac{1}{2}\operatorname{tr}(d\Omega)\Omega^{-1}(Z-n\Omega)\Omega^{-1} + n(d\mu)'\Omega^{-1}(\bar{y}-\mu) \\ &= \tfrac{1}{2}(\operatorname{vec} d\Omega)'(\Omega^{-1}\otimes\Omega^{-1})\operatorname{vec}(Z-n\Omega) + n(d\mu)'\Omega^{-1}(\bar{y}-\mu) \\ &= \tfrac{1}{2}(d v(\Omega))' D_m'(\Omega^{-1}\otimes\Omega^{-1})\operatorname{vec}(Z-n\Omega) + n(d\mu)'\Omega^{-1}(\bar{y}-\mu), \end{aligned} \tag{1}$$

where D_m is the duplication matrix (see section 3.8). The first-order conditions are

$$\Omega^{-1}(\bar{y}-\mu)=0, \qquad D_m'(\Omega^{-1}\otimes\Omega^{-1})\operatorname{vec}(Z-n\Omega)=0. \tag{2}$$

The first of these conditions implies $\hat{\mu}=\bar{y}$; the second can be written as

$$D_m'(\Omega^{-1}\otimes\Omega^{-1})D_m v(Z-n\Omega)=0 \tag{3}$$

since $Z-n\Omega$ is symmetric. Now, $D_m'(\Omega^{-1}\otimes\Omega^{-1})D_m$ is non-singular (see Theorem 3.13), so (3) implies $v(Z-n\Omega)=0$. Using again the symmetry of Z and Ω, we obtain

$$\hat{\Omega}=(1/n)\hat{Z}=(1/n)\sum_i (y_i-\bar{y})(y_i-\bar{y})'. \tag{4}$$

□

We shall call the above treatment of the symmetry condition (using the duplication matrix) *implicit*. In contrast, an *explicit* treatment of symmetry involves inclusion of the side condition $\Omega=\Omega'$. The next proof of Theorem 1 illustrates this approach.

Third proof of Theorem 1. Our starting point now is the Lagrangian function

$$\psi(\mu,\Omega) = -\tfrac{1}{2}mn\log 2\pi - \tfrac{1}{2}n\log|\Omega| - \tfrac{1}{2}\operatorname{tr}\Omega^{-1}Z - \operatorname{tr}L'(\Omega-\Omega'), \tag{5}$$

where L is an $m \times m$ matrix of Lagrange multipliers. Differentiating (5) yields

$$d\psi = \tfrac{1}{2}\operatorname{tr}(d\Omega)\Omega^{-1}(Z - n\Omega)\Omega^{-1} + \operatorname{tr}(L - L')d\Omega + n(d\mu)'\Omega^{-1}(\bar{y} - \mu), \qquad (6)$$

so that the first-order conditions are

$$\tfrac{1}{2}\Omega^{-1}(Z - n\Omega)\Omega^{-1} + L - L' = 0 \qquad (7)$$

$$\Omega^{-1}(\bar{y} - \mu) = 0 \qquad (8)$$

$$\Omega = \Omega'. \qquad (9)$$

From (8) follows $\hat{\mu} = \bar{y}$. Adding (7) to its transpose and using (9) yields $\Omega^{-1}(Z - n\Omega)\Omega^{-1} = 0$ and hence the desired result. □

5. THE TREATMENT OF POSITIVE DEFINITENESS

Finally we may impose both symmetry *and* positive definiteness on Ω by writing $\Omega = X'X$, X square. This leads to our final proof of Theorem 1.

Fourth proof of Theorem 1. The loglikelihood now becomes

$$\Lambda_n(\mu, X) = -\tfrac{1}{2}mn \log 2\pi - \tfrac{1}{2}n \log|X'X| - \tfrac{1}{2}\operatorname{tr}(X'X)^{-1}Z. \qquad (1)$$

Differentiating (1) yields

$$\begin{aligned} d\Lambda_n &= -\tfrac{1}{2}n \operatorname{tr}(X'X)^{-1}(dX'X) - \tfrac{1}{2}\operatorname{tr}(d(X'X)^{-1})Z - \tfrac{1}{2}\operatorname{tr}(X'X)^{-1}dZ \\ &= -n \operatorname{tr}(X'X)^{-1}X'dX + \tfrac{1}{2}\operatorname{tr}(X'X)^{-1}(dX'X)(X'X)^{-1}Z \\ &\quad + n(d\mu)'(X'X)^{-1}(\bar{y} - \mu) \\ &= -n \operatorname{tr} X^{-1}dX + \operatorname{tr}(X'X)^{-1}ZX^{-1}dX + n(d\mu)'(X'X)^{-1}(\bar{y} - \mu). \qquad (2) \end{aligned}$$

The first-order conditions are

$$nX^{-1} = (X'X)^{-1}ZX^{-1}, \qquad (X'X)^{-1}(\bar{y} - \mu) = 0 \qquad (3)$$

from which follows $\hat{\mu} = \bar{y}$ and $\hat{\Omega} = \hat{X}'\hat{X} = (1/n)\hat{Z}$. □

6. THE INFORMATION MATRIX

To obtain the information matrix we need to take the symmetry of Ω into account, either implicitly or explicitly. We prefer the implicit treatment using the duplication matrix.

Theorem 2

Let the random $m \times 1$ vectors $y_1, \ldots, y_n$ be independently and identically distributed such that

$$y_i \sim \mathcal{N}_m(\mu_0, \Omega_0) \qquad (i = 1, \ldots, n), \qquad (1)$$

where Ω_0 is positive definite, and let $n \geq m+1$. The information matrix for μ_0 and $\mathrm{v}(\Omega_0)$ is the $\frac{1}{2}m(m+3) \times \frac{1}{2}m(m+3)$ matrix

$$\mathscr{F}_n = n\begin{pmatrix} \Omega_0^{-1} & 0 \\ 0 & \frac{1}{2}D_m'(\Omega_0^{-1} \otimes \Omega_0^{-1})D_m \end{pmatrix}. \tag{2}$$

The asymptotic variance matrix of the ML estimators $\hat{\mu}$ and $\mathrm{v}(\hat{\Omega})$ is

$$\mathscr{F}^{-1} = \begin{pmatrix} \Omega_0 & 0 \\ 0 & 2D_m^+(\Omega_0 \otimes \Omega_0)D_m^{+\prime} \end{pmatrix}. \tag{3}$$

And the generalized asymptotic variance of $\mathrm{v}(\hat{\Omega})$ is

$$|2D_m^+(\Omega_0 \otimes \Omega_0)D_m^{+\prime}| = 2^m|\Omega_0|^{m+1}. \tag{4}$$

Proof. Since Ω is a *linear* function of $\mathrm{v}(\Omega)$, we have $\mathrm{d}^2\Omega = 0$ and hence the second differential of $\Lambda_n(\mu, \mathrm{v}(\Omega))$ is given by (3.8):

$$\begin{aligned} \mathrm{d}^2\Lambda_n(\mu, \mathrm{v}(\Omega)) = {} & \tfrac{1}{2}\mathrm{tr}(\mathrm{d}\Omega)(\mathrm{d}\Omega^{-1})(Z - n\Omega)\Omega^{-1} + \tfrac{1}{2}\mathrm{tr}(\mathrm{d}\Omega)\Omega^{-1}(Z - n\Omega)\mathrm{d}\Omega^{-1} \\ & + \tfrac{1}{2}\mathrm{tr}(\mathrm{d}\Omega)\Omega^{-1}(\mathrm{d}Z - n\mathrm{d}\Omega)\Omega^{-1} + n(\mathrm{d}\mu)'(\mathrm{d}\Omega^{-1})(\bar{y} - \mu) \\ & - n(\mathrm{d}\mu)'\Omega^{-1}\mathrm{d}\mu. \end{aligned} \tag{5}$$

Notice that we do not at this stage evaluate $\mathrm{d}^2\Lambda_n$ completely in terms of $\mathrm{d}\mu$ and $\mathrm{dv}(\Omega)$; this is unnecessary because, upon taking expectations, we find immediately

$$-\mathscr{E}\mathrm{d}^2\Lambda_n(\mu_0, \mathrm{v}(\Omega_0)) = \frac{n}{2}\mathrm{tr}(\mathrm{d}\Omega)\Omega_0^{-1}(\mathrm{d}\Omega)\Omega_0^{-1} + n(\mathrm{d}\mu)'\Omega_0^{-1}\mathrm{d}\mu, \tag{6}$$

since $\mathscr{E}\bar{y} = \mu_0$, $\mathscr{E}Z = n\Omega_0$ and $\mathscr{E}\mathrm{d}Z = 0$ (compare the passage from (3.8) to (3.9)). We now use the duplication matrix and obtain

$$\begin{aligned} -\mathscr{E}\mathrm{d}^2\Lambda_n(\mu_0, \mathrm{v}(\Omega_0)) & = \frac{n}{2}(\mathrm{vec}\,\mathrm{d}\Omega)'(\Omega_0^{-1} \otimes \Omega_0^{-1})\,\mathrm{vec}\,\mathrm{d}\Omega + n(\mathrm{d}\mu)'\Omega_0^{-1}\mathrm{d}\mu \\ & = \frac{n}{2}(\mathrm{dv}(\Omega))'D_m'(\Omega_0^{-1} \otimes \Omega_0^{-1})D_m\mathrm{dv}(\Omega) + n(\mathrm{d}\mu)'\Omega_0^{-1}\mathrm{d}\mu. \end{aligned} \tag{7}$$

Hence the information matrix for μ_0 and $\mathrm{v}(\Omega_0)$ is $\mathscr{F}_n = n\mathscr{F}$ with

$$\mathscr{F} = \begin{pmatrix} \Omega_0^{-1} & 0 \\ 0 & \frac{1}{2}D_m'(\Omega_0^{-1} \otimes \Omega_0^{-1})D_m \end{pmatrix}. \tag{8}$$

The asymptotic variance matrix of $\hat{\mu}$ and $\mathrm{v}(\hat{\Omega})$ is

$$\mathscr{F}^{-1} = \begin{pmatrix} \Omega_0 & 0 \\ 0 & 2D_m^+(\Omega_0 \otimes \Omega_0)D_m^{+\prime} \end{pmatrix}, \tag{9}$$

using Theorem 3.13(d). And the generalized asymptotic variance of $\mathrm{v}(\hat{\Omega})$ follows from (9) and Theorem 3.14(b). □

Exercises

1. Taking (5) as your starting point, show that

$$\begin{aligned}(1/n)\mathrm{d}^2\Lambda_n(\mu, \mathrm{v}(\Omega)) = &-(\mathrm{d}\mu)'\Omega^{-1}\mathrm{d}\mu - 2(\mathrm{d}\mu)'\Omega^{-1}(\mathrm{d}\Omega)\Omega^{-1}(\bar{y}-\mu)\\ &+\tfrac{1}{2}\mathrm{tr}(\mathrm{d}\Omega)\Omega^{-1}(\mathrm{d}\Omega)\Omega^{-1}\\ &-\mathrm{tr}(\mathrm{d}\Omega)\Omega^{-1}(\mathrm{d}\Omega)\Omega^{-1}(Z/n)\Omega^{-1}.\end{aligned}$$

2. Hence show that the Hessian matrix $\mathrm{H}_n(\mu, \mathrm{v}(\Omega))$ takes the form

$$-n\begin{pmatrix}\Omega^{-1} & [(\bar{y}-\mu)'\Omega^{-1}\otimes\Omega^{-1}]D_m\\ D_m'[\Omega^{-1}(\bar{y}-\mu)\otimes\Omega^{-1}] & \tfrac{1}{2}D_m'(\Omega^{-1}\otimes A)D_m\end{pmatrix}$$

with

$$A = \Omega^{-1}[(2/n)Z - \Omega]\Omega^{-1}.$$

7. ML ESTIMATION OF THE MULTIVARIATE NORMAL DISTRIBUTION WITH DISTINCT MEANS

Suppose now that we have not one but, say, p random samples, and let the j-th sample be from the $\mathcal{N}_m(\mu_{0j}, \Omega_0)$ distribution. We wish to estimate $\mu_{01}, \ldots, \mu_{0p}$ and the common variance matrix Ω_0.

Theorem 3

Let the random $m\times 1$ vectors y_{ij} $(i=1, \ldots, n_j;\ j=1, \ldots, p)$ be independently distributed such that

$$y_{ij} \sim \mathcal{N}_m(\mu_{0j}, \Omega_0) \qquad (i=1, \ldots, n_j;\ j=1, \ldots, p) \tag{1}$$

where Ω_0 is positive definite, and let

$$n = \sum_{j=1}^{p} n_j \geqslant m+p.$$

The ML estimators of $\mu_{01}, \ldots, \mu_{0p}$ and Ω_0 are

$$\hat{\mu}_j = (1/n_j)\sum_{i=1}^{n_j} y_{ij} \equiv \bar{y}_j \qquad (j=1, \ldots, p) \tag{2}$$

$$\hat{\Omega} = (1/n)\sum_{j=1}^{p}\sum_{i=1}^{n_j}(y_{ij}-\bar{y}_j)(y_{ij}-\bar{y}_j)'. \tag{3}$$

The information matrix for $\mu_{01}, \ldots, \mu_{0p}$ and $\mathrm{v}(\Omega_0)$ is

$$\mathscr{F}_n = n\begin{pmatrix}A\otimes\Omega_0^{-1} & 0\\ 0 & \tfrac{1}{2}D_m'(\Omega_0^{-1}\otimes\Omega_0^{-1})D_m\end{pmatrix}, \tag{4}$$

where A is a diagonal $p\times p$ matrix with diagonal elements n_j/n $(j=1, \ldots, p)$.

And the asymptotic variance matrix of the ML estimators $\hat{\mu}_1, \ldots, \hat{\mu}_p$ and $\mathrm{v}(\hat{\Omega})$ is

$$\mathscr{F}^{-1} = \begin{pmatrix} A^{-1} \otimes \Omega_0 & 0 \\ 0 & 2D_m^+(\Omega_0 \otimes \Omega_0)D_m^{+\prime} \end{pmatrix}. \tag{5}$$

Proof. The proof is left as an exercise for the reader. □

Exercise

1. Show that $\hat{\Omega}$ is positive definite (almost surely) if and only if $n-p \geqslant m$.

8. THE MULTIVARIATE LINEAR REGRESSION MODEL

Let us consider a system of linear regression equations

$$y_{ij} = x_i'\beta_{0j} + \varepsilon_{ij} \qquad (i=1, \ldots, n;\, j=1, \ldots, m), \tag{1}$$

where y_{ij} denotes the i-th observation on the j-th dependent variable, $x_1, \ldots, x_n$ are observations on the k regressors, $\beta_{01}, \ldots, \beta_{0m}$ are $k \times 1$ parameter vectors to be estimated, and ε_{ij} is a random disturbance term. We let $\varepsilon_i' = (\varepsilon_{i1}, \ldots, \varepsilon_{im})$ and assume that $\mathscr{E}\varepsilon_i = 0$ $(i=1, \ldots, n)$ and

$$\mathscr{E}\varepsilon_i\varepsilon_h' = \begin{cases} 0 & \text{if } i \neq h \\ \Omega_0 & \text{if } i = h. \end{cases} \tag{2}$$

Let $Y=(y_{ij})$ be the $n \times m$ matrix of the observations on the dependent variables and let

$$Y = (y_1, y_2, \ldots, y_n)' = (y_{(1)}, \ldots, y_{(m)}). \tag{3}$$

Similarly we define

$$X = (x_1, \ldots, x_n)', \qquad B_0 = (\beta_{01}, \ldots, \beta_{0m}) \tag{4}$$

of orders $n \times k$ and $k \times m$ respectively, and $\varepsilon_{(j)} = (\varepsilon_{1j}, \ldots, \varepsilon_{nj})'$. We can then write the system (1) either as

$$y_{(j)} = X\beta_{0j} + \varepsilon_{(j)} \qquad (j=1, \ldots, m) \tag{5}$$

or as

$$y_i' = x_i'B_0 + \varepsilon_i' \qquad (i=1, \ldots, n). \tag{6}$$

If the vectors $\varepsilon_{(1)}, \ldots, \varepsilon_{(m)}$ are uncorrelated, which is the case if Ω_0 is diagonal, we can estimate each β_{0j} separately. But in general this will not be the case and we have to estimate the whole system on efficiency grounds.

Theorem 4

Let the random $m \times 1$ vectors $y_1, \ldots, y_n$ be independently distributed such that

$$y_i \sim \mathscr{N}_m(B_0'x_i, \Omega_0) \qquad (i=1, \ldots, n), \tag{7}$$

where Ω_0 is positive definite and $X=(x_1, \ldots, x_n)'$ is a given (non-random) $n \times k$ matrix of full column rank k. Let $n \geq m+k$. The ML estimators of B_0 and Ω_0 are

$$\hat{B}=(X'X)^{-1}X'Y, \qquad \hat{\Omega}=(1/n)Y'MY, \tag{8}$$

where

$$Y=(y_1, \ldots, y_n)', \qquad M=I-X(X'X)^{-1}X'. \tag{9}$$

The information matrix for vec B_0 and $\mathrm{v}(\Omega_0)$ is

$$\mathscr{F}_n=n\begin{pmatrix}(1/n)\Omega_0^{-1}\otimes X'X & 0 \\ 0 & \frac{1}{2}D_m'(\Omega_0^{-1}\otimes\Omega_0^{-1})D_m\end{pmatrix}. \tag{10}$$

And, if $(1/n)X'X$ converges to a positive definite matrix Q when $n\to\infty$, the asymptotic variance matrix of vec $\hat{B}$ and $\mathrm{v}(\hat{\Omega})$ is

$$\mathscr{F}^{-1}=\begin{pmatrix}\Omega_0\otimes Q^{-1} & 0 \\ 0 & 2D_m^+(\Omega_0\otimes\Omega_0)D_m^{+\prime}\end{pmatrix}. \tag{11}$$

Proof. The loglikelihood is

$$\Lambda_n(B, \mathrm{v}(\Omega))=-\tfrac{1}{2}mn\log 2\pi-\tfrac{1}{2}n\log|\Omega|-\tfrac{1}{2}\mathrm{tr}\,\Omega^{-1}Z \tag{12}$$

where

$$Z=\sum_{i=1}^{n}(y_i-B'x_i)(y_i-B'x_i)'=(Y-XB)'(Y-XB), \tag{13}$$

and its first differential takes the form

$$\begin{aligned}\mathrm{d}\Lambda_n&=-\tfrac{1}{2}n\,\mathrm{tr}\,\Omega^{-1}\mathrm{d}\Omega+\tfrac{1}{2}\mathrm{tr}\,\Omega^{-1}(\mathrm{d}\Omega)\Omega^{-1}Z-\tfrac{1}{2}\mathrm{tr}\,\Omega^{-1}\mathrm{d}Z\\&=\tfrac{1}{2}\mathrm{tr}\,(\mathrm{d}\Omega)\Omega^{-1}(Z-n\Omega)\Omega^{-1}+\mathrm{tr}\,\Omega^{-1}(Y-XB)'X\mathrm{d}B.\end{aligned} \tag{14}$$

The first-order conditions are therefore

$$\Omega=(1/n)Z, \qquad (Y-XB)'X=0. \tag{15}$$

This leads to $\hat{B}=(X'X)^{-1}X'Y$, so that

$$\hat{\Omega}=(1/n)(Y-X\hat{B})'(Y-X\hat{B})=(1/n)(MY)'MY=(1/n)Y'MY. \tag{16}$$

The second differential is

$$\begin{aligned}\mathrm{d}^2\Lambda_n&=\mathrm{tr}\,(\mathrm{d}\Omega)(\mathrm{d}\Omega^{-1})(Z-n\Omega)\Omega^{-1}+\tfrac{1}{2}\mathrm{tr}\,(\mathrm{d}\Omega)\Omega^{-1}(\mathrm{d}Z-n\mathrm{d}\Omega)\Omega^{-1}\\&\quad+\mathrm{tr}\,(\mathrm{d}\Omega^{-1})(Y-XB)'X\mathrm{d}B-\mathrm{tr}\,\Omega^{-1}(\mathrm{d}B)'X'X\,\mathrm{d}B,\end{aligned} \tag{17}$$

and taking expectations we obtain

$$\begin{aligned}-\mathscr{E}\,\mathrm{d}^2\Lambda_n(B_0, \mathrm{v}(\Omega_0))&=\frac{n}{2}\mathrm{tr}\,(\mathrm{d}\Omega)\Omega_0^{-1}(\mathrm{d}\Omega)\Omega_0^{-1}+\mathrm{tr}\,\Omega_0^{-1}(\mathrm{d}B)'X'X\,\mathrm{d}B\\&=\frac{n}{2}(\mathrm{d}\mathrm{v}(\Omega))'D_m'(\Omega_0^{-1}\otimes\Omega_0^{-1})D_m\mathrm{d}\mathrm{v}(\Omega)\\&\quad+(\mathrm{d}\,\mathrm{vec}\,B)'(\Omega_0^{-1}\otimes X'X)\,\mathrm{d}\,\mathrm{vec}\,B.\end{aligned} \tag{18}$$

The information matrix and the inverse of its limit now follow easily from (18). □

Exercises

1. Use (17) to show that

$$\begin{aligned}(1/n)\mathrm{d}^2\Lambda_n(B, \mathrm{v}(\Omega)) = & -\operatorname{tr}\Omega^{-1}(\mathrm{d}B)'(X'X/n)\mathrm{d}B \\ & -2\operatorname{tr}(\mathrm{d}\Omega)\Omega^{-1}(\mathrm{d}B)'[X'(Y-XB)/n]\Omega^{-1} \\ & +\tfrac{1}{2}\operatorname{tr}(\mathrm{d}\Omega)\Omega^{-1}(\mathrm{d}\Omega)\Omega^{-1} \\ & -\operatorname{tr}(\mathrm{d}\Omega)\Omega^{-1}(\mathrm{d}\Omega)\Omega^{-1}(Z/n)\Omega^{-1}.\end{aligned}$$

2. Hence show that the Hessian matrix $\mathsf{H}_n(\operatorname{vec} B, \mathrm{v}(\Omega))$ takes the form

$$-n\begin{pmatrix}\Omega^{-1}\otimes(X'X/n) & (\Omega^{-1}\otimes(X'V/n)\Omega^{-1})D_m \\ D_m'(\Omega^{-1}\otimes\Omega^{-1}(V'X/n)) & \tfrac{1}{2}D_m'(\Omega^{-1}\otimes A)D_m\end{pmatrix}$$

with

$$V = Y - XB, \qquad A = \Omega^{-1}[(2/n)Z - \Omega]\Omega^{-1}.$$

Compare this result with the Hessian matrix obtained in exercise 6.2.

9. THE ERRORS-IN-VARIABLES MODEL

Consider the linear regression model

$$y_i = x_i'\beta_0 + \varepsilon_i \qquad (i = 1, \ldots, n), \tag{1}$$

where $x_1, \ldots, x_n$ are non-stochastic $k \times 1$ vectors. Assume that both y_i and x_i are measured with error, so that instead of observing y_i and x_i we observe y_i^* and x_i^* where

$$y_i^* = y_i + \zeta_i, \qquad x_i^* = x_i + \eta_i. \tag{2}$$

Then we have

$$\begin{pmatrix} y_i^* \\ x_i^* \end{pmatrix} = \begin{pmatrix} x_i'\beta_0 \\ x_i \end{pmatrix} + \begin{pmatrix} \varepsilon_i + \zeta_i \\ \eta_i \end{pmatrix} \tag{3}$$

or, for short,

$$z_i = \mu_{0i} + v_i \qquad (i = 1, \ldots, n). \tag{4}$$

If we assume that the distribution of $(v_1, \ldots, v_n)$ is completely known, then the problem is to estimate the vectors $x_1, \ldots, x_n$ and β_0. Letting $\alpha_0 = (-1, \beta_0')'$, we see that this is equivalent to estimating $\mu_{01}, \ldots, \mu_{0n}$ and α_0 subject to the constraints $\mu_{0i}'\alpha_0 = 0$ $(i = 1, \ldots, n)$. In this context the following result is of importance.

Theorem 5

Let the random $m \times 1$ vectors $y_1, \ldots, y_n$ be independently distributed such that

$$y_i \sim \mathcal{N}_m(\mu_{0i}, \Omega_0) \qquad (i = 1, \ldots, n), \tag{5}$$

where Ω_0 is positive definite *and known*, and the parameter vectors $\mu_{01}, \ldots, \mu_{0n}$ are subject to the constraint

$$\mu_{0i}'\alpha_0 = 0 \qquad (i=1, \ldots, n) \tag{6}$$

for some unknown α_0 in $\mathbb{R}^m$, normalized by $\alpha_0'\Omega_0\alpha_0 = 1$. The ML estimators of $\mu_{01}, \ldots, \mu_{0n}$ and α_0 are

$$\hat{\mu}_i = \Omega_0^{1/2}(I - uu')\Omega_0^{-1/2} y_i \qquad (i=1, \ldots, n), \tag{7}$$

$$\hat{\alpha} = \Omega_0^{-1/2} u, \tag{8}$$

where u is the normalized eigenvector ($u'u=1$) associated with the smallest eigenvalue of

$$\Omega_0^{-1/2}\left(\sum_{i=1}^{n} y_i y_i'\right)\Omega_0^{-1/2}.$$

Proof. Letting

$$Y = (y_1, \ldots, y_n)', \qquad M = (\mu_1, \ldots, \mu_n)', \tag{9}$$

we write the loglikelihood as

$$\Lambda_n(M, \alpha) = -\tfrac{1}{2}mn \log 2\pi - \tfrac{1}{2}n \log|\Omega_0| - \tfrac{1}{2}\operatorname{tr}(Y-M)\Omega_0^{-1}(Y-M)'. \tag{10}$$

We wish to maximize Λ_n subject to the constraints $M\alpha = 0$ and $\alpha'\Omega_0\alpha = 1$. Since Ω_0 is given, the problem becomes

$$\text{minimize} \qquad \tfrac{1}{2}\operatorname{tr}(Y-M)\Omega_0^{-1}(Y-M)' \tag{11}$$

$$\text{subject to} \qquad M\alpha = 0 \quad \text{and} \quad \alpha'\Omega_0\alpha = 1. \tag{12}$$

The Lagrangian function is

$$\psi(M, \alpha) = \tfrac{1}{2}\operatorname{tr}(Y-M)\Omega_0^{-1}(Y-M)' - l'M\alpha - \lambda(\alpha'\Omega_0\alpha - 1), \tag{13}$$

where l is a vector of Lagrange multipliers and λ is a (scalar) Lagrange multiplier. The first differential is

$$\begin{aligned} d\psi &= -\operatorname{tr}(Y-M)\Omega_0^{-1}(dM)' - l'(dM)\alpha - l'M d\alpha - 2\lambda\alpha'\Omega_0 d\alpha \\ &= -\operatorname{tr}[(Y-M)\Omega_0^{-1} + l\alpha'](dM)' - (l'M + 2\lambda\alpha'\Omega_0)d\alpha \end{aligned} \tag{14}$$

and the first-order conditions are thus

$$(Y-M)\Omega_0^{-1} = -l\alpha' \tag{15}$$

$$M'l = -2\lambda\Omega_0\alpha \tag{16}$$

$$M\alpha = 0 \tag{17}$$

$$\alpha'\Omega_0\alpha = 1. \tag{18}$$

As usual we first solve for the Lagrange multipliers. Post-multiplying (15) by $\Omega_0\alpha$ yields,

$$l = -Y\alpha, \tag{19}$$

using (17) and (18). Also, pre-multiplying (16) by α' yields

$$\lambda=0 \tag{20}$$

in view of (17) and (18). Inserting (19) and (20) into (15)–(18) gives

$$M=Y-Y\alpha\alpha'\Omega_0 \tag{21}$$

$$M'Y\alpha=0 \tag{22}$$

$$\alpha'\Omega_0\alpha=1. \tag{23}$$

(Note that $M\alpha=0$ is automatically satisfied.) Inserting (21) into (22) gives

$$(Y'Y-\eta\Omega_0)\alpha=0, \tag{24}$$

where $\eta=\alpha'Y'Y\alpha$, which we rewrite as

$$(\Omega_0^{-1/2}Y'Y\Omega_0^{-1/2}-\eta I)\Omega_0^{1/2}\alpha=0. \tag{25}$$

Given (21) and (23) we have

$$\operatorname{tr}(Y-M)\Omega_0^{-1}(Y-M)'=\alpha'Y'Y\alpha=\eta. \tag{26}$$

But this is the function we wish to minimize! Hence we take η as the *smallest* eigenvalue of $\Omega_0^{-1/2}Y'Y\Omega_0^{-1/2}$ and $\Omega_0^{1/2}\alpha$ as the associated normalized eigenvector. This yields (8), the ML estimator of α. The ML estimator of M then follows from (21). □

Exercise

1. If α_0 is normalized by $e'\alpha_0=-1$ (rather than by $\alpha_0'\Omega_0\alpha_0=1$), show that the ML estimators (7) and (8) become

$$\hat{\mu}_i=\Omega_0^{1/2}\left(I-\frac{uu'}{u'u}\right)\Omega_0^{-1/2}y_i, \qquad \hat{\alpha}=\Omega_0^{-1/2}u,$$

where u is the eigenvector (normalized by $e'\Omega_0^{-1/2}u=-1$) associated with the smallest eigenvalue of

$$\Omega_0^{-1/2}\left(\sum_i y_iy_i'\right)\Omega_0^{-1/2}.$$

10. THE NONLINEAR REGRESSION MODEL WITH NORMAL ERRORS

Let us now consider a system of n *nonlinear* regression equations with normal errors, which we write as

$$y\sim\mathcal{N}_n(\mu(\gamma_0),\Omega(\gamma_0)). \tag{1}$$

Here γ_0 denotes the true (but unknown) value of the parameter vector to be estimated. We assume that $\gamma_0 \in \Gamma$, an open subset of $\mathbb{R}^p$, and that p (the dimension of Γ) is independent of n. We also assume that $\Omega(\gamma)$ is positive definite for every $\gamma \in \Gamma$, and that μ and Ω are twice differentiable on Γ. We define the $p \times 1$ vector $l(\gamma) = (l_j(\gamma))$,

$$l_j(\gamma) = \tfrac{1}{2}\operatorname{tr}\left(\frac{\partial \Omega^{-1}(\gamma)}{\partial \gamma_j}\Omega(\gamma)\right) + u'(\gamma)\Omega^{-1}(\gamma)\frac{\partial \mu(\gamma)}{\partial \gamma_j} - \tfrac{1}{2}u'(\gamma)\frac{\partial \Omega^{-1}(\gamma)}{\partial \gamma_j}u(\gamma), \tag{2}$$

where $u(\gamma) = y - \mu(\gamma)$, and the $p \times p$ matrix $\mathscr{F}(\gamma) = (\mathscr{F}_{ij}(\gamma))$,

$$\mathscr{F}_{ij}(\gamma) = \left(\frac{\partial \mu(\gamma)}{\partial \gamma_i}\right)'\Omega^{-1}(\gamma)\frac{\partial \mu(\gamma)}{\partial \gamma_j} + \tfrac{1}{2}\operatorname{tr}\left(\frac{\partial \Omega^{-1}(\gamma)}{\partial \gamma_i}\Omega(\gamma)\frac{\partial \Omega^{-1}(\gamma)}{\partial \gamma_j}\Omega(\gamma)\right). \tag{3}$$

Theorem 6

The ML estimator of γ_0 in the nonlinear regression model (1) is obtained as a solution of the vector equation $l(\gamma) = 0$; the information matrix is $\mathscr{F}(\gamma_0)$; and the asymptotic variance matrix of the ML estimator $\hat{\gamma}$ is

$$\left(\lim_{n\to\infty} (1/n)\,\mathscr{F}(\gamma_0)\right)^{-1} \tag{4}$$

if the limit exists.

Proof. The loglikelihood takes the form

$$\Lambda(\gamma) = -(n/2)\log 2\pi - \tfrac{1}{2}\log|\Omega(\gamma)| - \tfrac{1}{2}u'\Omega^{-1}(\gamma)u, \tag{5}$$

where $u = u(\gamma) = y - \mu(\gamma)$. The first differential is

$$\begin{aligned} \mathrm{d}\Lambda(\gamma) &= -\tfrac{1}{2}\operatorname{tr}\Omega^{-1}\mathrm{d}\Omega - u'\Omega^{-1}\mathrm{d}u - \tfrac{1}{2}u'(\mathrm{d}\Omega^{-1})u \\ &= \tfrac{1}{2}\operatorname{tr}\Omega(\mathrm{d}\Omega^{-1}) + u'\Omega^{-1}\mathrm{d}\mu - \tfrac{1}{2}u'(\mathrm{d}\Omega^{-1})u. \end{aligned} \tag{6}$$

Hence $\partial\Lambda(\gamma)/\partial\gamma = l(\gamma)$, and the first-order conditions are given by $l(\gamma) = 0$. The second differential is

$$\begin{aligned} \mathrm{d}^2\Lambda(\gamma) &= \tfrac{1}{2}\operatorname{tr}(\mathrm{d}\Omega)(\mathrm{d}\Omega^{-1}) + \tfrac{1}{2}\operatorname{tr}\Omega(\mathrm{d}^2\Omega^{-1}) + (\mathrm{d}u)'\Omega^{-1}\mathrm{d}\mu \\ &\quad + u'\mathrm{d}(\Omega^{-1}\mathrm{d}\mu) - u'(\mathrm{d}\Omega^{-1})\mathrm{d}u - \tfrac{1}{2}u'(\mathrm{d}^2\Omega^{-1})u. \end{aligned} \tag{7}$$

Equation (7) can be further expanded, but this is not necessary here. Notice that $\mathrm{d}^2\Omega^{-1}$ (and $\mathrm{d}^2\mu$) does not vanish unless Ω^{-1} (and μ) is a linear (or affine) function of γ. Taking expectations at $\gamma = \gamma_0$, we obtain (letting $\Omega_0 = \Omega(\gamma_0)$)

$$\begin{aligned} -\mathscr{E}\,\mathrm{d}^2\Lambda(\gamma) &= \tfrac{1}{2}\operatorname{tr}\Omega_0(\mathrm{d}\Omega^{-1})\Omega_0(\mathrm{d}\Omega^{-1}) - \tfrac{1}{2}\operatorname{tr}(\Omega_0\mathrm{d}^2\Omega^{-1}) + (\mathrm{d}\mu)'\Omega_0^{-1}\mathrm{d}\mu \\ &\quad + \tfrac{1}{2}\operatorname{tr}(\Omega_0\mathrm{d}^2\Omega^{-1}) \\ &= \tfrac{1}{2}\operatorname{tr}\Omega_0(\mathrm{d}\Omega^{-1})\Omega_0(\mathrm{d}\Omega^{-1}) + (\mathrm{d}\mu)'\Omega_0^{-1}\mathrm{d}\mu, \end{aligned} \tag{8}$$

because $\mathscr{E}u_0=0$, $\mathscr{E}u_0u_0'=\Omega_0$. This shows that the information matrix is $\mathscr{F}(\gamma_0)$ and concludes the proof. □

Exercise

1. Use (7) to obtain the Hessian matrix $\mathsf{H}_n(\gamma)$.

11. A SPECIAL CASE: FUNCTIONAL INDEPENDENCE OF MEAN PARAMETERS AND VARIANCE PARAMETERS

Theorem 6 is rather general in that the same parameters may appear in both μ and Ω. We often encounter the special case where

$$\gamma=(\beta', \theta')' \tag{1}$$

and μ only depends on the β parameters while Ω only depends on the θ parameters.

Theorem 7

The ML estimators of $\beta_0=(\beta_{01}, \ldots, \beta_{0k})'$ and $\theta_0=(\theta_{01}, \ldots, \theta_{0m})'$ in the nonlinear regression model

$$y\sim\mathscr{N}_n(\mu(\beta_0), \Omega(\theta_0)) \tag{2}$$

are obtained by solving the equations

$$(y-\mu(\beta))'\Omega^{-1}(\theta)\frac{\partial\mu(\beta)}{\partial\beta_h}=0 \qquad (h=1, \ldots, k) \tag{3}$$

and

$$\operatorname{tr}\left(\frac{\partial\Omega^{-1}(\theta)}{\partial\theta_j}\Omega(\theta)\right)=(y-\mu(\beta))'\frac{\partial\Omega^{-1}(\theta)}{\partial\theta_j}(y-\mu(\beta)). \tag{4}$$

The information matrix for β_0 and θ_0 is $\mathscr{F}(\beta_0, \theta_0)$ where

$$\mathscr{F}(\beta, \theta)=$$

$$\begin{bmatrix} \left(\frac{\partial\mu(\beta)}{\partial\beta'}\right)'\Omega^{-1}(\theta)\left(\frac{\partial\mu(\beta)}{\partial\beta'}\right) & 0 \\ 0 & \frac{1}{2}\left(\frac{\partial\operatorname{vec}\Omega(\theta)}{\partial\theta'}\right)'(\Omega^{-1}(\theta)\otimes\Omega^{-1}(\theta))\left(\frac{\partial\operatorname{vec}\Omega(\theta)}{\partial\theta'}\right) \end{bmatrix}$$

and the asymptotic variance matrix of the ML estimators $\hat{\beta}$ and $\hat{\theta}$ is

$$\left(\lim_{n\to\infty}(1/n)\mathscr{F}(\beta_0, \theta_0)\right)^{-1},$$

if the limit exists.

Proof. Immediate from Theorem 6 and the expressions (10.2) and (10.3). □

Exercises

1. Under the conditions of Theorem 7 show that the asymptotic variance matrix of $\hat{\beta}$, denoted $\mathscr{V}_{as}(\hat{\beta})$, is
$$\mathscr{V}_{as}(\hat{\beta})=\left(\lim_{n\to\infty}(1/n)S_0'\Omega^{-1}(\theta_0)S_0\right)^{-1},$$
where S_0 denotes the $n\times k$ matrix of partial derivatives $\partial\mu(\beta)/\partial\beta'$ evaluated at β_0.
2. In particular, in the linear regression model $y\sim\mathscr{N}_n(X\beta_0, \Omega(\theta_0))$, show that
$$\mathscr{V}_{as}(\hat{\beta})=\left(\lim_{n\to\infty}(1/n)X'\Omega^{-1}(\theta_0)X\right)^{-1}.$$

12. GENERALIZATION OF THEOREM 6

In Theorem 6 we assumed that both μ and Ω depend on all the parameters in the system. In Theorem 7 we assumed that μ depends on some parameters β while Ω depends on some other parameters θ, and that μ does not depend on θ or Ω on β. The most general case, which we discuss in this section, assumes that β and θ may partially overlap. The following two theorems present the first-order conditions and the information matrix for this case.

Theorem 8

The ML estimators of $\beta_0=(\beta_{01},\ldots,\beta_{0k})'$, $\zeta_0=(\zeta_{01},\ldots,\zeta_{0l})'$ and $\theta_0=(\theta_{01},\ldots,\theta_{0m})'$ in the nonlinear regression model
$$y\sim\mathscr{N}_n(\mu(\beta_0,\zeta_0),\Omega(\theta_0,\zeta_0)) \tag{1}$$
are obtained by solving the equations
$$u'\Omega^{-1}\frac{\partial\mu}{\partial\beta_h}=0 \qquad (h=1,\ldots,k) \tag{2}$$
$$\tfrac{1}{2}\operatorname{tr}\left(\frac{\partial\Omega^{-1}}{\partial\zeta_i}\Omega\right)+u'\Omega^{-1}\frac{\partial\mu}{\partial\zeta_i}-\tfrac{1}{2}u'\frac{\partial\Omega^{-1}}{\partial\zeta_i}u=0 \qquad (i=1,\ldots,l) \tag{3}$$
$$\operatorname{tr}\left(\frac{\partial\Omega^{-1}}{\partial\theta_j}\Omega\right)=u'\frac{\partial\Omega^{-1}}{\partial\theta_j}u \qquad (j=1,\ldots,m), \tag{4}$$
where $u=y-\mu(\beta,\zeta)$.

Proof. Let $\gamma=(\beta',\zeta',\theta')'$. We know from Theorem 6 that we must solve the vector equation $l(\gamma)=0$, where the elements of l are given in (10.2). The results follows. □

Theorem 9

The information matrix for β_0, ζ_0 and θ_0 in the nonlinear regression model (1) is $\mathscr{F}(\beta_0, \zeta_0, \theta_0)$, where

$$\mathscr{F}(\beta, \zeta, \theta)=\begin{bmatrix} \mathscr{F}_{\beta\beta} & \mathscr{F}_{\beta\zeta} & 0 \\ \mathscr{F}_{\zeta\beta} & \mathscr{F}_{\zeta\zeta} & \mathscr{F}_{\zeta\theta} \\ 0 & \mathscr{F}_{\theta\zeta} & \mathscr{F}_{\theta\theta} \end{bmatrix} \tag{5}$$

and

$$\mathscr{F}_{\beta\beta}=(\mathsf{D}_\beta\mu)'\Omega^{-1}(\mathsf{D}_\beta\mu) \tag{6}$$

$$\mathscr{F}_{\beta\zeta}=(\mathsf{D}_\beta\mu)'\Omega^{-1}(\mathsf{D}_\zeta\mu) \tag{7}$$

$$\mathscr{F}_{\zeta\zeta}=(\mathsf{D}_\zeta\mu)'\Omega^{-1}(\mathsf{D}_\zeta\mu)+\tfrac{1}{2}(\mathsf{D}_\zeta \operatorname{vec}\Omega)'(\Omega^{-1}\otimes\Omega^{-1})(\mathsf{D}_\zeta \operatorname{vec}\Omega) \tag{8}$$

$$\mathscr{F}_{\zeta\theta}=\tfrac{1}{2}(\mathsf{D}_\zeta \operatorname{vec}\Omega)'(\Omega^{-1}\otimes\Omega^{-1})(\mathsf{D}_\theta \operatorname{vec}\Omega) \tag{9}$$

$$\mathscr{F}_{\theta\theta}=\tfrac{1}{2}(\mathsf{D}_\theta \operatorname{vec}\Omega)'(\Omega^{-1}\otimes\Omega^{-1})(\mathsf{D}_\theta \operatorname{vec}\Omega) \tag{10}$$

and where, as the notation indicates,

$$\mathsf{D}_\beta\mu=\frac{\partial\mu(\beta, \zeta)}{\partial\beta'}, \qquad \mathsf{D}_\zeta\mu=\frac{\partial\mu(\beta, \zeta)}{\partial\zeta'}, \tag{11}$$

$$\mathsf{D}_\theta \operatorname{vec}\Omega=\frac{\partial \operatorname{vec}\Omega(\theta, \zeta)}{\partial\theta'}, \quad \mathsf{D}_\zeta \operatorname{vec}\Omega=\frac{\partial \operatorname{vec}\Omega(\theta, \zeta)}{\partial\zeta'}. \tag{12}$$

Moreover, if $(1/n)\mathscr{F}(\beta_0, \zeta_0, \theta_0)$ tends to a finite positive definite matrix, say G, partitioned as

$$G=\begin{bmatrix} G_{\beta\beta} & G_{\beta\zeta} & 0 \\ G_{\zeta\beta} & G_{\zeta\zeta} & G_{\zeta\theta} \\ 0 & G_{\theta\zeta} & G_{\theta\theta} \end{bmatrix}, \tag{13}$$

then the asymptotic variance matrix of the ML estimators $\hat\beta$, $\hat\zeta$ and $\hat\theta$ is

$$G^{-1}=$$

$$\begin{bmatrix} G_{\beta\beta}^{-1}+G_{\beta\beta}^{-1}G_{\beta\zeta}Q^{-1}G_{\zeta\beta}G_{\beta\beta}^{-1} & -G_{\beta\beta}^{-1}G_{\beta\zeta}Q^{-1} & G_{\beta\beta}^{-1}G_{\beta\zeta}Q^{-1}G_{\zeta\theta}G_{\theta\theta}^{-1} \\ -Q^{-1}G_{\zeta\beta}G_{\beta\beta}^{-1} & Q^{-1} & -Q^{-1}G_{\zeta\theta}G_{\theta\theta}^{-1} \\ G_{\theta\theta}^{-1}G_{\theta\zeta}Q^{-1}G_{\zeta\beta}G_{\beta\beta}^{-1} & -G_{\theta\theta}^{-1}G_{\theta\zeta}Q^{-1} & G_{\theta\theta}^{-1}+G_{\theta\theta}^{-1}G_{\theta\zeta}Q^{-1}G_{\zeta\theta}G_{\theta\theta}^{-1} \end{bmatrix} \tag{14}$$

where

$$Q=G_{\zeta\zeta}-G_{\zeta\beta}G_{\beta\beta}^{-1}G_{\beta\zeta}-G_{\zeta\theta}G_{\theta\theta}^{-1}G_{\theta\zeta}. \tag{15}$$

Proof. **The structure of the information matrix follows from Theorem 6 and (10.3). The inverse of its limit follows from Theorem 1.3.** □

MISCELLANEOUS EXERCISES

1. Consider an m-dimensional system of demand equations

$$y_t = a + \Gamma f_t + v_t \qquad (t = 1, \dots, n),$$

where

$$f_t = (1/\imath'\Gamma\imath)(\imath' y_t)\imath + \quad Cz_t, \qquad C = I_m - (1/\imath'\Gamma\imath)\imath\imath'\Gamma,$$

$\imath$ is the $m \times 1$ summation vector $(1, 1, \dots, 1)'$, and Γ is diagonal. Let the $m \times 1$ vectors v_t be independently and identically distributed as $\mathcal{N}(0, \Omega)$. It is easy to see that $\imath'\Gamma f_t = \imath' y_t (t = 1, \dots, n)$ almost surely, and hence that $\imath' a = 0$, $\imath' v_t = 0$ $(t = 1, \dots, n)$ almost surely, and $\Omega\imath = 0$. Assume that $r(\Omega) = m - 1$ and denote the positive eigenvalues of Ω by $\lambda_1, \dots, \lambda_{m-1}$.

(a) Show that the loglikelihood of the sample is

$$\log L = \text{constant} - (n/2) \sum_{i=1}^{m-1} \log \lambda_i - (1/2)\text{tr } \Omega^+ V'V,$$

where $V = (v_1, \dots, v_n)'$. (The density of a singular normal distribution is given, e.g., in Mardia, Kent and Bibby, 1992, 41.)

(b) Show that the concentrated (with respect to Ω) loglikelihood is

$$\log L_c = \text{constant} - (n/2) \sum_{i=1}^{m-1} \log \mu_i,$$

where $\mu_1, \dots, \mu_{m-1}$ are the positive eigenvalues of $(1/n)V'V$. [*Hint*: Use Miscellaneous Exercise 8.7 to show that $\mathrm{d}\Omega^+ = -\Omega^+(\mathrm{d}\Omega)\Omega^+$, since Ω has locally constant rank.]

(c) Show that $\log L_c$ can be equivalently written as

$$\log L_c = \text{constant} - (n/2) \log |A|,$$

where

$$A = (1/n)V'V + (1/m)\imath\imath'.$$

[*Hint*: Use Exercise 1.11.3 and Theorem 3.5.]

(d) Show that the first-order condition with respect to a is given by

$$\sum_{t=1}^{n} (y_t - a - \Gamma f_t) = 0,$$

irrespective of whether we take account of the constraint $\imath' a = 0$.

(e) Show that the first-order condition with respect to $\gamma = \Gamma\imath$ is given by

$$\sum_{t=1}^{n} F_t C A^{-1} C' (y_t - a - \Gamma f_t) = 0,$$

where F_t is the diagonal matrix whose diagonal elements are the components of f_t (Barten 1969).

2. Let the random $p \times 1$ vectors $y_1, y_2, \ldots, y_n$ be independently distributed such that

$$y_t \sim \mathcal{N}_p(AB_0 c_t, \Omega_0) \qquad (t=1, \ldots, n)$$

where Ω_0 is positive definite, A is a known $p \times q$ matrix and $c_1, \ldots, c_n$ are known $k \times 1$ vectors. The matrices B_0 and Ω_0 are to be estimated. Let $C=(c_1, \ldots, c_n)$, $Y=(y_1, \ldots, y_n)'$ and denote the ML estimators of B_0 and Ω_0 by $\hat{B}$ and $\hat{\Omega}$. Assume that $r(C) \leqslant n-p$ and prove that

$$\hat{\Omega}=(1/n)(Y'-A\hat{B}C)(Y'-A\hat{B}C)'$$
$$A\hat{B}C=A(A'S^{-1}A)^+A'S^{-1}Y'C^+C,$$

where

$$S=(1/n)Y'(I-C^+C)Y$$

(cf. Von Rosen 1985).

BIBLIOGRAPHICAL NOTES

§2. The method of maximum likelihood originated with R. A. Fisher in the 1920s. Important contributions were made by H. Cramér, C. R. Rao and A. Wald. See Norden (1972, 1973) for a survey of maximum likelihood estimation. See also Cramer (1986).
§4–§6. Implicit versus explicit treatment of symmetry is discussed in more depth in Magnus (1988). The second proof of Theorem 1 and the proof of Theorem 2 follow Magnus and Neudecker (1980). The fourth proof of Theorem 1 follows Neudecker (1980a).
§8. See Zellner (1962) and Malinvaud (1966, chap. 6).
§9. See Madansky (1976), Malinvaud (1966, chap. 10) and Pollock (1979, chap. 8).
§10–§12. See Holly (1985) and Heijmans and Magnus (1986).

CHAPTER 16

Simultaneous equations

1. INTRODUCTION

In Chapter 13 we considered the simple linear regression model

$$y_t = x_t'\beta_0 + u_t \qquad (t=1, \ldots, n), \tag{1}$$

where y_t and u_t are scalar random variables and x_t and β_0 are $k \times 1$ vectors. In section 8 of Chapter 15 we generalized (1) into the *multivariate* linear regression model

$$y_t' = x_t'B_0 + u_t' \qquad (t=1, \ldots, n), \tag{2}$$

where y_t and u_t are random $m \times 1$ vectors, x_t is a $k \times 1$ vector and B_0 a $k \times m$ matrix.

In this chapter we consider a further generalization, where the model is specified by

$$y_t'\Gamma_0 + x_t'B_0 = u_t' \qquad (t=1, \ldots, n). \tag{3}$$

This model is known as the *simultaneous equations model.*

2. THE SIMULTANEOUS EQUATIONS MODEL

Thus, let economic theory specify a set of economic relations of the form

$$y_t'\Gamma_0 + x_t'B_0 = u_{0t}' \qquad (t=1, \ldots, n), \tag{1}$$

where y_t is an $m \times 1$ vector of observed endogenous variables, x_t is a $k \times 1$ vector of observed exogenous (non-random) variables and u_{0t} is an $m \times 1$ vector of unobserved random disturbances. The $m \times m$ matrix Γ_0 and the $k \times m$ matrix B_0 are unknown parameter matrices. We shall make the following assumption.

Assumption 1 (normality)

The vectors $\{u_{0t}, t=1, \ldots, n\}$ are independent and identically distributed as $\mathcal{N}(0, \Sigma_0)$ with Σ_0 a positive definite $m \times m$ matrix of unknown parameters.

Lemma 1

Given Assumption 1, the $m \times m$ matrix Γ_0 is non-singular.

Proof. Assume Γ_0 is singular and let a be an $m \times 1$ vector such that $\Gamma_0 a = 0$ and $a \neq 0$. Post-multiplying (1) by a then yields

$$x_t' B_0 a = u_{0t}' a. \tag{2}$$

Since the left-hand side of (2) is non-random, the variance of the random variable on the right-hand side must be zero. Hence $a' \Sigma_0 a = 0$, which contradicts the non-singularity of Σ_0. □

Given the non-singularity of Γ_0 we may post-multiply (1) with Γ_0^{-1}, thus obtaining the *reduced form*:

$$y_t' = x_t' \Pi_0 + v_{0t}' \qquad (t = 1, \ldots, n), \tag{3}$$

where

$$\Pi_0 = -B_0 \Gamma_0^{-1}, \qquad v_{0t}' = u_{0t}' \Gamma_0^{-1}. \tag{4}$$

Combining the observations we define

$$Y = (y_1, \ldots, y_n)', \qquad X = (x_1, \ldots, x_n)' \tag{5}$$

and, similarly,

$$U_0 = (u_{01}, \ldots, u_{0n})', \qquad V_0 = (v_{01}, \ldots, v_{0n})'. \tag{6}$$

Then we rewrite the structure (1) as

$$Y\Gamma_0 + XB_0 = U_0 \tag{7}$$

and the reduced form (3) as

$$Y = X\Pi_0 + V_0. \tag{8}$$

It is clear that the vectors $\{v_{0t}, t = 1, \ldots, n\}$ are independent and identically distributed as $\mathcal{N}(0, \Omega_0)$ where $\Omega_0 = \Gamma_0'^{-1} \Sigma_0 \Gamma_0^{-1}$. The loglikelihood function expressed in terms of the reduced form parameters (Π, Ω) follows from (15.8.12):

$$\Lambda_n(\Pi, \Omega) = -\tfrac{1}{2} mn \log 2\pi - \tfrac{1}{2} n \log |\Omega| - \tfrac{1}{2} \operatorname{tr} \Omega^{-1} W, \tag{9}$$

where

$$W = \sum_{t=1}^{n} (y_t' - x_t' \Pi)'(y_t' - x_t' \Pi) = (Y - X\Pi)'(Y - X\Pi). \tag{10}$$

Rewriting (9) in terms of (B, Γ, Σ), using $\Pi = -B\Gamma^{-1}$ and $\Omega = \Gamma'^{-1} \Sigma \Gamma^{-1}$, we obtain

$$\Lambda_n(B, \Gamma, \Sigma) = -\tfrac{1}{2} mn \log 2\pi + \tfrac{1}{2} n \log |\Gamma' \Gamma| - \tfrac{1}{2} n \log |\Sigma| - \tfrac{1}{2} \operatorname{tr} \Sigma^{-1} W^*, \tag{11}$$

where

$$W^* = \sum_{t=1}^{n} (y_t' \Gamma + x_t' B)'(y_t' \Gamma + x_t' B) = (Y\Gamma + XB)'(Y\Gamma + XB). \tag{12}$$

The essential feature of (11) is the presence of the Jacobian term $\frac{1}{2}\log|\Gamma'\Gamma|$ of the transformation from u_t to y_t.

There are two problems relating to the simultaneous equations model: the identification problem and the estimation problem. We shall discuss the identification problem first.

Exercise

1. In (11) we write $\frac{1}{2}\log|\Gamma'\Gamma|$ rather than $\log|\Gamma|$. Why?

3. THE IDENTIFICATION PROBLEM

It is clear that knowledge of the structural parameters $(B_0, \Gamma_0, \Sigma_0)$ implies knowledge of the reduced form parameters (Π_0, Ω_0), but that the converse is not true. It is also clear that a non-singular transformation of (2.1), say

$$y_t'\Gamma_0 G + x_t' B_0 G = u_{0t}' G, \qquad (1)$$

leads to the same loglikelihood (2.11) and the same reduced form parameters (Π_0, Ω_0). We say that $(B_0, \Gamma_0, \Sigma_0)$ and $(B_0 G, \Gamma_0 G, G'\Sigma_0 G)$ are *observationally equivalent*, and that therefore $(B_0, \Gamma_0, \Sigma_0)$ is not *identified*. The following definition makes these concepts precise.

Definition

Let $z = (z_1, \ldots, z_n)$ be a vector of random observations with continuous density function $h(z; \gamma_0)$ where γ_0 is a p-dimensional parameter vector lying in an open set $\Gamma \subset \mathbb{R}^p$. Let $\Lambda(\gamma; z)$ be the loglikelihood function. Then

(i) two parameter points γ and γ^* are *observationally equivalent* if $\Lambda(\gamma; z) = \Lambda(\gamma^*; z)$ for all z;
(ii) a parameter point γ in Γ is *(globally) identified* if there is no other point in Γ which is observationally equivalent;
(iii) a parameter point γ in Γ is *locally identified* if there exists an open neighbourhood $N(\gamma)$ of γ such that no other point of $N(\gamma)$ is observationally equivalent to γ.

The following assumption is essential for the reduced form parameter Π_0 to be identified.

Assumption 2 (rank)

The $n \times k$ matrix X has full column rank k.

Theorem 1

Consider the simultaneous equations model (2.1) under the normality assumption (Assumption 1) and rank condition (Assumption 2). Then, (i) the joint density of $(y_1, \ldots, y_n)$ depends on $(B_0, \Gamma_0, \Sigma_0)$ only through the reduced form parameters (Π_0, Ω_0); and (ii) Π_0 and Ω_0 are globally identified.

Proof. Since $Y=(y_1, \ldots, y_n)'$ is normally distributed, its density function depends only on its first two moments,

$$\mathscr{E} Y = X\Pi_0, \qquad \mathscr{V}(\operatorname{vec} Y) = \Omega_0 \otimes I_n. \tag{2}$$

Now, X has full column rank, so $X'X$ is non-singular and hence knowledge of these two moments is equivalent to knowledge of (Π_0, Ω_0). Thus the density of Y depends *only* on (Π_0, Ω_0). This proves (i). But it also shows that if we know the density of Y, we know the value of (Π_0, Ω_0), thus proving (ii). □

As a consequence of Theorem 1, a structural parameter of $(B_0, \Gamma_0, \Sigma_0)$ is identified if and only if its value can be deduced from the reduced form parameters (Π_0, Ω_0). Since without *a priori* restrictions on (B, Γ, Σ) *none* of the structural parameters are identified (Why not?), we introduce constraints

$$\psi_i(B, \Gamma, \Sigma) = 0 \qquad (i=1, \ldots, r). \tag{3}$$

The identifiability of the structure $(B_0, \Gamma_0, \Sigma_0)$ satisfying (3) then depends on the uniqueness of solutions of

$$\Pi_0 \Gamma + B = 0, \tag{4}$$

$$\Gamma' \Omega_0 \Gamma - \Sigma = 0, \tag{5}$$

$$\psi_i(B, \Gamma, \Sigma) = 0 \qquad (i=1, \ldots, r). \tag{6}$$

4. IDENTIFICATION WITH LINEAR CONSTRAINTS ON B AND Γ ONLY

In this section we shall assume that all prior information is in the form of linear restrictions on B and Γ, apart from the obvious symmetry constraint on Σ. We shall prove our next theorem.

Theorem 2

Consider the simultaneous equations model (2.1) under the normality assumption (Assumption 1) and rank condition (Assumption 2). Assume further that prior information is available in the form of linear restrictions on B and Γ:

$$R_1 \operatorname{vec} B + R_2 \operatorname{vec} \Gamma = r. \tag{1}$$

Then $(B_0, \Gamma_0, \Sigma_0)$ is globally identified if and only if the matrix

$$R_1(I_m \otimes B_0) + R_2(I_m \otimes \Gamma_0) \tag{2}$$

has full column rank m^2.

Proof. The identifiability of the structure $(B_0, \Gamma_0, \Sigma_0)$ depends on the uniqueness of solutions of

$$\Pi_0\Gamma + B = 0 \tag{3}$$

$$\Gamma'\Omega_0\Gamma - \Sigma = 0 \tag{4}$$

$$R_1 \operatorname{vec} B + R_2 \operatorname{vec} \Gamma - r = 0 \tag{5}$$

$$\Sigma = \Sigma'. \tag{6}$$

Now (6) is redundant since it is implied by (4). From (3) we obtain

$$\operatorname{vec} B = -(I_m \otimes \Pi_0) \operatorname{vec} \Gamma \tag{7}$$

and, from (4), $\Sigma = \Gamma'\Omega_0\Gamma$. Inserting (7) into (5) we see that the identifiability hinges on the uniqueness of solutions of the linear equation

$$[R_2 - R_1(I_m \otimes \Pi_0)] \operatorname{vec} \Gamma = r. \tag{8}$$

By Theorem 2.12, equation (8) has a unique solution for vec Γ if and only if the matrix $R_2 - R_1(I_m \otimes \Pi_0)$ has full column rank m^2. Post-multiplying this matrix by the non-singular matrix $I_m \otimes \Gamma_0$ we obtain (2). □

5. IDENTIFICATION WITH LINEAR CONSTRAINTS ON B, Γ AND Σ

In Theorem 2 we obtained a *global* result, but this is only possible if the constraint functions are linear in B and Γ and independent of Σ. The reason is that, even with linear constraints on B, Γ and Σ, our problem becomes one of solving a system of *nonlinear* equations, for which in general only *local* results can be obtained.

Theorem 3

Consider the simultaneous equations model (2.1) under the normality assumption (Assumption 1) and rank condition (Assumption 2). Assume further that prior information is available in the form of linear restrictions on B, Γ and Σ:

$$R_1 \operatorname{vec} B + R_2 \operatorname{vec} \Gamma + R_3 \mathrm{v}(\Sigma) = r. \tag{1}$$

Then $(B_0, \Gamma_0, \Sigma_0)$ is locally identified if the matrix

$$W = R_1(I_m \otimes B_0) + R_2(I_m \otimes \Gamma_0) + 2R_3 D_m^+(I_m \otimes \Sigma_0) \tag{2}$$

has full column rank m^2.

Remark. If we define the parameter set P as the set of all (B, Γ, Σ) such that Γ is non-singular and Σ is positive definite, and the restricted parameter set P' as the subset of P satisfying the restriction

$$R_1 \operatorname{vec} B + R_2 \operatorname{vec} \Gamma + R_3 \mathrm{v}(\Sigma) = r, \tag{3}$$

then condition (2), which is sufficient for the local identification of $(B_0, \Gamma_0, \Sigma_0)$, becomes a necessary condition as well if it is assumed that there exists an open neighbourhood of $(B_0, \Gamma_0, \Sigma_0)$ in the restricted parameter set P' in which the matrix

$$W(B, \Gamma, \Sigma) = R_1(I_m \otimes B) + R_2(I_m \otimes \Gamma) + 2R_3 D_m^+(I_m \otimes \Sigma) \tag{4}$$

has constant rank.

Proof. The identifiability of $(B_0, \Gamma_0, \Sigma_0)$ depends on the uniqueness of solutions of

$$\Pi_0 \Gamma + B = 0 \tag{5}$$

$$\Gamma' \Omega_0 \Gamma - \Sigma = 0 \tag{6}$$

$$R_1 \operatorname{vec} B + R_2 \operatorname{vec} \Gamma + R_3 v(\Sigma) - r = 0. \tag{7}$$

The symmetry of Σ follows again from the symmetry of Ω_0 and (6). Equations (5)–(7) form a system of *nonlinear* equations (because of (6)) in B, Γ and $v(\Sigma)$. Differentiating (5)–(7) gives

$$\Pi_0 \mathrm{d}\Gamma + \mathrm{d}B = 0 \tag{8}$$

$$(\mathrm{d}\Gamma)' \Omega_0 \Gamma + \Gamma' \Omega_0 (\mathrm{d}\Gamma) - \mathrm{d}\Sigma = 0 \tag{9}$$

$$R_1 \, \mathrm{d} \operatorname{vec} B + R_2 \, \mathrm{d} \operatorname{vec} \Gamma + R_3 \, \mathrm{d}v(\Sigma) = 0 \tag{10}$$

and hence, upon taking vecs in (8) and (9),

$$(I_m \otimes \Pi_0) \, \mathrm{d} \operatorname{vec} \Gamma + \mathrm{d} \operatorname{vec} B = 0 \tag{11}$$

$$(\Gamma' \Omega_0 \otimes I_m) \, \mathrm{d} \operatorname{vec} \Gamma' + (I_m \otimes \Gamma' \Omega_0) \, \mathrm{d} \operatorname{vec} \Gamma - \mathrm{d} \operatorname{vec} \Sigma = 0 \tag{12}$$

$$R_1 \, \mathrm{d} \operatorname{vec} B + R_2 \, \mathrm{d} \operatorname{vec} \Gamma + R_3 \, \mathrm{d}v(\Sigma) = 0. \tag{13}$$

Writing $\operatorname{vec} \Sigma = D_m v(\Sigma)$, $\operatorname{vec} \Gamma' = K_m \operatorname{vec} \Gamma$ and using Theorem 3.9(a), (12) becomes

$$(I_{m^2} + K_m)(I_m \otimes \Gamma' \Omega_0) \, \mathrm{d} \operatorname{vec} \Gamma - D_m \, \mathrm{d}v(\Sigma) = 0. \tag{14}$$

From (11), (14) and (13) we obtain the Jacobian matrix

$$J(\Gamma) = \begin{bmatrix} I_m \otimes \Pi_0 & I_{mk} & 0 \\ (I_{m^2} + K_m)(I_m \otimes \Gamma' \Omega_0) & 0 & -D_m \\ R_2 & R_1 & R_3 \end{bmatrix}, \tag{15}$$

where we notice that J depends on Γ, but not on B and Σ. (This follows of course from the fact that the only nonlinearity in (5)–(7) is in Γ.) A sufficient condition for $(B_0, \Gamma_0, \Sigma_0)$ to be locally identifiable is that J evaluated at Γ_0 has full column rank. (This follows essentially from the implicit function theorem.) But, when evaluated at Γ_0, we can write

$$J(\Gamma_0)=\begin{bmatrix} 0 & I_{mk} & 0 \\ 0 & 0 & -D_m \\ W & R_1 & R_3 \end{bmatrix}\begin{bmatrix} (I_m\otimes\Gamma_0)^{-1} & 0 & 0 \\ I_m\otimes\Pi_0 & I_{mk} & 0 \\ -2D_m^+(I_m\otimes\Gamma'\Omega_0) & 0 & I_{1/2m(m+1)} \end{bmatrix}, \tag{16}$$

using the fact that $D_mD_m^+=\frac{1}{2}(I_{m^2}+K_m)$, see Theorem 3.12(b). The second partitioned matrix in (16) is non-singular. Hence $J(\Gamma_0)$ has full column rank if and only if the first partitioned matrix in (16) has full column rank; this, in turn, is the case if and only if W has full column rank. □

6. NONLINEAR CONSTRAINTS

Exactly the same techniques as used in establishing Theorem 3 (linear constraints) enable us to establish Theorem 4 (nonlinear constraints).

Theorem 4

Consider the simultaneous equations model (2.1) under the normality assumption (Assumption 1) and rank condition (Assumption 2). Assume that prior information is available in the form of nonlinear continuously differentiable restrictions on B, Γ and Σ:

$$f(B,\Gamma,\mathrm{v}(\Sigma))=0. \tag{1}$$

Then $(B_0, \Gamma_0, \Sigma_0)$ is locally identified if the matrix

$$W=R_1(I_m\otimes B_0)+R_2(I_m\otimes\Gamma_0)+2R_3D_m^+(I_m\otimes\Sigma_0) \tag{2}$$

has full column rank m^2 where the matrices

$$R_1=\frac{\partial f}{\partial(\operatorname{vec} B)'},\qquad R_2=\frac{\partial f}{\partial(\operatorname{vec}\Gamma)'},\qquad R_3=\frac{\partial f}{\partial(\mathrm{v}(\Sigma))'} \tag{3}$$

are evaluated at $(B_0, \Gamma_0, \mathrm{v}(\Sigma_0))$.

Proof. The proof is left as an exercise. □

7. FULL-INFORMATION MAXIMUM LIKELIHOOD (FIML): THE INFORMATION MATRIX (GENERAL CASE)

We now turn to the problem of estimating simultaneous equations models, assuming that sufficient restrictions are present for identification. Maximum likelihood estimation of the structural parameters $(B_0, \Gamma_0, \Sigma_0)$ calls for maximization of the loglikelihood function (2.11) subject to the *a priori* and identifying constraints. This method of estimation is known as full-information maximum likelihood (FIML). Finding the FIML estimates involves nonlinear optimization and can be computationally burdensome. We shall first find the information matrix for the rather general case where every element of B, Γ and Σ can be expressed as a known function of some parameter θ.

Theorem 5

Consider a random sample of size n from the process defined by the simultaneous equations model (2.1) under the normality assumption (Assumption 1) and the rank condition (Assumption 2). Assume that (B, Γ, Σ) satisfies certain *a priori* (nonlinear) twice differentiable constraints

$$B=B(\theta), \qquad \Gamma=\Gamma(\theta), \qquad \Sigma=\Sigma(\theta), \tag{1}$$

where θ is an unknown parameter vector. The true value of θ is denoted θ_0, so that $B_0=B(\theta_0)$, $\Gamma_0=\Gamma(\theta_0)$ and $\Sigma_0=\Sigma(\theta_0)$. Let $\Lambda_n(\theta)$ be the loglikelihood, so that

$$\begin{aligned}\Lambda_n(\theta)=&-(mn/2)\log 2\pi+(n/2)\log|\Gamma'\Gamma|-(n/2)\log|\Sigma|\\ &-\tfrac{1}{2}\operatorname{tr}\Sigma^{-1}(Y\Gamma+XB)'(Y\Gamma+XB).\end{aligned} \tag{2}$$

Then the information matrix $\mathscr{F}_n(\theta_0)$, determined by

$$-\mathscr{E}\mathrm{d}^2\Lambda_n(\theta_0)=(\mathrm{d}\theta)'\mathscr{F}_n(\theta_0)\mathrm{d}\theta, \tag{3}$$

is given by

$$\mathscr{F}_n(\theta_0)=$$

$$n\begin{pmatrix}\Delta_1\\ \Delta_2\end{pmatrix}'\begin{pmatrix}K_{m+k,m}[\Gamma_0^{-1}J\otimes(\Gamma_0^{-1}J)']+P_n\otimes\Sigma_0^{-1} & -(\Gamma_0^{-1}J)'\otimes\Sigma_0^{-1}\\ -\Gamma_0^{-1}J\otimes\Sigma_0^{-1} & \tfrac{1}{2}\Sigma_0^{-1}\otimes\Sigma_0^{-1}\end{pmatrix}\begin{pmatrix}\Delta_1\\ \Delta_2\end{pmatrix}, \tag{4}$$

where

$$\Delta_1=\partial\operatorname{vec}A'/\partial\theta', \qquad \Delta_2=\partial\operatorname{vec}\Sigma/\partial\theta', \tag{5}$$

$$P_n=\begin{pmatrix}\Pi_0'Q_n\Pi_0+\Omega_0 & \Pi_0'Q_n\\ Q_n\Pi_0 & Q_n\end{pmatrix}, \qquad Q_n=(1/n)X'X, \tag{6}$$

$$\Pi_0=-B_0\Gamma_0^{-1}, \qquad \Omega_0=\Gamma_0'^{-1}\Sigma_0\Gamma_0^{-1}, \qquad A'=(\Gamma':B'), \qquad J=(I_m:0), \tag{7}$$

and Δ_1 and Δ_2 are evaluated at θ_0.

Proof. We rewrite the loglikelihood as

$$\Lambda_n(\theta)=\text{constant}+\tfrac{1}{2}n\log|\Gamma'\Gamma|-\tfrac{1}{2}n\log|\Sigma|-\tfrac{1}{2}\operatorname{tr}\Sigma^{-1}A'Z'ZA, \tag{8}$$

where $Z=(Y:X)$. The first differential is

$$\begin{aligned}\mathrm{d}\Lambda_n=&\,n\operatorname{tr}\Gamma^{-1}\mathrm{d}\Gamma-\tfrac{1}{2}n\operatorname{tr}\Sigma^{-1}\mathrm{d}\Sigma-\operatorname{tr}\Sigma^{-1}A'Z'Z\mathrm{d}A\\ &+\tfrac{1}{2}\operatorname{tr}\Sigma^{-1}(\mathrm{d}\Sigma)\Sigma^{-1}A'Z'ZA,\end{aligned} \tag{9}$$

and the second differential

$$\begin{aligned}\mathrm{d}^2\Lambda_n = & -n\,\mathrm{tr}\,(\Gamma^{-1}\mathrm{d}\Gamma)^2 + n\,\mathrm{tr}\,\Gamma^{-1}\mathrm{d}^2\Gamma + \tfrac{1}{2}n\,\mathrm{tr}\,(\Sigma^{-1}\mathrm{d}\Sigma)^2 \\ & -\tfrac{1}{2}n\,\mathrm{tr}\,\Sigma^{-1}\mathrm{d}^2\Sigma - \mathrm{tr}\,\Sigma^{-1}(\mathrm{d}A)'Z'Z\mathrm{d}A \\ & +2\,\mathrm{tr}\,\Sigma^{-1}(\mathrm{d}\Sigma)\Sigma^{-1}A'Z'Z\mathrm{d}A - \mathrm{tr}\,\Sigma^{-1}A'Z'Z\mathrm{d}^2A \\ & -\mathrm{tr}\,A'Z'ZA(\Sigma^{-1}\mathrm{d}\Sigma)^2\Sigma^{-1} + \tfrac{1}{2}\mathrm{tr}\,\Sigma^{-1}A'Z'ZA\Sigma^{-1}\mathrm{d}^2\Sigma. \end{aligned} \tag{10}$$

It is easily verified that

$$(1/n)\mathscr{E}(Z'Z) = P_n \tag{11}$$

$$(1/n)\mathscr{E}(\Sigma_0^{-1}A_0'Z'Z) = \Gamma_0^{-1}J \tag{12}$$

and

$$(1/n)\mathscr{E}(A_0'Z'ZA_0) = \Sigma_0. \tag{13}$$

Using these results we obtain

$$\begin{aligned} -(1/n)\mathscr{E}\mathrm{d}^2\Lambda_n(\theta_0) = & \,\mathrm{tr}(\Gamma_0^{-1}\mathrm{d}\Gamma)^2 - \mathrm{tr}\,\Gamma_0^{-1}\mathrm{d}^2\Gamma - \tfrac{1}{2}\mathrm{tr}(\Sigma_0^{-1}\mathrm{d}\Sigma)^2 \\ & +\tfrac{1}{2}\mathrm{tr}\,\Sigma_0^{-1}\mathrm{d}^2\Sigma + \mathrm{tr}\,\Sigma_0^{-1}(\mathrm{d}A)'P_n\mathrm{d}A - 2\mathrm{tr}\Sigma_0^{-1}(\mathrm{d}\Sigma)\Gamma_0^{-1}J\mathrm{d}A \\ & +\mathrm{tr}\,\Gamma_0^{-1}J\mathrm{d}^2A + \mathrm{tr}(\Sigma_0^{-1}\mathrm{d}\Sigma)^2 - \tfrac{1}{2}\mathrm{tr}\,\Sigma_0^{-1}\mathrm{d}^2\Sigma \\ = & \,\mathrm{tr}(\Gamma_0^{-1}\mathrm{d}\Gamma)^2 + \mathrm{tr}\,\Sigma_0^{-1}(\mathrm{d}A)'P_n\mathrm{d}A - 2\,\mathrm{tr}\,\Sigma_0^{-1}(\mathrm{d}\Sigma)\Gamma_0^{-1}J\mathrm{d}A \\ & +\tfrac{1}{2}\mathrm{tr}\,(\Sigma_0^{-1}\mathrm{d}\Sigma)^2 \\ = & \,(\mathrm{d}\,\mathrm{vec}\,A')'[K_{m+k,m}(\Gamma_0^{-1}J\otimes J'\Gamma_0'^{-1}) + P_n\otimes\Sigma_0^{-1}]\mathrm{d}\,\mathrm{vec}\,A' \\ & -2(\mathrm{d}\,\mathrm{vec}\,\Sigma)'(\Gamma_0^{-1}J\otimes\Sigma_0^{-1})\mathrm{d}\,\mathrm{vec}\,A' + \tfrac{1}{2}(\mathrm{d}\,\mathrm{vec}\,\Sigma)'(\Sigma_0^{-1}\otimes\Sigma_0^{-1})\mathrm{d}\,\mathrm{vec}\,\Sigma. \end{aligned} \tag{14}$$

Finally, since $\mathrm{d}\,\mathrm{vec}\,A' = \Delta_1\mathrm{d}\theta$ and $\mathrm{d}\,\mathrm{vec}\,\Sigma = \Delta_2\mathrm{d}\theta$, the result follows. □

8. FULL-INFORMATION MAXIMUM LIKELIHOOD (FIML): THE ASYMPTOTIC VARIANCE MATRIX (SPECIAL CASE)

Theorem 5 provides us with the information matrix of the FIML estimator $\hat{\theta}$, assuming that B, Γ and Σ can all be expressed as (nonlinear) functions of a parameter vector θ. Our real interest, however, lies not so much in the information matrix as in the inverse of its limit, known as the asymptotic variance matrix. But to make further progress we need to assume more about the functions B, Γ and Σ. Therefore we shall assume that B and Γ depend on some parameter, say ζ, functionally independent of $\mathrm{v}(\Sigma)$. If Σ is also constrained, say $\Sigma=\Sigma(\sigma)$, where σ and ζ are independent, the results are less appealing (see exercise 3).

Theorem 6

Consider a random sample of size n from the process defined by the simultaneous equations model (2.1) under the normality assumption (Assumption 1) and the rank condition (Assumption 2). Assume that B and Γ satisfy certain *a priori* (nonlinear) twice differentiable constraints:

$$B=B(\zeta), \qquad \Gamma=\Gamma(\zeta), \tag{1}$$

where ζ is an unknown parameter vector, functionally independent of $v(\Sigma)$. Then the information matrix $\mathscr{F}_n(\zeta_0, v(\Sigma_0))$ is given by

$$\mathscr{F}_n(\zeta_0, v(\Sigma_0))=n\begin{pmatrix}\Delta'[K_{m+k,m}(C\otimes C')+P_n\otimes\Sigma_0^{-1}]\Delta & -\Delta'(C'\otimes\Sigma_0^{-1})D_m\\ -D_m'(C\otimes\Sigma_0^{-1})\Delta & \frac{1}{2}D_m'(\Sigma_0^{-1}\otimes\Sigma_0^{-1})D_m\end{pmatrix} \tag{2}$$

where

$$\Delta=(\Delta_\gamma':\Delta_\beta')', \qquad \Delta_\beta=\frac{\partial\,\mathrm{vec}\,B'}{\partial\zeta'}, \qquad \Delta_\gamma=\frac{\partial\,\mathrm{vec}\,\Gamma'}{\partial\zeta'}, \tag{3}$$

all evaluated at ζ_0, $C=\Gamma_0^{-1}J$ and J and P_n are defined in Theorem 5.

Moreover, if $Q_n=(1/n)X'X$ tends to a positive definite limit Q as $n\to\infty$, so that P_n tends to a positive semidefinite limit, say P, then the asymptotic variance matrix of the ML estimators $\hat{\zeta}$ and $v(\hat{\Sigma})$ is

$$\begin{pmatrix}V^{-1} & 2V^{-1}\Delta_\gamma'E_0'D_m^{+\prime}\\ 2D_m^+E_0\Delta_\gamma V^{-1} & 2D_m^+(\Sigma_0\otimes\Sigma_0+2E_0\Delta_\gamma V^{-1}\Delta_\gamma'E_0')D_m^{+\prime}\end{pmatrix} \tag{4}$$

with

$$V=\Delta'[(P-C'\Sigma_0C)\otimes\Sigma_0^{-1}]\Delta, \qquad E_0=\Sigma_0\Gamma_0^{-1}\otimes I_m. \tag{5}$$

Proof. We apply Theorem 5. Let $\theta=(\zeta', v(\Sigma)')'$. Then

$$\Delta_1=\begin{pmatrix}\partial\,\mathrm{vec}\,\Gamma'/\partial\theta'\\ \partial\,\mathrm{vec}\,B'/\partial\theta'\end{pmatrix}=\begin{pmatrix}\Delta_\gamma & 0\\ \Delta_\beta & 0\end{pmatrix}=(\Delta:0) \tag{6}$$

and

$$\Delta_2=\partial\,\mathrm{vec}\,\Sigma/\partial\theta'=(0:D_m). \tag{7}$$

Thus, (2) follows from (7.4). The asymptotic variance matrix is obtained as the inverse of

$$\mathscr{F}=\begin{pmatrix}\mathscr{F}_{11} & \mathscr{F}_{12}\\ \mathscr{F}_{21} & \mathscr{F}_{22}\end{pmatrix} \tag{8}$$

where

$$\mathscr{F}_{11}=\Delta'[K_{m+k,m}(C\otimes C')+P\otimes\Sigma_0^{-1}]\Delta \tag{9}$$

$$\mathscr{F}_{12}=-\Delta'(C'\otimes\Sigma_0^{-1})D_m \tag{10}$$

$$\mathscr{F}_{22}=\tfrac{1}{2}D_m'(\Sigma_0^{-1}\otimes\Sigma_0^{-1})D_m. \tag{11}$$

We have

$$\mathscr{F}^{-1}=\begin{pmatrix}W^{-1} & -W^{-1}\mathscr{F}_{12}\mathscr{F}_{22}^{-1}\\ -\mathscr{F}_{22}^{-1}\mathscr{F}_{21}W^{-1} & \mathscr{F}_{22}^{-1}+\mathscr{F}_{22}^{-1}\mathscr{F}_{21}W^{-1}\mathscr{F}_{12}\mathscr{F}_{22}^{-1}\end{pmatrix} \tag{12}$$

with

$$W=\mathscr{F}_{11}-\mathscr{F}_{12}\mathscr{F}_{22}^{-1}\mathscr{F}_{21}. \tag{13}$$

From (10) and (11) we obtain first

$$\begin{aligned}\mathscr{F}_{12}\mathscr{F}_{22}^{-1} &= -2\Delta'(C'\otimes\Sigma_0^{-1})D_m D_m^+(\Sigma_0\otimes\Sigma_0)D_m^{+\prime}\\ &= -2\Delta'(C'\Sigma_0\otimes I_m)D_m^{+\prime},\end{aligned} \tag{14}$$

using Theorem 3.13(b) and (2.2.4). Hence

$$\begin{aligned}\mathscr{F}_{12}\mathscr{F}_{22}^{-1}\mathscr{F}_{21} &= 2\Delta'(C'\Sigma_0\otimes I_m)D_m^{+\prime}D_m'(C\otimes\Sigma_0^{-1})\Delta\\ &= \Delta'(C'\Sigma_0\otimes I_m)(I_{m^2}+K_m)(C\otimes\Sigma_0^{-1})\Delta\\ &= \Delta'[C'\Sigma_0 C\otimes\Sigma_0^{-1}+K_{m+k,m}(C\otimes C')]\Delta,\end{aligned} \tag{15}$$

using Theorems 3.12(b) and 3.9(a). Inserting (9) and (15) in (13) yields

$$W=\Delta'[(P-C'\Sigma_0 C)\otimes\Sigma_0^{-1}]\Delta=V. \tag{16}$$

To obtain the remaining terms of $\mathscr{F}^{-1}$ we partition $C=(\Gamma_0^{-1}:0)$ and rewrite (14) as

$$\begin{aligned}\mathscr{F}_{12}\mathscr{F}_{22}^{-1} &= -2(\Delta_\gamma' : \Delta_\beta')\begin{pmatrix}\Gamma_0'^{-1}\Sigma_0\otimes I_m\\ 0\end{pmatrix}D_m^{+\prime}\\ &= -2\Delta_\gamma'(\Gamma_0'^{-1}\Sigma_0\otimes I_m)D_m^{+\prime}\\ &= -2\Delta_\gamma' E_0' D_m^{+\prime}.\end{aligned} \tag{17}$$

Hence

$$-W^{-1}\mathscr{F}_{12}\mathscr{F}_{22}^{-1}=2V^{-1}\Delta_\gamma' E_0' D_m^{+\prime} \tag{18}$$

and

$$\begin{aligned}&\mathscr{F}_{22}^{-1}+\mathscr{F}_{22}^{-1}\mathscr{F}_{21}W^{-1}\mathscr{F}_{12}\mathscr{F}_{22}^{-1}\\ &\quad=2D_m^+(\Sigma_0\otimes\Sigma_0)D_m^{+\prime}+4D_m^+E_0\Delta_\gamma V^{-1}\Delta_\gamma' E_0' D_m^{+\prime}.\end{aligned} \tag{19}$$

□

Exercises

1. In the special case of Theorem 6 where Γ_0 is a known matrix of constants and $B=B(\zeta)$, show that the asymptotic variance matrix of $\hat{\zeta}$ and $\mathrm{v}(\hat{\Sigma})$ is

$$\mathscr{F}^{-1}=\begin{pmatrix}[\Delta_\beta'(Q\otimes\Sigma_0^{-1})\Delta_\beta]^{-1} & 0\\ 0 & 2D_m^+(\Sigma_0\otimes\Sigma_0)D_m^{+\prime}\end{pmatrix}.$$

2. How does this result relate to (8.11) in Theorem 15.4.
3. Assume, in addition to the set-up of Theorem 6, that Σ is diagonal and let σ be the $m\times 1$ vector of its diagonal elements. Obtain the asymptotic variance matrix of $(\hat{\zeta},\hat{\sigma})$. In particular, show that

$$\mathscr{V}_{as}(\hat{\zeta})=\left[\Delta'\left(K_{m+k,m}(C\otimes C')+P\otimes\Sigma_0^{-1}-2\sum_{i=1}^{m}(C'E_{ii}C\otimes E_{ii})\right)\Delta\right]^{-1}$$

where E_{ii} is an $m\times m$ matrix with a one in the i-th diagonal position and zeros elsewhere.

9. LIMITED-INFORMATION MAXIMUM LIKELIHOOD (LIML): THE FIRST-ORDER CONDITIONS

In contrast to the FIML method of estimation, the limited-information maximum likelihood (LIML) method estimates the parameters of a single structural equation, say the first, subject only to those constraints that involve solely the coefficients of the equation being estimated. We shall only consider the standard case where all constraints are of the exclusion type. Then LIML can be represented as a special case of FIML where every equation (apart from the first) is just identified. Thus we write

$$y = Y\gamma_0 + X_1\beta_0 + u_0 \tag{1}$$

$$Y = X_1\Pi_{01} + X_2\Pi_{02} + V_0. \tag{2}$$

The matrices Π_{01} and Π_{02} are unrestricted. The LIML estimates of β_0 and γ_0 in equation (1) are then defined as the ML estimates of β_0 and γ_0 in the system (1)+(2).

We shall first obtain the first-order conditions.

Theorem 7

Consider a single equation from a simultaneous equations system,

$$y = Y\gamma_0 + X_1\beta_0 + u_0, \tag{3}$$

completed by the reduced form of Y,

$$Y = X_1\Pi_{01} + X_2\Pi_{02} + V_0, \tag{4}$$

where y $(n\times 1)$ and Y $(n\times m)$ contain the observations on the endogenous variables, X_1 $(n\times k_1)$ and X_2 $(n\times k_2)$ are exogenous (non-random) and u_0 $(n\times 1)$ and V_0 $(n\times m)$ are random disturbances. We assume that the n rows of $(u_0 : V_0)$ are independent and identically distributed as $\mathcal{N}(0, \Psi_0)$, where Ψ_0 is a positive definite $(m+1)\times(m+1)$ matrix partitioned as

$$\Psi_0 = \begin{pmatrix} \sigma_0^2 & \theta_0' \\ \theta_0 & \Omega_0 \end{pmatrix}. \tag{5}$$

There are $m+k_1+m(k_1+k_2)+\frac{1}{2}(m+1)(m+2)$ parameters to be estimated, namely $\gamma_0\,(m\times 1)$, $\beta_0\,(k_1\times 1)$, $\Pi_{01}\,(k_1\times m)$, $\Pi_{02}\,(k_2\times m)$, σ_0^2, $\theta_0\,(m\times 1)$ and $\mathrm{v}(\Omega_0)(\frac{1}{2}m(m+1)\times 1)$. We define

$$X = (X_1 : X_2), \qquad Z = (Y : X_1), \tag{6}$$

$$\Pi = (\Pi_1' : \Pi_2')', \qquad \alpha = (\gamma' : \beta')'. \tag{7}$$

If $\hat{u}$ and $\hat{V}$ are solutions of the equations

$$u = [I - Z(Z'(I - VV^+)Z)^{-1}Z'(I - VV^+)]y \tag{8}$$

$$V = [I - X(X'(I - uu^+)X)^{-1}X'(I - uu^+)]Y, \tag{9}$$

where $(X:u)$ and $(Z:V)$ are assumed to have full column rank, then the ML estimators of α_0, Π_0 and Ψ_0 are

$$\hat{\alpha} = (Z'(I - \hat{V}\hat{V}^+)Z)^{-1}Z'(I - \hat{V}\hat{V}^+)y \tag{10}$$

$$\hat{\Pi} = (X'(I - \hat{u}\hat{u}^+)X)^{-1}X'(I - \hat{u}\hat{u}^+)Y \tag{11}$$

$$\hat{\Psi} = \frac{1}{n}\begin{pmatrix} \hat{u}\hat{u} & \hat{u}'\hat{V} \\ \hat{V}'\hat{u} & \hat{V}'\hat{V} \end{pmatrix}. \tag{12}$$

Remark. To solve equations (8) and (9) we can use the following iterative scheme. Choose $u^{(0)}=0$ as the starting value. Then compute

$$V^{(1)} = V(u^{(0)}) = (I - X(X'X)^{-1}X')Y \tag{13}$$

and $u^{(1)} = u(V^{(1)})$, $V^{(2)} = V(u^{(1)})$, and so on. If this scheme converges, a solution has been found.

Proof. Given (6) and (7) we may rewrite (3) and (4) as

$$y = Z\alpha_0 + u_0, \qquad Y = X\Pi_0 + V_0. \tag{14}$$

We define $W = (u:V)$, where

$$u = u(\alpha) = y - Z\alpha, \qquad V = V(\Pi) = Y - X\Pi. \tag{15}$$

Then we can write the loglikelihood function as

$$\Lambda(\alpha, \pi, \psi) = \text{constant} - \tfrac{1}{2}n \log|\Psi| - \tfrac{1}{2}\operatorname{tr} W\Psi^{-1}W', \tag{16}$$

where $\pi = \operatorname{vec}\Pi'$ and $\psi = \mathrm{v}(\Psi)$. The first differential is

$$\mathrm{d}\Lambda = -\tfrac{1}{2}n \operatorname{tr}\Psi^{-1}\mathrm{d}\Psi + \tfrac{1}{2}\operatorname{tr} W\Psi^{-1}(\mathrm{d}\Psi)\Psi^{-1}W' - \operatorname{tr} W\Psi^{-1}(\mathrm{d}W)'. \tag{17}$$

Since $\mathrm{d}W = -(Z\mathrm{d}\alpha,\ X\mathrm{d}\Pi)$ and

$$\Psi^{-1} = \frac{1}{\eta^2}\begin{pmatrix} 1 & -\theta'\Omega^{-1} \\ -\Omega^{-1}\theta & \eta^2\Omega^{-1} + \Omega^{-1}\theta\theta'\Omega^{-1} \end{pmatrix} \tag{18}$$

where $\eta^2 = \sigma^2 - \theta'\Omega^{-1}\theta$, we obtain

$$\begin{aligned} \operatorname{tr} W\Psi^{-1}(\mathrm{d}W)' = -(1/\eta^2)[&(\mathrm{d}\alpha)'Z'u - \theta'\Omega^{-1}(\mathrm{d}\Pi)'X'u - (\mathrm{d}\alpha)'Z'V\Omega^{-1}\theta \\ &+ \operatorname{tr} V(\eta^2\Omega^{-1} + \Omega^{-1}\theta\theta'\Omega^{-1})(\mathrm{d}\Pi)'X'] \end{aligned} \tag{19}$$

and hence

$$\begin{aligned} \mathrm{d}\Lambda = {} & \tfrac{1}{2}\operatorname{tr}(\Psi^{-1}W'W\Psi^{-1} - n\Psi^{-1})\mathrm{d}\Psi + (1/\eta^2)(\mathrm{d}\alpha)'(Z'u - Z'V\Omega^{-1}\theta) \\ & - (1/\eta^2)\operatorname{tr}[X'u\theta'\Omega^{-1} - X'V(\eta^2\Omega^{-1} + \Omega^{-1}\theta\theta'\Omega^{-1})](\mathrm{d}\Pi)'. \end{aligned} \tag{20}$$

Hence the first-order conditions are

$$\Psi = (1/n)W'W \tag{21}$$

$$Z'u = Z'V\Omega^{-1}\theta \tag{22}$$

$$X'V(\eta^2\Omega^{-1} + \Omega^{-1}\theta\theta'\Omega^{-1}) = X'u\theta'\Omega^{-1}. \tag{23}$$

Post-multiplying (23) by $\Omega-(1/\sigma^2)\theta\theta'$ yields

$$\sigma^2 X'V = X'u\theta'. \tag{24}$$

Inserting $\sigma^2=u'u/n$, $\Omega=V'V/n$ and $\theta=V'u/n$ in (22) and (24) gives

$$Z'u = Z'V(V'V)^{-1}V'u \tag{25}$$

$$X'V = X'(1/u'u)uu'V \tag{26}$$

and hence, since $u=y-Z\alpha$ and $V=Y-X\Pi$,

$$Z'(I-VV^+)Z\alpha = Z'(I-VV^+)y \tag{27}$$

$$X'(I-uu^+)X\Pi = X'(I-uu^+)Y. \tag{28}$$

Since $(X:u)$ and $(Z:V)$ have full column rank, the matrices $Z'(I-VV^+)Z$ and $X'(I-uu^+)X$ are non-singular. This gives

$$Z\alpha = Z(Z'(I-VV^+)Z)^{-1}Z'(I-VV^+)y \tag{29}$$

$$X\Pi = X(X'(I-uu^+)X)^{-1}X'(I-uu^+)Y. \tag{30}$$

Hence, we can express u in terms of V and V in terms of u as follows:

$$u = [I - Z(Z'(I-VV^+)Z)^{-1}Z'(I-VV^+)]y \tag{31}$$

$$V = [I - X(X'(I-uu^+)X)^{-1}X'(I-uu^+)]Y. \tag{32}$$

Given a solution $(\hat{u}, \hat{V})$ of these equations, we obtain $\hat{\alpha}$ from (29), $\hat{\Pi}$ from (30) and $\hat{\Psi}$ from (21). □

10. LIMITED-INFORMATION MAXIMUM LIKELIHOOD (LIML): THE INFORMATION MATRIX

Having obtained the first-order conditions for LIML estimation, we proceed to derive the information matrix.

Theorem 8

Consider a single equation from a simultaneous equations system,

$$y = Y\gamma_0 + X_1\beta_0 + u_0, \tag{1}$$

completed by the reduced form of Y,

$$Y = X_1\Pi_{01} + X_2\Pi_{02} + V_0. \tag{2}$$

Under the conditions of Theorem 7 and letting $\pi=\operatorname{vec}\Pi'$ and $\psi=\operatorname{v}(\Psi)$, the information matrix in terms of the parametrization (α,π,ψ) is

$$\mathcal{F}_n(\alpha_0,\pi_0,\psi_0) = n\begin{bmatrix}\mathcal{F}_{\alpha\alpha} & \mathcal{F}_{\alpha\pi} & \mathcal{F}_{\alpha\psi}\\ \mathcal{F}_{\pi\alpha} & \mathcal{F}_{\pi\pi} & 0\\ \mathcal{F}_{\psi\alpha} & 0 & \mathcal{F}_{\psi\psi}\end{bmatrix} \tag{3}$$

with

$$\mathscr{F}_{\alpha\alpha}=(1/\eta_0^2)A_{zz} \tag{4}$$

$$\mathscr{F}_{\alpha\pi}=-(1/\eta_0^2)(A_{zx}\otimes\theta_0'\Omega_0^{-1})=\mathscr{F}'_{\pi\alpha} \tag{5}$$

$$\mathscr{F}_{\alpha\psi}=(e'\Psi_0^{-1}\otimes S')D_{m+1}=\mathscr{F}'_{\psi\alpha} \tag{6}$$

$$\mathscr{F}_{\pi\pi}=(1/\eta_0^2)[(1/n)X'X\otimes(\eta_0\Omega_0^{-1}+\Omega_0^{-1}\theta_0\theta_0'\Omega_0^{-1})] \tag{7}$$

$$\mathscr{F}_{\psi\psi}=\tfrac{1}{2}D'_{m+1}(\Psi_0^{-1}\otimes\Psi_0^{-1})D_{m+1}, \tag{8}$$

where A_{zz} and A_{zx} are defined as

$$A_{zz}=\frac{1}{n}\begin{pmatrix}\Pi_0'X'X\Pi_0+n\Omega_0 & \Pi_0'X'X_1\\ X_1'X\Pi_0 & X_1'X_1\end{pmatrix},\qquad A_{zx}=\frac{1}{n}\begin{pmatrix}\Pi_0'X'X\\ X_1'X\end{pmatrix} \tag{9}$$

of orders $(m+k_1)\times(m+k_1)$ and $(m+k_1)\times(k_2+k_1)$ respectively, $e=(1,0,\ldots,0)'$ of order $(m+1)\times 1$, $\eta_0^2=\sigma_0^2-\theta_0'\Omega_0^{-1}\theta_0$, and S is the $(m+1)\times(m+k_1)$ selection matrix

$$S=\begin{pmatrix}0 & 0\\ I_m & 0\end{pmatrix}. \tag{10}$$

Proof. Recall from (9.17) that the first differential of the loglikelihood function $\Lambda(\alpha,\pi,\psi)$ is

$$\mathrm{d}\Lambda=-\tfrac{1}{2}n\operatorname{tr}\Psi^{-1}\mathrm{d}\Psi+\tfrac{1}{2}\operatorname{tr}W\Psi^{-1}(\mathrm{d}\Psi)\Psi^{-1}W'-\operatorname{tr}W\Psi^{-1}(\mathrm{d}W)', \tag{11}$$

where $W=(u:V)$, $u=y-Z\alpha$, $V=Y-X\Pi$. Hence, the second differential is

$$\begin{aligned}\mathrm{d}^2\Lambda=&\tfrac{1}{2}n\operatorname{tr}\Psi^{-1}(\mathrm{d}\Psi)\Psi^{-1}\mathrm{d}\Psi-\operatorname{tr}W\Psi^{-1}(\mathrm{d}\Psi)\Psi^{-1}(\mathrm{d}\Psi)\Psi^{-1}W'\\&+2\operatorname{tr}W\Psi^{-1}(\mathrm{d}\Psi)\Psi^{-1}(\mathrm{d}W)'-\operatorname{tr}(\mathrm{d}W)\Psi^{-1}(\mathrm{d}W)',\end{aligned} \tag{12}$$

using the (obvious) facts that both Ψ and W are *linear* in the parameters, so that $\mathrm{d}^2\Psi=0$ and $\mathrm{d}^2W=0$. Let

$$W_0=(u_0:V_0)=(w_{01},\ldots,w_{0n})'. \tag{13}$$

Then $\{w_{0t}, t=1,\ldots,n\}$ are independent and identically distributed as $\mathcal{N}(0,\Psi_0)$. Hence

$$(1/n)\mathscr{E}W_0'W_0=(1/n)\mathscr{E}\sum_{t=1}^{n}w_{0t}w_{0t}'=\Psi_0, \tag{14}$$

and also, since $\mathrm{d}W=-(Z\mathrm{d}\alpha:X\mathrm{d}\Pi)$,

$$(1/n)\mathscr{E}(\mathrm{d}W)'W_0=\begin{pmatrix}-(\mathrm{d}\alpha)'S'\Psi_0\\ 0\end{pmatrix} \tag{15}$$

and

$$(1/n)\mathscr{E}(\mathrm{d}W)'(\mathrm{d}W)=\begin{pmatrix}(\mathrm{d}\alpha)'A_{zz}(\mathrm{d}\alpha) & (\mathrm{d}\alpha)'A_{zx}(\mathrm{d}\Pi)\\ (\mathrm{d}\Pi)'A_{zx}'(\mathrm{d}\alpha) & (1/n)(\mathrm{d}\Pi)'X'X(\mathrm{d}\Pi)\end{pmatrix}. \tag{16}$$

Now, writing the inverse of Ψ as in (9.18), we obtain

$$\begin{aligned}-(1/n)\mathscr{E}\,\mathrm{d}^2\Lambda(\alpha_0,\pi_0,\psi_0) &= \tfrac{1}{2}\operatorname{tr}\Psi_0^{-1}(\mathrm{d}\Psi)\Psi_0^{-1}\mathrm{d}\Psi+2(\mathrm{d}\alpha)'S'(\mathrm{d}\Psi)\Psi_0^{-1}e\\ &\quad+(1/\eta_0^2)(\mathrm{d}\alpha)'A_{zz}(\mathrm{d}\alpha)-(2/\eta_0^2)\theta_0'\Omega_0^{-1}(\mathrm{d}\Pi)'A_{zx}'(\mathrm{d}\alpha)\\ &\quad+(1/\eta_0^2)\operatorname{tr}(\eta_0^2\Omega_0^{-1}+\Omega_0^{-1}\theta_0\theta_0'\Omega_0^{-1})\\ &\quad(\mathrm{d}\Pi)'[(1/n)X'X](\mathrm{d}\Pi)\\ &=\tfrac{1}{2}(\mathrm{d}v(\Psi))'D_{m+1}'(\Psi_0^{-1}\otimes\Psi_0^{-1})D_{m+1}\mathrm{d}v(\Psi)\\ &\quad+2(\mathrm{d}\alpha)'(e'\Psi_0^{-1}\otimes S')D_{m+1}\mathrm{d}v(\Psi)\\ &\quad+(1/\eta_0^2)(\mathrm{d}\alpha)'A_{zz}(\mathrm{d}\alpha)\\ &\quad-(2/\eta_0^2)(\mathrm{d}\alpha)'(A_{zx}\otimes\theta_0'\Omega_0^{-1})\mathrm{d}\operatorname{vec}\Pi'\\ &\quad+(1/\eta_0^2)(\mathrm{d}\operatorname{vec}\Pi')'\{[(1/n)X'X]\\ &\quad\otimes(\eta_0^2\Omega_0^{-1}+\Omega_0^{-1}\theta_0\theta_0'\Omega_0^{-1})\}\,\mathrm{d}\operatorname{vec}\Pi',\end{aligned}\tag{17}$$

and the result follows. □

11. LIMITED-INFORMATION MAXIMUM LIKELIHOOD (LIML): THE ASYMPTOTIC VARIANCE MATRIX

Again, the derivation of the information matrix in Theorem 8 is only an intermediary result. Our real interest lies in the asymptotic variance matrix, which we shall now derive.

Theorem 9

Consider a single equation from a simultaneous equations system,

$$y=Y\gamma_0+X_1\beta_0+u_0 \tag{1}$$

completed by the reduced form of Y,

$$Y=X_1\Pi_{01}+X_2\Pi_{02}+V_0. \tag{2}$$

Assume, in addition to the conditions of Theorem 7, that Π_{02} has full column rank m and that $(1/n)X'X$ tends to a positive definite $(k_1+k_2)\times(k_1+k_2)$ matrix Q as $n\to\infty$. Then, letting $\pi_1=\operatorname{vec}\Pi_1'$, $\pi_2=\operatorname{vec}\Pi_2'$ and $\omega=\mathrm{v}(\Omega)$, the asymptotic variance matrix of the ML estimators $(\hat\beta,\hat\gamma,\hat\pi_1,\hat\pi_2,\hat\sigma^2,\hat\theta,\mathrm{v}(\hat\Omega))$ is

$$\mathscr{F}^{-1}=\begin{array}{c} \begin{array}{ccc}\beta\ \ \gamma & \pi_1\ \ \pi_2 & \sigma^2\ \ \theta\ \ \omega\end{array}\\ \begin{bmatrix}\mathscr{F}^{\alpha\alpha} & \mathscr{F}^{\alpha\pi} & \mathscr{F}^{\alpha\psi}\\ \mathscr{F}^{\pi\alpha} & \mathscr{F}^{\pi\pi} & \mathscr{F}^{\pi\psi}\\ \mathscr{F}^{\psi\alpha} & \mathscr{F}^{\psi\pi} & \mathscr{F}^{\psi\psi}\end{bmatrix}\end{array} \tag{3}$$

with

$$\mathscr{F}^{\alpha\alpha}=\sigma_0^2\begin{pmatrix} P_1^{-1} & -P_1^{-1}P_2 \\ -P_2'P_1^{-1} & Q_{11}^{-1}+P_2'P_1^{-1}P_2 \end{pmatrix} \tag{4}$$

$$\mathscr{F}^{\alpha\pi}=\begin{pmatrix} -P_1^{-1}\Pi_{02}'Q_{21}Q_{11}^{-1}\otimes\theta_0' & P_1^{-1}\Pi_{02}'\otimes\theta_0' \\ (Q_{11}^{-1}+P_2'P_1^{-1}\Pi_{02}'Q_{21}Q_{11}^{-1})\otimes\theta_0' & -P_2'P_1^{-1}\Pi_{02}'\otimes\theta_0' \end{pmatrix} \tag{5}$$

$$\mathscr{F}^{\alpha\psi}=\sigma_0^2\begin{pmatrix} -2P_1^{-1}\theta_0 & -P_1^{-1}\Omega_0 & 0 \\ 2P_2'P_1^{-1}\theta_0 & P_2'P_1^{-1}\Omega_0 & 0 \end{pmatrix} \tag{6}$$

$$\mathscr{F}^{\pi\pi}=\begin{pmatrix} Q^{11} & Q^{12} \\ Q^{21} & Q^{22} \end{pmatrix}\otimes\Omega_0-(1/\sigma_0^2)\begin{pmatrix} H_*^{11} & H_*^{12} \\ H_*^{21} & H_*^{22} \end{pmatrix}\otimes\theta_0\theta_0' \tag{7}$$

$$\mathscr{F}^{\pi\psi}=\begin{pmatrix} 2Q_{11}^{-1}Q_{12}\Pi_{02}P_1^{-1}\theta_0\otimes\theta_0 & Q_{11}^{-1}Q_{12}\Pi_{02}P_1^{-1}\Omega_0\otimes\theta_0 & 0 \\ -2\Pi_{02}P_1^{-1}\theta_0\otimes\theta_0 & -\Pi_{02}P_1^{-1}\Omega_0\otimes\theta_0 & 0 \end{pmatrix} \tag{8}$$

$$\mathscr{F}^{\psi\psi}=\begin{bmatrix} 2\sigma_0^4+4\sigma_0^2\theta_0'P_1^{-1}\theta_0 & 2\sigma_0^2\theta_0'(I+P_1^{-1}\Omega_0) & 2(\theta_0'\otimes\theta_0')D_m^{+\prime} \\ 2\sigma_0^2(I+\Omega_0P_1^{-1})\theta_0 & \sigma_0^2(\Omega_0+\Omega_0P_1^{-1}\Omega_0)+\theta_0\theta_0' & 2(\theta_0'\otimes\Omega_0)D_m^{+\prime} \\ 2D_m^+(\theta_0\otimes\theta_0) & 2D_m^+(\theta_0\otimes\Omega_0) & 2D_m^+(\Omega_0\otimes\Omega_0)D_m^{+\prime} \end{bmatrix}, \tag{9}$$

where

$$Q=\begin{pmatrix} Q_{11} & Q_{12} \\ Q_{21} & Q_{22} \end{pmatrix}=(Q_1:Q_2), \tag{10}$$

$$Q^{-1}=\begin{pmatrix} Q^{11} & Q^{12} \\ Q^{21} & Q^{22} \end{pmatrix}=\begin{pmatrix} Q_{11}^{-1}+Q_{11}^{-1}Q_{12}G^{-1}Q_{21}Q_{11}^{-1} & -Q_{11}^{-1}Q_{12}G^{-1} \\ -G^{-1}Q_{21}Q_{11}^{-1} & G^{-1} \end{pmatrix}, \tag{11}$$

$$P_1=\Pi_{02}'G\Pi_{02}, \qquad P_2=\Pi_{01}'+\Pi_{02}'Q_{21}Q_{11}^{-1}, \tag{12}$$

$$G=Q_{22}-Q_{21}Q_{11}^{-1}Q_{12}, \qquad H=G^{-1}-\Pi_{02}P_1^{-1}\Pi_{02}', \tag{13}$$

and

$$H_*=\begin{pmatrix} H_*^{11} & H_*^{12} \\ H_*^{21} & H_*^{22} \end{pmatrix}=\begin{pmatrix} Q_{11}^{-1}Q_{12}HQ_{21}Q_{11}^{-1} & -Q_{11}^{-1}Q_{12}H \\ -HQ_{21}Q_{11}^{-1} & H \end{pmatrix}. \tag{14}$$

Proof. Theorem 8 gives the information matrix. The *asymptotic* information matrix, denoted as $\mathscr{F}$, is obtained as the limit of $(1/n)\,\mathscr{F}_n$ for $n\to\infty$. We find

$$\mathscr{F}=\begin{bmatrix} \mathscr{F}_{\alpha\alpha} & \mathscr{F}_{\alpha\pi} & \mathscr{F}_{\alpha\psi} \\ \mathscr{F}_{\pi\alpha} & \mathscr{F}_{\pi\pi} & 0 \\ \mathscr{F}_{\psi\alpha} & 0 & \mathscr{F}_{\psi\psi} \end{bmatrix} \tag{15}$$

with

$$\mathscr{F}_{\alpha\alpha}=(1/\eta_0^2)A_{zz} \tag{16}$$

$$\mathscr{F}_{\alpha\pi} = -(1/\eta_0^2)(A_{zx}\otimes\theta_0'\Omega_0^{-1}) = \mathscr{F}'_{\pi\alpha} \tag{17}$$

$$\mathscr{F}_{\alpha\psi} = (e'\Psi_0^{-1}\otimes S')D_{m+1} = \mathscr{F}'_{\psi\alpha} \tag{18}$$

$$\mathscr{F}_{\pi\pi} = (1/\eta_0^2)[Q\otimes(\eta_0^2\Omega_0^{-1}+\Omega_0^{-1}\theta_0\theta_0'\Omega_0^{-1})] \tag{19}$$

$$\mathscr{F}_{\psi\psi} = \tfrac{1}{2}D'_{m+1}(\Psi_0^{-1}\otimes\Psi_0^{-1})D_{m+1}, \tag{20}$$

where A_{zz} and A_{zx} are now defined as the limits of (10.9):

$$A_{zz} = \begin{pmatrix} \Pi_0'Q\Pi_0+\Omega_0 & \Pi_0'Q_1 \\ Q_1'\Pi_0 & Q_{11} \end{pmatrix}, \qquad A_{zx} = \begin{pmatrix} \Pi_0'Q \\ Q_1' \end{pmatrix}, \tag{21}$$

and e, η_0^2 and S are defined in Theorem 8.

It follows from Theorem 1.3 that

$$\mathscr{F}^{-1} = \begin{bmatrix} \mathscr{F}^{\alpha\alpha} & \mathscr{F}^{\alpha\pi} & \mathscr{F}^{\alpha\psi} \\ \mathscr{F}^{\pi\alpha} & \mathscr{F}^{\pi\pi} & \mathscr{F}^{\pi\psi} \\ \mathscr{F}^{\psi\alpha} & \mathscr{F}^{\psi\pi} & \mathscr{F}^{\psi\psi} \end{bmatrix} \tag{22}$$

with

$$\mathscr{F}^{\alpha\alpha} = (\mathscr{F}_{\alpha\alpha} - \mathscr{F}_{\alpha\pi}\mathscr{F}_{\pi\pi}^{-1}\mathscr{F}_{\pi\alpha} - \mathscr{F}_{\alpha\psi}\mathscr{F}_{\psi\psi}^{-1}\mathscr{F}_{\psi\alpha})^{-1} \tag{23}$$

$$\mathscr{F}^{\alpha\pi} = -\mathscr{F}^{\alpha\alpha}\mathscr{F}_{\alpha\pi}\mathscr{F}_{\pi\pi}^{-1} \tag{24}$$

$$\mathscr{F}^{\alpha\psi} = -\mathscr{F}^{\alpha\alpha}\mathscr{F}_{\alpha\psi}\mathscr{F}_{\psi\psi}^{-1} \tag{25}$$

$$\mathscr{F}^{\pi\pi} = \mathscr{F}_{\pi\pi}^{-1} + \mathscr{F}_{\pi\pi}^{-1}\mathscr{F}_{\pi\alpha}\mathscr{F}^{\alpha\alpha}\mathscr{F}_{\alpha\pi}\mathscr{F}_{\pi\pi}^{-1} \tag{26}$$

$$\mathscr{F}^{\pi\psi} = \mathscr{F}_{\pi\pi}^{-1}\mathscr{F}_{\pi\alpha}\mathscr{F}^{\alpha\alpha}\mathscr{F}_{\alpha\psi}\mathscr{F}_{\psi\psi}^{-1} \tag{27}$$

$$\mathscr{F}^{\psi\psi} = \mathscr{F}_{\psi\psi}^{-1} + \mathscr{F}_{\psi\psi}^{-1}\mathscr{F}_{\psi\alpha}\mathscr{F}^{\alpha\alpha}\mathscr{F}_{\alpha\psi}\mathscr{F}_{\psi\psi}^{-1}. \tag{28}$$

To evaluate $\mathscr{F}^{\alpha\alpha}$, which is the asymptotic variance matrix of $\hat{\alpha}$, we need some intermediary results:

$$\mathscr{F}_{\pi\pi}^{-1} = Q^{-1}\otimes[\Omega_0 - (1/\sigma_0^2)\theta_0\theta_0'] \tag{29}$$

$$\mathscr{F}_{\alpha\pi}\mathscr{F}_{\pi\pi}^{-1} = -A_{zx}Q^{-1}\otimes(1/\sigma_0^2)\theta_0' \tag{30}$$

$$\mathscr{F}_{\alpha\pi}\mathscr{F}_{\pi\pi}^{-1}\mathscr{F}_{\pi\alpha} = [(1/\eta_0^2)-(1/\sigma_0^2)]A_{zx}Q^{-1}A'_{zx}, \tag{31}$$

and also, using Theorems 3.13(d), 3.13(b), 3.12(b) and 3.9(a),

$$\mathscr{F}_{\psi\psi}^{-1} = 2D_{m+1}^{+}(\Psi_0\otimes\Psi_0)D_{m+1}^{+\prime} \tag{32}$$

$$\mathscr{F}_{\alpha\psi}\mathscr{F}_{\psi\psi}^{-1} = 2(e'\otimes S'\Psi_0)D_{m+1}^{+\prime} \tag{33}$$

$$\mathscr{F}_{\alpha\psi}\mathscr{F}_{\psi\psi}^{-1}\mathscr{F}_{\psi\alpha} = (e'\Psi_0^{-1}e)S'\Psi_0S = (1/\eta_0^2)S'\Psi_0S, \tag{34}$$

since $S'e=0$. Hence

$$\begin{aligned}\mathscr{F}^{\alpha\alpha} &= [(1/\eta_0^2)(A_{zz}-A_{zx}Q^{-1}A'_{zx}-S'\Psi_0S)+(1/\sigma_0^2)A_{zx}Q^{-1}A'_{zx}]^{-1} \\ &= \sigma_0^2(A_{zx}Q^{-1}A'_{zx})^{-1}.\end{aligned} \tag{35}$$

It is not difficult to partition the expression for $\mathcal{F}^{\alpha\alpha}$ in (35). Since

$$A_{zx}Q^{-1}=\begin{pmatrix}\Pi'_{01} & \Pi'_{02}\\ I_{k_1} & 0\end{pmatrix}, \tag{36}$$

we have

$$A_{zx}Q^{-1}A'_{zx}=\begin{pmatrix}\Pi'_0Q\Pi_0 & \Pi'_0Q_1\\ Q'_1\Pi_0 & Q_{11}\end{pmatrix} \tag{37}$$

and

$$(A_{zx}Q^{-1}A'_{zx})^{-1}=\begin{pmatrix}P_1^{-1} & -P_1^{-1}P_2\\ -P'_2P_1^{-1} & Q_{11}^{-1}+P'_2P_1^{-1}P_2\end{pmatrix} \tag{38}$$

with P_1 and P_2 defined in (12). Hence

$$\mathcal{F}^{\alpha\alpha}=\sigma_0^2\begin{pmatrix}P_1^{-1} & -P_1^{-1}P_2\\ -P'_2P_1^{-1} & Q_{11}^{-1}+P'_2P_1^{-1}P_2\end{pmatrix}. \tag{39}$$

We now proceed to obtain expressions for the other blocks of $\mathcal{F}^{-1}$. We have

$$\begin{aligned}\mathcal{F}^{\alpha\pi}&=\begin{pmatrix}P_1^{-1} & -P_1^{-1}P_2\\ -P'_2P_1^{-1} & Q_{11}^{-1}+P'_2P_1^{-1}P_2\end{pmatrix}\begin{pmatrix}\Pi'_{01} & \Pi'_{02}\\ I_{k_1} & 0\end{pmatrix}\otimes\theta'_0\\ &=\begin{pmatrix}-P_1^{-1}\Pi'_{02}Q_{21}Q_{11}^{-1}\otimes\theta'_0 & P_1^{-1}\Pi'_{02}\otimes\theta'_0\\ (Q_{11}^{-1}+P'_2P_1^{-1}\Pi'_{02}Q_{21}Q_{11}^{-1})\otimes\theta'_0 & -P'_2P_1^{-1}\Pi'_{02}\otimes\theta'_0\end{pmatrix},\end{aligned} \tag{40}$$

$$\begin{aligned}\mathcal{F}^{\alpha\psi}&=-2\mathcal{F}^{\alpha\alpha}(e'\otimes S'\Psi_0)D_{m+1}^{+\prime}\\ &=-2\mathcal{F}^{\alpha\alpha}\begin{pmatrix}\theta_0 & \Omega_0 & 0\\ 0 & 0 & 0\end{pmatrix}D_{m+1}^{+\prime}\\ &=-2\sigma_0^2\begin{pmatrix}P_1^{-1} & -P_1^{-1}P_2\\ -P'_2P_1^{-1} & Q_{11}^{-1}+P'_2P_1^{-1}P_2\end{pmatrix}\begin{pmatrix}\theta_0 & \frac{1}{2}\Omega_0 & 0\\ 0 & 0 & 0\end{pmatrix}\\ &=-2\sigma_0^2\begin{pmatrix}P_1^{-1}\theta_0 & \frac{1}{2}P_1^{-1}\Omega_0 & 0\\ -P'_2P_1^{-1}\theta_0 & -\frac{1}{2}P'_2P_1^{-1}\Omega_0 & 0\end{pmatrix},\end{aligned} \tag{41}$$

using Theorem 3.17. Further

$$\begin{aligned}\mathcal{F}^{\pi\pi}&=Q^{-1}\otimes[\Omega_0-(1/\sigma_0^2)\theta_0\theta'_0]+Q^{-1}A'_{zx}\mathcal{F}^{\alpha\alpha}A_{zx}Q^{-1}\otimes(1/\sigma_0^4)\theta_0\theta'_0\\ &=Q^{-1}\otimes\Omega_0-(1/\sigma_0^2)[Q^{-1}-(1/\sigma_0^2)Q^{-1}A'_{zx}\mathcal{F}^{\alpha\alpha}A_{zx}Q^{-1}]\otimes\theta_0\theta'_0.\end{aligned} \tag{42}$$

With Q and G as defined in (10) and (13) one easily verifies that

$$Q^{-1}=\begin{pmatrix}Q_{11}^{-1}+Q_{11}^{-1}Q_{12}G^{-1}Q_{21}Q_{11}^{-1} & -Q_{11}^{-1}Q_{12}G^{-1}\\ -G^{-1}Q_{21}Q_{11}^{-1} & G^{-1}\end{pmatrix}. \tag{43}$$

Also,

$$\begin{aligned}&Q^{-1}A'_{zx}\mathcal{F}^{\alpha\alpha}A_{zx}Q^{-1}\\ &=\sigma_0^2\begin{pmatrix}Q_{11}^{-1}+Q_{11}^{-1}Q_{12}\Pi_{02}P_1^{-1}\Pi'_{02}Q_{21}Q_{11}^{-1} & -Q_{11}^{-1}Q_{12}\Pi_{02}P_1^{-1}\Pi'_{02}\\ -\Pi_{02}P_1^{-1}\Pi'_{02}Q_{21}Q_{11}^{-1} & \Pi_{02}P_1^{-1}\Pi'_{02}\end{pmatrix}.\end{aligned} \tag{44}$$

Hence

$$Q^{-1}-(1/\sigma_0^2)Q^{-1}A'_{zx}\mathscr{F}^{\alpha\alpha}A_{zx}Q^{-1}=\begin{pmatrix} Q_{11}^{-1}Q_{12}HQ_{21}Q_{11}^{-1} & -Q_{11}^{-1}Q_{12}H \\ -HQ_{21}Q_{11}^{-1} & H \end{pmatrix} \tag{45}$$

where H is defined in (13).

Inserting (43) and (45) in (42) gives (7).

Next,

$$\begin{aligned}\mathscr{F}^{\pi\psi} &= -\mathscr{F}_{\pi\pi}^{-1}\mathscr{F}_{\pi\alpha}\mathscr{F}^{\alpha\psi} \\ &= (Q^{-1}A'_{zx}\otimes\theta_0)\begin{pmatrix} -2P_1^{-1}\theta_0 & -P_1^{-1}\Omega_0 & 0 \\ 2P'_2P_1^{-1}\theta_0 & P'_2P_1^{-1}\Omega_0 & 0\end{pmatrix} \\ &= \begin{pmatrix} \Pi_{01}\otimes\theta_0 & I\otimes\theta_0 \\ \Pi_{02}\otimes\theta_0 & 0\end{pmatrix}\begin{pmatrix} -2P_1^{-1}\theta_0 & -P_1^{-1}\Omega_0 & 0 \\ 2P'_2P_1^{-1}\theta_0 & P'_2P_1^{-1}\Omega_0 & 0\end{pmatrix} \\ &= \begin{pmatrix} 2Q_{11}^{-1}Q_{12}\Pi_{02}P_1^{-1}\theta_0\otimes\theta_0 & Q_{11}^{-1}Q_{12}\Pi_{02}P_1^{-1}\Omega_0\otimes\theta_0 & 0 \\ -2\Pi_{02}P_1^{-1}\theta_0\otimes\theta_0 & -\Pi_{02}P_1^{-1}\Omega_0\otimes\theta_0 & 0\end{pmatrix}\end{aligned} \tag{46}$$

and finally

$$\begin{aligned}\mathscr{F}^{\psi\psi} &= 2D_{m+1}^{+}(\Psi_0\otimes\Psi_0)D_{m+1}^{+\prime}+4D_{m+1}^{+}(e\otimes\Psi_0S)\mathscr{F}^{\alpha\alpha}(e'\otimes S'\Psi_0)D_{m+1}^{+\prime} \\ &= 2D_{m+1}^{+}(\Psi_0\otimes\Psi_0)D_{m+1}^{+\prime}+4D_{m+1}^{+}(ee'\otimes\Psi_0S\mathscr{F}^{\alpha\alpha}S'\Psi_0)D_{m+1}^{+\prime}.\end{aligned} \tag{47}$$

Using Theorem 3.15 we find

$$D_{m+1}^{+}(\Psi_0\otimes\Psi_0)D_{m+1}^{+\prime}=\begin{bmatrix} \sigma_0^4 & \sigma_0^2\theta'_0 & (\theta'_0\otimes\theta'_0)D_m^{+\prime} \\ \sigma_0^2\theta_0 & \tfrac{1}{2}(\sigma_0^2\Omega_0+\theta_0\theta'_0) & (\theta'_0\otimes\Omega_0)D_m^{+\prime} \\ D_m^{+}(\theta_0\otimes\theta_0) & D_m^{+}(\theta_0\otimes\Omega_0) & D_m^{+}(\Omega_0\otimes\Omega_0)D_m^{+\prime}\end{bmatrix} \tag{48}$$

and

$$D_{m+1}^{+}(ee'\otimes\Psi_0S\mathscr{F}^{\alpha\alpha}S'\Psi_0)D_{m+1}^{+\prime}=\sigma_0^2\begin{bmatrix} \theta'_0P_1^{-1}\theta_0 & \tfrac{1}{2}\theta'_0P_1^{-1}\Omega_0 & 0 \\ \tfrac{1}{2}\Omega_0P_1^{-1}\theta_0 & \tfrac{1}{4}\Omega_0P_1^{-1}\Omega_0 & 0 \\ 0 & 0 & 0\end{bmatrix}, \tag{49}$$

because

$$\Psi_0S\mathscr{F}^{\alpha\alpha}S'\Psi_0=\sigma_0^2\begin{pmatrix} \theta'_0P_1^{-1}\theta_0 & \theta'_0P_1^{-1}\Omega_0 \\ \Omega_0P_1^{-1}\theta_0 & \Omega_0P_1^{-1}\Omega_0\end{pmatrix}. \tag{50}$$

This concludes the proof. □

Exercises

1. Show that

$$A_{zx}=\lim_{n\to\infty}(1/n)\mathscr{E}Z'X.$$

2. Hence prove that

$$\mathscr{V}_{as}(\hat{\alpha}) = \sigma_0^2\left(\lim_{n\to\infty}(1/n)(\mathscr{E}Z'X)(X'X)^{-1}(\mathscr{E}X'Z)\right)^{-1}$$

(see Holly and Magnus 1988).
3. Let $\theta_0 = (\theta'_{01}, \theta'_{02})'$. What is the interpretation of the hypothesis $\theta_{02} = 0$?
4. Show that

$$\mathscr{V}_{as}(\hat{\theta}_2) = \sigma_0^2(\Omega'_{02} + \Omega'_{02}P_1^{-1}\Omega_{02}) + \theta_{02}\theta'_{02},$$

where $\Omega_0 = (\Omega_{01} : \Omega_{02})$ is partitioned conformably to θ_0 (see Smith 1985). How would you test the hypothesis $\theta_{02} = 0$?
5. Show that H_* is positive semidefinite.
6. Hence show that

$$Q^{-1} \otimes [\Omega_0 - (1/\sigma_0^2)\theta_0\theta'_0] \leqslant \mathscr{V}_{as}(\text{vec}\,\hat{\Pi}') \leqslant Q^{-1} \otimes \Omega_0$$

where $B \leqslant A$ means that $A - B$ is positive semidefinite.

BIBLIOGRAPHICAL NOTES

§3–§6. The identification problem is thoroughly discussed by Fisher (1966). See also Koopmans *et al.* (1950), Malinvaud (1966, chap. 18) and Rothenberg (1971). The remark in §5 is based on th. 5.A.2 in Fisher (1966). See also Hsiao (1983).
§7–§8. See also Koopmans *et al.* (1950) and Rothenberg and Leenders (1964).
§9. The fact that LIML can be represented as a special case of FIML where every equation (apart from the first) is just identified is discussed by Godfrey and Wickens (1982).
§10–§11. See Smith (1985) and Holly and Magnus (1988).

CHAPTER 17

Topics in psychometrics

1. INTRODUCTION

In this chapter we shall explore some of the optimization problems that occur in psychometrics. Most of these are concerned with the eigenstructure of variance matrices, that is, with their eigenvalues and eigenvectors. The theorems in this chapter fall into four categories. Thus, sections 2–7 deal with *principal components analysis*. Here, a set of p scalar random variables $x_1, \ldots, x_p$ is transformed linearly and orthogonally into an equal number of new random variables $v_1, \ldots, v_p$. The transformation is such that the new variables are uncorrelated. The first principal component v_1 is the normalized linear combination of the x variables with maximum variance; the second principal component v_2 is the normalized linear combination having maximum variance out of all linear combinations uncorrelated with v_1; and so on. One hopes that the first few components account for a large proportion of the variance of the x variables. Another way of looking at principal components analysis is to approximate the variance matrix of x, say Ω, which is assumed known, 'as well as possible' by another positive semidefinite matrix of lower rank. If Ω is *not* known we use an estimate S of Ω based on a sample of x, and try to approximate S rather than Ω.

Instead of approximating S, which depends on the observation matrix X (containing the sample values of x), we can also attempt to approximate X directly. For example, we could approximate X be a lower-rank matrix, say $\tilde{X}$. Employing a singular value decomposition we can write $\tilde{X} = ZA'$, where A is semi-orthogonal. Hence, $X = ZA' + E$, where Z and A have to be determined subject to A being semi-orthogonal such that $\operatorname{tr} E'E$ is minimized. This method of approximating X is called one-mode *component analysis* and is discussed in section 8. Generalizations to two-mode and multimode component analysis are also discussed (sections 10 and 11).

In contrast to principal components analysis, which is primarily concerned with explaining the variance structure, *factor analysis* attempts to explain the covariances of the variables x in terms of a smaller number of non-observables, called 'factors'. This typically leads to the model

$$x = Ay + \mu + \varepsilon, \tag{1}$$

where y and ε are unobservable and independent. One usually assumes that $y \sim \mathcal{N}(0, I_m)$, $\varepsilon \sim \mathcal{N}(0, \Phi)$, where Φ is diagonal. The variance matrix of x is then $AA' + \Phi$, and the problem is to estimate A and Φ from the data. Interesting optimization problems arise in this context and are discussed in sections 12–15.

A final section deals with *canonical correlations*. Here, again, the idea is to reduce the number of variables without sacrificing too much information. Whereas principal components analysis regards the variables as arising from a *single* set, canonical correlation analysis assumes that the variables fall naturally into *two* sets. Instead of studying the two complete sets, the aim is to select only a few uncorrelated linear combinations of the two sets of variables, which are pairwise highly correlated.

2. POPULATION PRINCIPAL COMPONENTS

Let x be a $p \times 1$ random vector with mean μ and positive definite variance matrix Ω. It is assumed that Ω is known. Let $\lambda_1 \geqslant \lambda_2 \geqslant \ldots \geqslant \lambda_p > 0$ be the eigenvalues of Ω and let $T = (t_1, t_2, \ldots, t_p)$ be a $p \times p$ orthogonal matrix such that

$$T'\Omega T = \Lambda = \text{diag}(\lambda_1, \lambda_2, \ldots, \lambda_p). \tag{1}$$

If the eigenvalues $\lambda_1, \ldots, \lambda_p$ are distinct, then T is unique apart from possible sign reversals of its columns. If multiple eigenvalues occur, T is not unique. The i-th column of T is, of course, an eigenvector of Ω associated with the eigenvalue λ_i.

We now define the $p \times 1$ vector of transformed random variables

$$v = T'x \tag{2}$$

as the *vector of principal components* of x. The i-th element of v, say v_i, is called the *i-th principal component.*

Theorem 1

The principal components $v_1, v_2, \ldots, v_p$ are uncorrelated, and $\mathscr{V}(v_i) = \lambda_i$, $i = 1, \ldots, p$.

Proof. We have

$$\mathscr{V}(v) = \mathscr{V}(T'x) = T'\mathscr{V}(x)T$$
$$= T'\Omega T = \Lambda,$$

and the result follows. □

3. OPTIMALITY OF PRINCIPAL COMPONENTS

The principal components have the following optimality property.

Theorem 2

The first principal component v_1 is the normalized linear combination of $x_1, \ldots, x_p$ with maximum variance. That is,

$$\max_{a'a=1} \mathscr{V}(a'x) = \mathscr{V}(v_1) = \lambda_1. \tag{1}$$

The second principal component v_2 is the normalized linear combination of $x_1, \ldots, x_p$ with maximum variance subject to being uncorrelated to v_1. That is,

$$\max_{\substack{a'a=1 \\ t_1'a=0}} \mathscr{V}(a'x) = \mathscr{V}(v_2) = \lambda_2, \tag{2}$$

where t_1 denotes the first column of T. In general, for $i = 1, 2, \ldots, p$, the i-th principal component v_i is the normalized linear combination of $x_1, \ldots, x_p$ with maximum variance subject to being uncorrelated to $v_1, v_2, \ldots, v_{i-1}$. That is,

$$\max_{\substack{a'a=1 \\ T_{i-1}'a=0}} \mathscr{V}(a'x) = \mathscr{V}(v_i) = \lambda_i, \tag{3}$$

where T_{i-1} denotes the $p \times (i-1)$ matrix consisting of the first $i-1$ columns of T.

Proof. We want to find a linear combination of the elements of x, say $a'x$ such that $\mathscr{V}(a'x)$ is maximal subject to the conditions $a'a = 1$ (normalization) and $\mathscr{C}(a'x, v_j) = 0$, $j = 1, 2, \ldots, i-1$. Noting that

$$\mathscr{V}(a'x) = a'\Omega a \tag{4}$$

and also that

$$\begin{aligned} \mathscr{C}(a'x, v_j) &= \mathscr{C}(a'x, t_j'x) \\ &= a'\Omega t_j = \lambda_j a' t_j, \end{aligned} \tag{5}$$

the problem boils down to

$$\text{maximize} \quad \frac{a'\Omega a}{a'a} \tag{6}$$

$$\text{subject to} \quad t_j'a = 0 \qquad (j = 1, \ldots, i-1). \tag{7}$$

From Theorem 11.6 we know that the constrained maximum is λ_i and is obtained for $a = t_i$. □

Notice that the principal components are unique (apart from sign) if and only if all eigenvalues are distinct. But Theorem 2 holds irrespective of multiplicities among the eigenvalues.

Since principal components analysis attempts to 'explain' the variability in x, we need some measure of the amount of total variation in x that has been explained by the first r principal components. One such measure is

$$\mu_r = \frac{\mathscr{V}(v_1)+ \ldots + \mathscr{V}(v_r)}{\mathscr{V}(x_1)+ \ldots + \mathscr{V}(x_p)}. \tag{8}$$

It is clear that

$$\mu_r = \frac{\lambda_1 + \lambda_2 + \ldots + \lambda_r}{\lambda_1 + \lambda_2 + \ldots + \lambda_p}, \tag{9}$$

and hence that $0 < \mu_r \leqslant 1$ and $\mu_p = 1$.

Principal components analysis is only useful when, for a relatively small value of r, μ_r is close to one; in that case a small number of principal components explain most of the variation in x.

4. A RELATED RESULT

Another way of looking at the problem of explaining the variation in x is to try and find a matrix V of specified rank $r \leqslant p$ which provides the 'best' approximation of Ω. It turns out that the optimal V is a matrix whose r non-zero eigenvalues are the r largest eigenvalues of Ω.

Theorem 3

Let Ω be a given positive definite $p \times p$ matrix and let $1 \leqslant r \leqslant p$. Let ϕ be a real-valued function defined by

$$\phi(V) = \operatorname{tr}(\Omega - V)^2 \tag{1}$$

where V is positive semidefinite of rank r. The minimum of ϕ is obtained for

$$V = \sum_{i=1}^{r} \lambda_i t_i t_i' \tag{2}$$

where $\lambda_1, \ldots, \lambda_r$ are the r *largest* eigenvalues of Ω and $t_1, \ldots, t_r$ are corresponding orthonormal eigenvectors. The minimum value of ϕ is the sum of the squares of the $p-r$ *smallest* eigenvalues of Ω.

Proof. In order to force positive semidefiniteness on V, we write $V = AA'$ where A is a $p \times r$ matrix of full column rank r. Let

$$\phi(A) = \operatorname{tr}(\Omega - AA')^2. \tag{3}$$

Then we must minimize ϕ with respect to A. The first differential is

$$\begin{aligned} \mathrm{d}\phi(A) &= -2\operatorname{tr}(\Omega - AA')\mathrm{d}(AA') \\ &= -4\operatorname{tr} A'(\Omega - AA')\mathrm{d}A. \end{aligned} \tag{4}$$

The first-order condition is thus

$$\Omega A = A(A'A). \tag{5}$$

As $A'A$ is symmetric it can be diagonalized. Thus, if $\mu_1, \mu_2, \ldots, \mu_r$ denote the eigenvalues of $A'A$, then there exists an orthogonal $r \times r$ matrix S such that

$$S'A'AS = M = \operatorname{diag}(\mu_1, \mu_2, \ldots, \mu_r). \tag{6}$$

Defining $Q = ASM^{-1/2}$, we can now rewrite (5) as

$$\Omega Q = QM, \qquad Q'Q = I_r. \tag{7}$$

Hence, every eigenvalue of $A'A$ is an eigenvalue of Ω, and Q is a corresponding matrix of orthonormal eigenvectors.

Given (5) and (6) the objective function ϕ can be rewritten as

$$\phi(A) = \operatorname{tr} \Omega^2 - \operatorname{tr} M^2. \tag{8}$$

For a minimum we thus put $\mu_1, \ldots, \mu_r$ equal to $\lambda_1, \ldots, \lambda_r$, the r *largest* eigenvalues of Ω. Then

$$V = AA' = QM^{1/2}S'SM^{1/2}Q' = QMQ' = \sum_{i=1}^{r} \lambda_i t_i t_i'. \tag{9}$$

□

Exercises

1. Show that the explained variation in x as defined in (3.8) is given by $\mu_r = \operatorname{tr} V / \operatorname{tr} \Omega$.
2. Show that if, in Theorem 3, we only require V to be symmetric (rather than positive semidefinite), we obtain the same result.

5. SAMPLE PRINCIPAL COMPONENTS

In applied research the variance matrix Ω is usually not known and must be estimated. To this end we consider a random sample $x_1, x_2, \ldots, x_n$ of size $n > p$ from the distribution of the random $p \times 1$ vector x. We let

$$\mathscr{E}x = \mu, \qquad \mathscr{V}(x) = \Omega, \tag{1}$$

where both μ and Ω are unknown (but are assumed to be finite). In addition we assume that Ω is positive definite, and denote its eigenvalues by $\lambda_1 \geq \lambda_2 \geq \ldots \geq \lambda_p > 0$.

The observations in the sample can be combined into the $n \times p$ observation matrix

$$X = \begin{bmatrix} x_{11} & \cdots & x_{1p} \\ \vdots & & \vdots \\ x_{n1} & \cdots & x_{np} \end{bmatrix} = (x_1, \ldots, x_n)'. \tag{2}$$

The *sample variance** of x, denoted S, is

$$S=(1/n)X'MX=(1/n)\sum_{i=1}^{n}(x_i-\bar{x})(x_i-\bar{x})', \tag{3}$$

where

$$\bar{x}=(1/n)\sum_{i=1}^{n}x_i, \qquad M=I_n-(1/n)\imath\imath', \qquad \imath=(1, 1, \ldots, 1)'.$$

We denote the eigenvalues of S by $l_1>l_2>\ldots>l_p$, and notice that these are distinct with probability one even when the eigenvalues of Ω are not all distinct. Let $Q=(q_1, q_2, \ldots, q_p)$ be a $p\times p$ orthogonal matrix such that

$$Q'SQ=L\equiv\text{diag}(l_1, l_2, \ldots, l_p). \tag{4}$$

We then define the $p\times 1$ vector

$$\hat{v}=Q'x \tag{5}$$

as the vector of *sample principal components* of x, and its i-th element $\hat{v}_i$ as the *i-th sample principal component.*

Recall that $T=(t_1, \ldots, t_p)$ denotes a $p\times p$ orthogonal matrix such that

$$T'\Omega T=\Lambda=\text{diag}(\lambda_1, \ldots, \lambda_p). \tag{6}$$

We would expect that the matrices S, Q and L from the sample provide good estimates for the corresponding population matrices Ω, T and Λ. That this is indeed the case follows from the next theorem.

Theorem 4 (Anderson)

If x follows a p-dimensional normal distribution, then S is the ML estimator for Ω. If, in addition, the eigenvalues of Ω are all distinct, then the ML estimators for λ_i and t_i are l_i and q_i respectively ($i=1, \ldots, p$).

Remark. If the eigenvalues of both Ω and S are distinct (as in the second part of Theorem 4), then the eigenvectors t_i and q_i ($i=1, \ldots, p$) are unique apart from their sign. We can resolve this indeterminacy by requiring that the first non-zero element in each column of T and Q is positive.

Exercise

1. If Ω is singular, show that $r(X)\leq r(\Omega)+1$. Conclude that X cannot have full rank p and S must be singular, if $r(\Omega)\leq p-2$.

* The sample variance is more commonly defined as $S^*=[n/(n-1)]S$, which has the advantage of being an unbiased estimator for Ω. We prefer to work with S as given in (3) because, given normality, it is the ML estimator for Ω.

6. OPTIMALITY OF SAMPLE PRINCIPAL COMPONENTS

In direct analogy with population principal components, the sample principal components have the following optimality property.

Theorem 5

The first sample principal component $\hat{v}_1$ is the normalized linear combination of x, say $a_1'x$, with maximum sample variance. That is, the vector a_1 maximizes $a_1'Sa_1$ subject to the constraint $a_1'a_1=1$. In general, for $i=1, 2, \ldots, p$, the i-th sample principal component $\hat{v}_i$ is the normalized linear combination of x, say $a_i'x$, with maximum sample variance subject to having zero sample correlation with $\hat{v}_1, \ldots, \hat{v}_{i-1}$. That is, the vector a_i maximizes $a_i'Sa_i$ subject to the constraints $a_i'a_i=1$ and $q_j'a_i=0, j=1, \ldots, i-1$.

7. SAMPLE ANALOGUE OF THEOREM 3

Precisely as in section 4, the problem can also be viewed as one of approximating the sample variance matrix S, of rank p, by a matrix V of given rank $r \leqslant p$.

Theorem 6

The positive semidefinite $p \times p$ matrix V of given rank $r \leqslant p$ which provides the best approximation to $S \equiv (1/n)X'MX$ in the sense that it minimizes $\operatorname{tr}(S-V)^2$ is given by

$$V=\sum_{i=1}^{r} l_i q_i q_i'.$$

8. ONE-MODE COMPONENT ANALYSIS

Let X be the $n \times p$ observation matrix and $M=I_n-(1/n)\imath\imath'$. As in (5.3) we express the sample variance matrix S as

$$S=(1/n)X'MX. \tag{1}$$

In Theorem 6 we found the best approximation to S by a matrix V of given rank. Of course, instead of approximating S we can also approximate X by a matrix of given (lower) rank. This is attempted in component analysis.

In the one-mode component model we try to approximate the p columns of $X=(x^1, \ldots, x^p)$ by linear combinations of a smaller number of vectors $z^1, \ldots, z^r$. In other words, we write

$$x^j=\sum_{h=1}^{r} \alpha_{jh} z^h + e^j \qquad (j=1, \ldots, p) \tag{2}$$

and try to make the residuals e^j 'as small as possible' by suitable choices of $\{z^h\}$ and $\{\alpha_{jh}\}$. In matrix notation (2) becomes

$$X = ZA' + E. \tag{3}$$

The $n \times r$ matrix Z is known as the *core matrix*. Without loss of generality we may assume $A'A = I_r$ (see exercise 1). Even with this constraint on A there is some indeterminacy in (3). We can post-multiply Z with an orthogonal matrix R and pre-multiply A' with R' without changing ZA' or the constraint $A'A = I_r$.

Let us introduce the set of matrices

$$\mathcal{O}_{p\times r} = \{A : A \in \mathbb{R}^{p\times r}, A'A = I_r\}. \tag{4}$$

This is the set of all semi-orthogonal $p \times r$ matrices, also known as the *Stiefel manifold*.

With this notation we can now prove Theorem 7.

Theorem 7 (Eckart and Young)

Let X be a given $n \times p$ matrix and let ϕ be a real-valued function defined by

$$\phi(A, Z) = \operatorname{tr}(X - ZA')(X - ZA')' \tag{5}$$

where $A \in \mathcal{O}_{p\times r}$ and $Z \in \mathbb{R}^{n\times r}$. The minimum of ϕ is obtained when A is a $p \times r$ matrix of orthonormal eigenvectors associated with the r *largest* eigenvalues of $X'X$ and $Z = XA$. The 'best' approximation $\tilde{X}$ (of rank r) to X is then $\tilde{X} = XAA'$. The constrained minimum of ϕ is the sum of the $p - r$ *smallest* eigenvalues of $X'X$.

Proof. Define the Lagrangian function

$$\psi(A, Z) = \tfrac{1}{2}\operatorname{tr}(X - ZA')(X - ZA')' - \tfrac{1}{2}\operatorname{tr} L(A'A - I), \tag{6}$$

where L is a *symmetric* $r \times r$ matrix of Lagrange multipliers. Differentiating ψ we obtain

$$\begin{aligned} \mathrm{d}\psi &= \operatorname{tr}(X - ZA')\mathrm{d}(X - ZA')' - \tfrac{1}{2}\operatorname{tr} L[(\mathrm{d}A)'A + A'\mathrm{d}A] \\ &= -\operatorname{tr}(X - ZA')A(\mathrm{d}Z)' - \operatorname{tr}(X - ZA')(\mathrm{d}A)Z' - \operatorname{tr} LA'\mathrm{d}A \\ &= -\operatorname{tr}(X - ZA')A(\mathrm{d}Z)' - \operatorname{tr}(Z'X - Z'ZA' + LA')\mathrm{d}A. \end{aligned} \tag{7}$$

The first-order conditions are

$$(X - ZA')A = 0 \tag{8}$$

$$Z'X - Z'ZA' + LA' = 0 \tag{9}$$

$$A'A = I. \tag{10}$$

From (8) and (10) we find

$$Z = XA. \tag{11}$$

Post-multiplying both sides of (9) by A gives

$$L = Z'Z - Z'XA = 0, \tag{12}$$

in view of (10) and (11). Hence (9) can be rewritten as

$$(X'X)A = A(A'X'XA). \tag{13}$$

Now, let P be an orthogonal $r \times r$ matrix such that

$$P'A'X'XAP = \Lambda_1, \tag{14}$$

where Λ_1 is a diagonal $r \times r$ matrix containing the eigenvalues of $A'X'XA$ on its diagonal. Let $T_1 = AP$. Then (13) can be written

$$X'XT_1 = T_1\Lambda_1. \tag{15}$$

Hence, T_1 is a semi-orthogonal $p \times r$ matrix that diagonalizes $X'X$, and the r diagonal elements in Λ_1 are eigenvalues of $X'X$.

Given $Z = XA$, we have

$$(X - ZA')(X - ZA')' = X(I - AA')X' \tag{16}$$

and thus

$$\operatorname{tr}(X - ZA')(X - ZA')' = \operatorname{tr}X'X - \operatorname{tr}\Lambda_1. \tag{17}$$

To minimize (17), we must maximize tr Λ_1; hence Λ_1 contains the r *largest* eigenvalues of $X'X$, and T_1 contains eigenvectors associated with these r eigenvalues. The 'best' approximation to X is then

$$ZA' = XAA' = XT_1T_1', \tag{18}$$

so that an optimal choice is $A = T_1, Z = XT_1$. From (17) it is clear that the value of the constrained minimum is the sum of the $p-r$ smallest eigenvalues of $X'X$.

□

We notice that the 'best' approximation to X, say $\tilde{X}$, is given by (18): $\tilde{X} = XAA'$. It is important to observe that $\tilde{X}$ is part of a singular value decomposition of X, namely the part corresponding to the r largest eigenvalues of $X'X$. To see this, assume that $r(X) = p$ and that the eigenvalues of $X'X$ are given by $\lambda_1 \geqslant \lambda_2 \geqslant \cdots \geqslant \lambda_p > 0$. Let $\Lambda = \operatorname{diag}(\lambda_1, \ldots, \lambda_p)$ and let

$$X = S\Lambda^{1/2}T' \tag{19}$$

be a singular value decomposition of X, with $S'S = T'T = I_p$. Let

$$\Lambda_1 = \operatorname{diag}(\lambda_1, \ldots, \lambda_r), \Lambda_2 = \operatorname{diag}(\lambda_{r+1}, \ldots, \lambda_p) \tag{20}$$

and partition S and T accordingly as

$$S = (S_1 : S_2), T = (T_1 : T_2). \tag{21}$$

Then,

$$X = S_1\Lambda_1^{1/2}T_1' + S_2\Lambda_2^{1/2}T_2'. \tag{22}$$

From (19)–(21) we see that $X'XT_1 = T_1\Lambda_1$, in accordance with (15). The approximation $\tilde{X}$ can then be written as

$$\tilde{X} = XAA' = XT_1T_1' = (S_1\Lambda_1^{1/2}T_1' + S_2\Lambda_2^{1/2}T_2')T_1T_1' = S_1\Lambda_1^{1/2}T_1'. \tag{23}$$

This result will be helpful in the treatment of two-mode component analysis in section 10. Notice that when $r(ZA') = r(X)$, then $\tilde{X} = X$ (see also Exercise 3).

Exercises

1. Suppose $r(A) = r' \leqslant r$. Use the singular-value decomposition of A to show that $ZA' = Z^* A^{*\prime}$, where $A^{*\prime} A^* = I_{r'}$. Conclude that, in (3), we may assume $A'A = I_r$.
2. Consider the optimization problem

$$\begin{array}{ll} \text{minimize} & \phi(X) \\ \text{subject to} & F(X) = 0. \end{array}$$

If $F(X)$ is symmetric for all X, prove that the Lagrangian function is

$$\psi(X) = \phi(X) - \operatorname{tr} LF(X)$$

where L is *symmetric*.

3. If X has rank $\leqslant r$ show that

$$\min \operatorname{tr} (X - ZA')(X - ZA')' = 0$$

over all A in $\mathcal{O}_{p \times r}$ and Z in $\mathbb{R}^{n \times r}$.

9. RELATIONSHIP BETWEEN ONE-MODE COMPONENT ANALYSIS AND SAMPLE PRINCIPAL COMPONENTS

In the one-mode component model we attempted to approximate the $n \times p$ matrix X by ZA' satisfying $A'A = I_r$. The solution, from Theorem 7, is

$$ZA' = XT_1 T_1' \tag{1}$$

where T_1 is a $p \times r$ matrix of eigenvectors associated with the r largest eigenvalues of $X'X$.

If, instead of X, we approximate MX by ZA' under the constraint $A'A = I_r$, we find in precisely the same way

$$ZA' = MXT_1 T_1', \tag{2}$$

but now T_1 is a $p \times r$ matrix of eigenvectors associated with the r largest eigenvalues of $(MX)'(MX) = X'MX$. This suggests that a suitable approximation to $X'MX$ is provided by

$$(ZA')'ZA' = T_1 T_1' X' MXT_1 T_1' = T_1 \Lambda_1 T_1' \tag{3}$$

where Λ_1 is an $r \times r$ matrix containing the r largest eigenvalues of $X'MX$. Now, (3) is precisely the approximation obtained in Theorem 6. Thus one-mode component analysis and sample principal components are tightly connected.

10. TWO-MODE COMPONENT ANALYSIS

Suppose that our data set consists of a 27×6 matrix X containing the scores given by $n=27$ individuals to each of $p=6$ television commercials. A one-mode component analysis would attempt to reduce p from 6 to 2 (say). There is no reason, however, why we should not also reduce n, say from 27 to 4. This is attempted in two-mode component analysis, where the purpose is to find matrices A, B and Z such that

$$X=BZA'+E \tag{1}$$

with $A'A = I_{r_1}$ and $B'B = I_{r_2}$, and 'minimal' residual matrix E. (In our example $r_1 = 2, r_2 = 4$.) When $r_1 = r_2$ the result follows directly from Theorem 7 and we obtain Theorem 8.

Theorem 8

Let X be a given $n \times p$ matrix and let ϕ be a real-valued function defined by

$$\phi(A, B, Z)=\text{tr}(X-BZA')(X-BZA')' \tag{2}$$

where $A \in \mathcal{O}_{p\times r}$, $B \in \mathcal{O}_{n\times r}$ and $Z \in \mathbb{R}^{r\times r}$. The minimum of ϕ is obtained when A, B and Z satisfy

$$A = T_1, \qquad B = S_1, \qquad Z = \Lambda_1^{1/2}, \tag{3}$$

where Λ_1 is a diagonal $r \times r$ matrix containing the r *largest* eigenvalues of XX' (and of $X'X$), S_1 is an $n \times r$ matrix of orthonormal eigenvectors of XX' associated with these r eigenvalues,

$$XX'S_1 = S_1\Lambda_1, \tag{4}$$

and T_1 is a $p \times r$ matrix of orthonormal eigenvectors of $X'X$ defined by

$$T_1 = X'S_1\Lambda_1^{-1/2}. \tag{5}$$

The constrained minimum of ϕ is the sum of the $p - r$ *smallest* eigenvalues of XX'.

Proof. Immediate from Theorem 7 and the discussion following its proof. □

In the more general case where $r_1 \neq r_2$ the solution is essentially the same. A better approximation does not exist. Suppose $r_2 > r_1$. Then we can extend B with $r_2 - r_1$ additional columns such that $B'B = I_{r_2}$, and we can extend Z with $r_2 - r_1$ additional rows of zeros. The approximation is still the same: $BZA' = S_1\Lambda_1^{1/2}T_1'$. Adding columns to B turns out to be useless; it does not lead to a better approximation to X, since the rank of BZA' remains r_1.

11. MULTIMODE COMPONENT ANALYSIS

Continuing our example of section 10, suppose that we now have an enlarged data set consisting of a three-dimensional matrix X of order $27 \times 6 \times 5$ containing scores by $p_1 = 27$ individuals to each of $p_2 = 6$ television commercials; each commercial is shown $p_3 = 5$ times to every individual. A three-mode component analysis would attempt to reduce p_1, p_2 and p_3 to, say, $r_1 = 6$, $r_2 = 2$, $r_3 = 3$. Since, in principle, there is no limit to the number of modes we might be interested in, let us consider the s-mode model. First, however, we reconsider the two-mode case

$$X = BZA' + E. \tag{1}$$

We rewrite (1) as

$$x = (A \otimes B)z + e \tag{2}$$

where $x = \text{vec}\, X$, $z = \text{vec}\, Z$ and $e = \text{vec}\, E$. This suggests the following formulation for the s-mode component case:

$$x = (A_1 \otimes A_2 \otimes \ldots \otimes A_s)z + e, \tag{3}$$

where A_i is a $p_i \times r_i$ matrix satisfying $A_i' A_i = I_{r_i}$ $(i = 1, \ldots, s)$. The data vector x and the 'core' vector z can be considered as stacked versions of s-dimensional matrices X and Z. The elements in x are identified by s indices with the i-th index assuming the values $1, 2, \ldots, p_i$. The elements are arranged in such a way that the first index runs slowly and the last index runs fast. The elements in z are also identified by s indices; the i-th index runs from 1 to r_i.

The mathematical problem is to choose A_i $(i = 1, \ldots, s)$ and z in such a way that the residual e is 'as small as possible'.

Theorem 9

Let $p_1, p_2, \ldots, p_s$ and $r_1, r_2, \ldots, r_s$ be given integers, $p_i, r_i \geq 1$, and put

$$p = \prod_{i=1}^{s} p_i \quad \text{and} \quad r = \prod_{i=1}^{s} r_i.$$

Let x be a given $p \times 1$ vector and let ϕ be a real-valued function defined by

$$\phi(A, z) = (x - Az)'(x - Az) \tag{4}$$

where $A = A_1 \otimes A_2 \otimes \cdots \otimes A_s$, $A_i \in \mathcal{O}_{p_i \times r_i}$ $(i = 1, \cdots, s)$ and $z \in \mathbb{R}^r$. The minimum of ϕ is obtained when $A_1, \ldots, A_s$ and z satisfy

$$A_i = T_i \quad (i = 1, \ldots, s), \qquad z = (T_1 \otimes \ldots \otimes T_s)'x, \tag{5}$$

where T_i is a $p_i \times r_i$ matrix of orthonormal eigenvectors associated with the r_i largest eigenvalues of $X_i' T_{(i)} T_{(i)}' X_i$. Here $T_{(i)}$ denotes the $(p/p_i) \times (r/r_i)$ matrix

$$T_{(i)} = T_1 \otimes \ldots \otimes T_{i-1} \otimes T_{i+1} \otimes \ldots \otimes T_s, \tag{6}$$

and X_i is the $(p/p_i)\times p_i$ matrix defined by

$$\operatorname{vec} X_i' = Q_i x \qquad (i=1,\ldots,s) \tag{7}$$

where

$$Q_i = I_{\alpha_{i-1}} \otimes K_{\beta_{s-i},p_i} \qquad (i=1,\ldots,s) \tag{8}$$

with

$$\alpha_0=1, \qquad \alpha_1=p_1, \qquad \alpha_2=p_1p_2, \ \ldots, \ \alpha_s=p \tag{9}$$

and

$$\beta_0=1, \qquad \beta_1=p_s, \qquad \beta_2=p_sp_{s-1}, \ \ldots, \ \beta_s=p. \tag{10}$$

The minimum value of ϕ is $x'x-z'z$.

Remark. The solution has to be obtained iteratively. Take $A_2^{(0)},\ldots,A_s^{(0)}$ as starting values for $A_2,\ldots,A_s$. Compute $A_{(1)}^{(0)} = A_2^{(0)} \otimes \ldots \otimes A_s^{(0)}$. Then form a first approximate of A_1, say $A_1^{(1)}$, as the $p_1 \times r_1$ matrix of orthonormal eigenvectors associated with the r_1 largest eigenvalues of $X_1'A_{(1)}^{(0)}A_{(1)}^{(0)\prime}X_1$. Next, use $A_1^{(1)}$ and $A_3^{(0)},\ldots,A_s^{(0)}$ to compute $A_{(2)}^{(0)} = A_1^{(1)} \otimes A_3^{(0)} \otimes \ldots \otimes A_s^{(0)}$, and form $A_2^{(1)}$, the first approximate of A_2, in a similar manner. Having computed $A_1^{(1)},\ldots,A_s^{(1)}$, we form a new approximate of A_1, say $A_1^{(2)}$. This process is continued until convergence.

Proof. Analogous to the $p\times p$ matrices Q_i we define the $r\times r$ matrices

$$R_i = I_{\gamma_{i-1}} \otimes K_{\delta_{s-i},r_i} \qquad (i=1,\ldots,s) \tag{11}$$

where

$$\gamma_0=1, \qquad \gamma_1=r_1, \qquad \gamma_2=r_1r_2, \ \ldots, \ \gamma_s=r \tag{12}$$

and

$$\delta_0=1, \qquad \delta_1=r_s, \qquad \delta_2=r_sr_{s-1}, \ \ldots, \ \delta_s=r. \tag{13}$$

We also define the $(r/r_i)\times r_i$ matrices Z_i by

$$\operatorname{vec} Z_i' = R_i z \qquad (i=1,\ldots,s), \tag{14}$$

and notice that

$$Q_i(A_1 \otimes A_2 \otimes \ldots \otimes A_s)R_i' = A_{(i)} \otimes A_i, \tag{15}$$

where $A_{(i)}$ is defined in the same way as $T_{(i)}$.

Now, let ψ be the Lagrangian function

$$\psi(A,Z) = \tfrac{1}{2}(x-Az)'(x-Az) - \tfrac{1}{2}\sum_{i=1}^{s} \operatorname{tr} L_i(A_i'A_i - I), \tag{16}$$

where L_i $(i=1,\ldots,s)$ is a symmetric $r_i\times r_i$ matrix of Lagrange multipliers. We have

$$\mathrm{d}\psi = -(x-Az)'(\mathrm{d}A)z - (x-Az)'A\,\mathrm{d}z - \sum_{i=1}^{s} \operatorname{tr} L_iA_i'\mathrm{d}A_i. \tag{17}$$

Since $A = Q_i'(A_{(i)} \otimes A_i)R_i$ for $i=1,\ldots,s$, we obtain

$$\mathrm{d}A = \sum_{i=1}^{s} Q_i'(A_{(i)} \otimes \mathrm{d}A_i)R_i \tag{18}$$

and hence

$$(x-Az)'(\mathrm{d}A)z=\sum_{i=1}^{s}(x-Az)'Q_i'(A_{(i)}\otimes \mathrm{d}A_i)R_iz$$
$$=\sum_{i=1}^{s}(\operatorname{vec}X_i'-Q_iAR_i'\operatorname{vec}Z_i')'(A_{(i)}\otimes \mathrm{d}A_i)\operatorname{vec}Z_i'$$
$$=\sum_{i=1}^{s}(\operatorname{vec}(X_i'-A_iZ_i'A_{(i)}'))'(A_{(i)}\otimes \mathrm{d}A_i)\operatorname{vec}Z_i'$$
$$=\sum_{i=1}^{s}\operatorname{tr}Z_i'A_{(i)}'(X_i-A_{(i)}Z_iA_i')\mathrm{d}A_i. \tag{19}$$

Inserting (19) in (17) we thus find

$$\mathrm{d}\psi=-(x-Az)'A\mathrm{d}z-\sum_{i=1}^{s}\operatorname{tr}[Z_i'A_{(i)}'(X_i-A_{(i)}Z_iA_i')+L_iA_i']\mathrm{d}A_i,$$

from which follow the first-order conditions

$$A'(x-Az)=0 \tag{20}$$

$$Z_i'A_{(i)}'X_i-Z_i'A_{(i)}'A_{(i)}Z_iA_i'+L_iA_i'=0 \qquad (i=1,\ldots,s) \tag{21}$$

$$A_i'A_i=I_{r_i} \qquad (i=1,\ldots,s). \tag{22}$$

We find again

$$z=A'x, \tag{23}$$

from which it follows that $Z_i=A_{(i)}'X_iA_i$. Hence $L_i=0$, and (21) can be simplified to

$$(X_i'A_{(i)}A_{(i)}'X_i)A_i=A_i(A_i'X_i'A_{(i)}A_{(i)}'X_iA_i). \tag{24}$$

For $i=1,\ldots,s$, let S_i be an orthogonal $r_i\times r_i$ matrix such that

$$S_i'A_i'X_i'A_{(i)}A_{(i)}'X_iA_iS_i=\Lambda_i \qquad \text{(diagonal)}. \tag{25}$$

Then (24) can be written as

$$(X_i'A_{(i)}A_{(i)}'X_i)(A_iS_i)=(A_iS_i)\Lambda_i. \tag{26}$$

We notice that

$$\operatorname{tr}\Lambda_i=\operatorname{tr}Z_i'Z_i=z'z=\lambda, \quad \text{say}, \tag{27}$$

is the same for all i. Then

$$(x-Az)'(x-Az)=x'x-\lambda. \tag{28}$$

To minimize (28), we must maximize λ; hence Λ_i contains the r_i largest eigenvalues of $X_i'A_{(i)}A_{(i)}'X_i$, and $AS_i=T_i$. Then, by (23),

$$Az=AA'x=(A_1A_1'\otimes\ldots\otimes A_sA_s')x$$
$$=(T_1T_1'\otimes\ldots\otimes T_sT_s')x, \tag{29}$$

and an optimal choice is $A_i=T_i$ $(i=1,\ldots,s)$ and $z=(T_1\otimes\ldots\otimes T_s)'x$. □

Exercise

1. Show that the matrices Q_i and R_i defined in (8) and (11) satisfy:

$$Q_1 = K_{p/p_1, p_1}, \qquad Q_s = I_p$$

and

$$R_1 = K_{r/r_1, r_1}, \qquad R_s = I_r.$$

12. FACTOR ANALYSIS

Let x be an observable $p \times 1$ random vector with $\mathscr{E}x = \mu$ and $\mathscr{V}(x) = \Omega$. The factor analysis model assumes that the observations are generated by the structure

$$x = Ay + \mu + \varepsilon, \tag{1}$$

where y is an $m \times 1$ vector of non-observable random variables called 'common factors', A is a $p \times m$ matrix of unknown parameters called 'factor loadings', and ε is a $p \times 1$ vector of non-observable random errors. It is assumed that $y \sim \mathscr{N}(0, I_m)$, $\varepsilon \sim \mathscr{N}(0, \Phi)$, where Φ is diagonal positive definite, and that y and ε are independent. Given these assumptions we find that $x \sim \mathscr{N}(\mu, \Omega)$ with

$$\Omega = AA' + \Phi. \tag{2}$$

There is clearly a problem of identifying A from AA', because if $A^* = AT$ is an orthogonal transformation of A, then $A^*A^{*\prime} = AA'$. We shall see later (section 15) how this ambiguity can be solved.

Suppose that a random sample of $n > p$ observations $x_1, \ldots, x_n$ of x is obtained. The loglikelihood is

$$\Lambda_n(\mu, A, \Phi) = -\tfrac{1}{2}np \log 2\pi - \tfrac{1}{2}n \log|\Omega| - \tfrac{1}{2}\sum_{i=1}^{n} (x_i - \mu)'\Omega^{-1}(x_i - \mu). \tag{3}$$

Maximizing Λ with respect to μ yields

$$\hat{\mu} = (1/n)\sum_{i=1}^{n} x_i.$$

Substituting $\hat{\mu}$ for μ in (3) yields the so-called concentrated loglikelihood

$$\Lambda_n^c(A, \Phi) = -\tfrac{1}{2}np \log 2\pi - \tfrac{1}{2}n(\log|\Omega| + \operatorname{tr}\Omega^{-1}S), \tag{4}$$

with

$$S = (1/n)\sum_{i=1}^{n} (x_i - \bar{x})(x_i - \bar{x})'. \tag{5}$$

Clearly maximizing (4) is equivalent to minimizing $\log|\Omega| + \operatorname{tr}\Omega^{-1}S$ with respect to A and Φ. The following theorem assumes Φ known, and thus minimizes with respect to A only.

Theorem 10

Let S and Φ be two given positive definite $p\times p$ matrices, Φ diagonal, and let $1\leqslant m\leqslant p$. Let ϕ be a real-valued function defined by

$$\phi(A)=\log|AA'+\Phi|+\operatorname{tr}(AA'+\Phi)^{-1}S, \tag{6}$$

where $A\in\mathbb{R}^{p\times m}$. The minimum of ϕ is obtained when

$$A=\Phi^{1/2}T(\Lambda-I_m)^{1/2}, \tag{7}$$

where Λ is a diagonal $m\times m$ matrix containing the m *largest* eigenvalues of $\Phi^{-1/2}S\Phi^{-1/2}$ and T is a $p\times m$ matrix of corresponding orthonormal eigenvectors. The minimum value of ϕ is

$$p+\log|S|+\sum_{i=m+1}^{p}(\lambda_i-\log\lambda_i-1), \tag{8}$$

where $\lambda_{m+1},\ldots,\lambda_p$ denote the $p-m$ *smallest* eigenvalues of $\Phi^{-1/2}S\Phi^{-1/2}$.

Proof. Define

$$\Omega=AA'+\Phi,\qquad C=\Omega^{-1}-\Omega^{-1}S\Omega^{-1}. \tag{9}$$

Then $\phi=\log|\Omega|+\operatorname{tr}\Omega^{-1}S$ and hence

$$\begin{aligned}\mathrm{d}\phi&=\operatorname{tr}\Omega^{-1}\mathrm{d}\Omega-\operatorname{tr}\Omega^{-1}(\mathrm{d}\Omega)\Omega^{-1}S=\operatorname{tr}C\mathrm{d}\Omega\\&=\operatorname{tr}C((\mathrm{d}A)A'+A(\mathrm{d}A)')=2\operatorname{tr}A'C\mathrm{d}A.\end{aligned} \tag{10}$$

The first-order condition is

$$CA=0, \tag{11}$$

or, equivalently,

$$A=S\Omega^{-1}A. \tag{12}$$

From (12) we obtain

$$\begin{aligned}AA'\Phi^{-1}A&=S\Omega^{-1}AA'\Phi^{-1}A\\&=S\Omega^{-1}(\Omega-\Phi)\Phi^{-1}A\\&=S\Phi^{-1}A-S\Omega^{-1}A=S\Phi^{-1}A-A.\end{aligned} \tag{13}$$

Hence

$$S\Phi^{-1}A=A(I_m+A'\Phi^{-1}A). \tag{14}$$

Assume that $r(A)=m'\leqslant m$, and let Q be a semi-orthogonal $m\times m'$ matrix $(Q'Q=I_{m'})$ such that

$$A'\Phi^{-1}AQ=QM, \tag{15}$$

where M is diagonal and contains the m' non-zero eigenvalues of $A'\Phi^{-1}A$. Then (14) can be written as

$$S\Phi^{-1}AQ=AQ(I+M) \tag{16}$$

from which we obtain

$$(\Phi^{-1/2}S\Phi^{-1/2})\tilde{T}=\tilde{T}(I+M), \tag{17}$$

where $\tilde{T} \equiv \Phi^{-1/2}AQM^{-1/2}$ is a semi-orthogonal $p \times m'$ matrix.

Our next step is to rewrite $\Omega = AA' + \Phi$ as

$$\Omega = \Phi^{1/2}(I + \Phi^{-1/2}AA'\Phi^{-1/2})\Phi^{1/2}, \tag{18}$$

so that the determinant and inverse of Ω can be expressed as

$$|\Omega| = |\Phi|\,|I + A'\Phi^{-1}A| \tag{19}$$

and

$$\Omega^{-1} = \Phi^{-1} - \Phi^{-1}A(I + A'\Phi^{-1}A)^{-1}A'\Phi^{-1}. \tag{20}$$

Then, using (14),

$$\begin{aligned}\Omega^{-1}S &= \Phi^{-1}S - \Phi^{-1}A(I + A'\Phi^{-1}A)^{-1}(I + A'\Phi^{-1}A)A' \\ &= \Phi^{-1}S - \Phi^{-1}AA'. \end{aligned} \tag{21}$$

Given the first-order condition, we thus have

$$\begin{aligned}\phi &= \log|\Omega| + \operatorname{tr}\Omega^{-1}S \\ &= \log|\Phi| + \log|I + A'\Phi^{-1}A| + \operatorname{tr}\Phi^{-1}S - \operatorname{tr}A'\Phi^{-1}A \\ &= p + \log|S| + [\operatorname{tr}(\Phi^{-1/2}S\Phi^{-1/2}) - \log|\Phi^{-1/2}S\Phi^{-1/2}| - p] \\ &\quad - [\operatorname{tr}(I_m + A'\Phi^{-1}A) - \log|I_m + A'\Phi^{-1}A| - m] \\ &= p + \log|S| + \sum_{i=1}^{p}(\lambda_i - \log\lambda_i - 1) - \sum_{j=1}^{m}(\nu_j - \log\nu_j - 1), \end{aligned} \tag{22}$$

where $\lambda_1 \geq \lambda_2 \geq \ldots \geq \lambda_p$ are the eigenvalues of $\Phi^{-1/2}S\Phi^{-1/2}$ and $\nu_1 \geq \nu_2 \geq \ldots \geq \nu_m$ are the eigenvalues of $I_m + A'\Phi^{-1}A$. From (15) and (17) we see that $\nu_1, \ldots, \nu_{m'}$ are also eigenvalues of $\Phi^{-1/2}S\Phi^{-1/2}$ and that the remaining eigenvalues $\nu_{m'+1}, \ldots, \nu_m$ are all one. Since we wish to minimize ϕ, we make $\nu_1, \ldots, \nu_{m'}$ as large as possible, that is equal to the m' *largest* eigenvalues of $\Phi^{-1/2}S\Phi^{-1/2}$. Thus,

$$\nu_i = \begin{cases} \lambda_i & (i = 1, \ldots, m') \\ 1 & (i = m'+1, \ldots, m). \end{cases} \tag{23}$$

Given (23), (22) reduces to

$$\phi = p + \log|S| + \sum_{i=m'+1}^{p}(\lambda_i - \log\lambda_i - 1), \tag{24}$$

which, in turn, is minimized when m' is taken as large as possible; that is, $m' = m$.

Given $m' = m$, Q is orthogonal, $T = \tilde{T} = \Phi^{-1/2}AQM^{-1/2}$ and $\Lambda = I + M$. Hence

$$AA' = \Phi^{1/2}T(\Lambda - I)T'\Phi^{1/2} \tag{25}$$

and A can be chosen as $A = \Phi^{1/2}T(\Lambda - I)^{1/2}$. □

Notice that the optimal choice for A is such that $A'\Phi^{-1}A$ is a diagonal matrix, even though this was not imposed.

13. A ZIGZAG ROUTINE

Theorem 10 provides the basis for (at least) two procedures by which ML estimates of A *and* Φ in the factor model can be found. The first procedure is to minimize the concentrated function (12.8) with respect to the p diagonal elements of Φ. The second procedure is based on the first-order conditions obtained from minimizing the function

$$\psi(A,\Phi)=\log|AA'+\Phi|+\operatorname{tr}(AA'+\Phi)^{-1}S. \tag{1}$$

The function ψ is the same as the function ϕ defined in (12.6) except that ϕ is a function of A *given* Φ, while ψ is a function of A *and* Φ.

In this section we investigate the second procedure. The first procedure is discussed in section 14.

From (12.12) we see that the first-order condition of ψ with respect to A is given by

$$A=S\Omega^{-1}A, \tag{2}$$

where $\Omega=AA'+\Phi$. To obtain the first-order condition with respect to Φ, we differentiate ψ holding A constant. This yields

$$\begin{aligned}\mathrm{d}\psi&=\operatorname{tr}\Omega^{-1}\mathrm{d}\Omega-\operatorname{tr}\Omega^{-1}(\mathrm{d}\Omega)\Omega^{-1}S\\&=\operatorname{tr}\Omega^{-1}\mathrm{d}\Phi-\operatorname{tr}\Omega^{-1}(\mathrm{d}\Phi)\Omega^{-1}S\\&=\operatorname{tr}(\Omega^{-1}-\Omega^{-1}S\Omega^{-1})\mathrm{d}\Phi.\end{aligned} \tag{3}$$

Since Φ is diagonal, the first-order condition with respect to Φ is

$$\mathrm{dg}(\Omega^{-1}-\Omega^{-1}S\Omega^{-1})=0. \tag{4}$$

Pre- and post-multiplying (4) by Φ we obtain the equivalent condition

$$\mathrm{dg}(\Phi\Omega^{-1}\Phi)=\mathrm{dg}(\Phi\Omega^{-1}S\Omega^{-1}\Phi). \tag{5}$$

(The equivalence follows from the fact that Φ is diagonal and non-singular.) Now, given the first-order condition for A in (2), and writing $\Omega-AA'$ for Φ, we have

$$\begin{aligned}S\Omega^{-1}\Phi&=S\Omega^{-1}(\Omega-AA')=S-S\Omega^{-1}AA'\\&=S-AA'=S+\Phi-\Omega,\end{aligned} \tag{6}$$

so that

$$\begin{aligned}\Phi\Omega^{-1}S\Omega^{-1}\Phi&=\Phi\Omega^{-1}(S+\Phi-\Omega)\\&=\Phi\Omega^{-1}S+\Phi\Omega^{-1}\Phi-\Phi\\&=\Phi\Omega^{-1}\Phi+S-\Omega,\end{aligned} \tag{7}$$

using the fact that $\Phi\Omega^{-1}S=S\Omega^{-1}\Phi$. Hence, given (2), (5) is equivalent to

$$\mathrm{dg}(\Omega)=\mathrm{dg}(S), \tag{8}$$

that is,

$$\Phi=\mathrm{dg}(S-AA'). \tag{9}$$

Thus, Theorem 10 provides an explicit solution for A as a function of Φ, and (9) gives Φ as an explicit function of A. A zigzag routine suggests itself: choose an appropriate starting value for Φ, then calculate AA' from (12.25), then Φ from (9), etc. If convergence occurs (which is not guaranteed), then the resulting values for Φ and AA' correspond to a (local) minimum of ψ.

From (12.25) and (9) we summarize this iterative procedure as

$$\phi_i^{(k+1)} = s_{ii} - \phi_i^{(k)} \sum_{j=1}^{m} (\lambda_j^{(k)} - 1)(t_{ij}^{(k)})^2 \qquad (i = 1, \ldots, p) \tag{10}$$

for $k=0, 1, 2, \ldots$. Here s_{ii} denotes the i-th diagonal element of S, $\lambda_j^{(k)}$ the j-th largest eigenvalue of $(\Phi^{(k)})^{-1/2} S (\Phi^{(k)})^{-1/2}$, and $(t_{1j}^{(k)}, \ldots, t_{pj}^{(k)})'$ the corresponding eigenvector.

What is an appropriate starting value for Φ? From (9) we see that $0 < \phi_i < s_{ii}$ $(i=1, \ldots, p)$. This suggests that we choose our starting value as

$$\Phi^{(0)} = \alpha \mathrm{dg}(S) \tag{11}$$

for some α satisfying $0 < \alpha < 1$. Calculating A from (12.7) given $\Phi = \Phi^{(0)}$ leads to

$$A^{(1)} = (\mathrm{dg} S)^{1/2} T(\Lambda - \alpha I_m)^{1/2}, \tag{12}$$

where Λ is a diagonal $m \times m$ matrix containing the m largest eigenvalues of $S^* \equiv (\mathrm{dg}\, S)^{-1/2} S\, (\mathrm{dg}\, S)^{-1/2}$ and T is a $p \times m$ matrix of corresponding orthonormal eigenvectors. This shows that α must be chosen smaller than each of the m largest eigenvalues of S^*.

14. A NEWTON–RAPHSON ROUTINE

Instead of using the first-order conditions to set up a zigzag procedure, we can also use the Newton–Raphson method in order to find the values of $\phi_1, \ldots, \phi_p$ that minimize the concentrated function (12.8). The Newton–Raphson method requires knowledge of the first- and second-order derivatives of this function, and these are provided by the following theorem.

Theorem 11

Let S be a given positive definite $p \times p$ matrix and let $1 \leqslant m \leqslant p-1$. Let γ be a real-valued function defined by

$$\gamma(\phi_1, \ldots, \phi_p) = \sum_{i=m+1}^{p} (\lambda_i - \log \lambda_i - 1), \tag{1}$$

where $\lambda_{m+1}, \ldots, \lambda_p$ denote the $p-m$ smallest eigenvalues of $\Phi^{-1/2} S \Phi^{-1/2}$ and $\Phi = \mathrm{diag}(\phi_1, \ldots, \phi_p)$ is diagonal positive definite of order $p \times p$. At points $(\phi_1, \ldots, \phi_p)$ where $\lambda_{m+1}, \ldots, \lambda_p$ are all distinct eigenvalues of $\Phi^{-1/2} S \Phi^{-1/2}$, the gradient of γ is the $p \times 1$ vector

$$g(\phi) = -\Phi^{-1}\left(\sum_{i=m+1}^{p} (\lambda_i - 1) u_i \odot u_i\right) \tag{2}$$

and the Hessian is the $p \times p$ matrix

$$G(\phi) = \Phi^{-1}\left(\sum_{i=m+1}^{p} u_i u_i' \odot B^i\right)\Phi^{-1}, \tag{3}$$

where

$$B^i = (2\lambda_i - 1) u_i u_i' + 2\lambda_i(\lambda_i - 1)(\lambda_i I - \Phi^{-1/2} S \Phi^{-1/2})^+, \tag{4}$$

and u_i ($i = m+1, \ldots, p$) denotes the orthonormal eigenvector of $\Phi^{-1/2} S \Phi^{-1/2}$ associated with λ_i.

Remark. The symbol $\odot$ denotes the Hadamard product: $A \odot B = (a_{ij} b_{ij})$, see section 3.6.

Proof. Let $\phi = (\phi_1, \ldots, \phi_p)$ and $S^*(\phi) = \Phi^{-1/2} S \Phi^{-1/2}$. Let ϕ_0 be a given point in $\mathbb{R}^p_+$ (the positive orthant of $\mathbb{R}^p$) and $S_0^* = S^*(\phi_0)$. Let

$$\lambda_1 \geqslant \lambda_2 \geqslant \ldots \geqslant \lambda_m > \lambda_{m+1} > \ldots > \lambda_p \tag{5}$$

denote the eigenvalues of S_0^* and let $u_1, \ldots, u_p$ be the corresponding eigenvectors. (Notice that the $p - m$ smallest eigenvalues of S_0^* are assumed distinct.) Then, according to Theorem 8.7, there is a neighbourhood, say $N(\phi_0)$, where differentiable eigenvalue functions $\lambda^{(i)}$ and eigenvector functions $u^{(i)}$ ($i = m+1, \ldots, p$) exist satisfying

$$S^* u^{(i)} = \lambda^{(i)} u^{(i)}, \qquad u^{(i)\prime} u^{(i)} = 1 \tag{6}$$

and

$$u^{(i)}(\phi_0) = u_i, \qquad \lambda^{(i)}(\phi_0) = \lambda_i. \tag{7}$$

Furthermore, at $\phi = \phi_0$,

$$\mathrm{d}\lambda^{(i)} = u_i'(\mathrm{d}S^*) u_i \tag{8}$$

and

$$\mathrm{d}^2\lambda^{(i)} = 2u_i'(\mathrm{d}S^*) T_i^+ (\mathrm{d}S^*) u_i + u_i'(\mathrm{d}^2 S^*) u_i, \tag{9}$$

where $T_i = \lambda_i I - S_0^*$. (See also Theorem 8.10.)

In the present case, $S^* = \Phi^{-1/2} S \Phi^{-1/2}$ and hence

$$\mathrm{d}S^* = -\tfrac{1}{2}(\Phi^{-1}(\mathrm{d}\Phi) S^* + S^*(\mathrm{d}\Phi)\Phi^{-1}) \tag{10}$$

and

$$\mathrm{d}^2 S^* = \tfrac{3}{4}(\Phi^{-1}(\mathrm{d}\Phi)\Phi^{-1}(\mathrm{d}\Phi) S^* + S^*(\mathrm{d}\Phi)\Phi^{-1}(\mathrm{d}\Phi)\Phi^{-1}) + \tfrac{1}{2}\Phi^{-1}(\mathrm{d}\Phi) S^*(\mathrm{d}\Phi)\Phi^{-1}. \tag{11}$$

Inserting (10) into (8) yields

$$\mathrm{d}\lambda^{(i)} = -\lambda_i u_i'\Phi^{-1}(\mathrm{d}\Phi)u_i. \tag{12}$$

Similarly, inserting (10) and (11) into (9) yields

$$\begin{aligned}
\mathrm{d}^2\lambda^{(i)} &= \tfrac{1}{2}\lambda_i^2 u_i'(\mathrm{d}\Phi)\Phi^{-1}T_i^+\Phi^{-1}(\mathrm{d}\Phi)u_i \\
&\quad + \lambda_i u_i'\Phi^{-1}(\mathrm{d}\Phi)S_0^*T_i^+\Phi^{-1}(\mathrm{d}\Phi)u_i \\
&\quad + \tfrac{1}{2}u_i'\Phi^{-1}(\mathrm{d}\Phi)S_0^*T_i^+S_0^*(\mathrm{d}\Phi)\Phi^{-1}u_i \\
&\quad + \tfrac{3}{2}\lambda_i u_i'\Phi^{-1}(\mathrm{d}\Phi)\Phi^{-1}(\mathrm{d}\Phi)u_i \\
&\quad + \tfrac{1}{2}u_i'\Phi^{-1}(\mathrm{d}\Phi)S_0^*(\mathrm{d}\Phi)\Phi^{-1}u_i \\
&= \tfrac{1}{2}u_i'(\mathrm{d}\Phi)\Phi^{-1}C_i\Phi^{-1}(\mathrm{d}\Phi)u_i,
\end{aligned} \tag{13}$$

where

$$C_i = \lambda_i^2 T_i^+ + 2\lambda_i S_0^* T_i^+ + S_0^* T_i^+ S_0^* + 3\lambda_i I + S_0^*. \tag{14}$$

Now, since

$$T_i^+ = \sum_{j\neq i}(\lambda_i - \lambda_j)^{-1}u_j u_j' \qquad \text{and} \qquad S_0^* = \sum_j \lambda_j u_j u_j',$$

we have

$$S_0^* T_i^+ = \sum_{j\neq i}[\lambda_j/(\lambda_i-\lambda_j)]u_j u_j' \qquad \text{and} \qquad S_0^* T_i^+ S_0^* = \sum_{j\neq i}[\lambda_j^2/(\lambda_i-\lambda_j)]u_j u_j'.$$

Hence we obtain

$$C_i = 4\lambda_i(u_i u_i' + \lambda_i T_i^+). \tag{15}$$

We can now take the differentials of

$$\gamma = \sum_{i=m+1}^{p}(\lambda_i - \log\lambda_i - 1).$$

We have

$$\mathrm{d}\gamma = \sum_{i=m+1}^{p}(1-\lambda_i^{-1})\mathrm{d}\lambda^{(i)} \tag{16}$$

and

$$\mathrm{d}^2\gamma = \sum_{i=m+1}^{p}[(\lambda_i^{-1}\mathrm{d}\lambda^{(i)})^2 + (1-\lambda_i^{-1})\mathrm{d}^2\lambda^{(i)}]. \tag{17}$$

Inserting (12) in (16) gives

$$\mathrm{d}\gamma = -\sum_{i=m+1}^{p}(\lambda_i - 1)u_i'\Phi^{-1}(\mathrm{d}\Phi)u_i. \tag{18}$$

Inserting (12) and (13) in (17) gives

$$\begin{aligned} d^2\gamma &= \sum_{i=m+1}^{p} u_i'(d\Phi)\Phi^{-1}[u_iu_i' + \tfrac{1}{2}(1-\lambda_i^{-1})C_i]\Phi^{-1}(d\Phi)u_i \\ &= \sum_{i=m+1}^{p} u_i'(d\Phi)\Phi^{-1}[(2\lambda_i - 1)u_iu_i' + 2\lambda_i(\lambda_i - 1)T_i^+]\Phi^{-1}(d\Phi)u_i \\ &= \sum_{i=m+1}^{p} u_i'(d\Phi)\Phi^{-1}B^i\Phi^{-1}(d\Phi)u_i, \end{aligned} \tag{19}$$

in view of (15). The first-order partial derivatives are thus

$$\frac{\partial\gamma}{\partial\phi_h} = -\phi_h^{-1}\sum_{i=m+1}^{p}(\lambda_i - 1)u_{ih}^2 \qquad (h = 1, \ldots, p), \tag{20}$$

where u_{ih} denotes the h-th component of u_i; the second-order partial derivatives are

$$\frac{\partial^2\gamma}{\partial\phi_h\partial\phi_k} = (\phi_h\phi_k)^{-1}\sum_{i=m+1}^{p} u_{ih}u_{ik}B_{hk}^i \qquad (h, k = 1, \ldots, p) \tag{21}$$

and the result follows. □

Given knowledge of the gradient $g(\phi)$ and the Hessian $G(\phi)$ from (2) and (3), the Newton–Raphson method proceeds as follows. First choose a starting value $\phi^{(0)}$. Then, for $k = 0, 1, 2, \ldots$, compute

$$\phi^{(k+1)} = \phi^{(k)} - [G(\phi^{(k)})]^{-1}g(\phi^{(k)}). \tag{22}$$

This method appears to work well in practice, and yields the values $\phi_1, \ldots, \phi_p$ which minimize (1). Given these values we can compute A from (12.7), thus completing the solution.

There is, however, one proviso. In Theorem 11 we require that the $p-m$ smallest eigenvalues of $\Phi^{-1/2}S\Phi^{-1/2}$ are all distinct. But, by rewriting (12.2) as

$$\Phi^{-1/2}\Omega\Phi^{-1/2} = I + \Phi^{-1/2}AA'\Phi^{-1/2}, \tag{23}$$

we see that the $p-m$ smallest eigenvalues of $\Phi^{-1/2}\Omega\Phi^{-1/2}$ are all one. Therefore, if the sample size increases, the $p-m$ smallest eigenvalues of $\Phi^{-1/2}S\Phi^{-1/2}$ will all converge to one. For large samples an optimization method based on Theorem 11 may therefore not give reliable results.

15. KAISER'S VARIMAX METHOD

The factorization $\Omega = AA' + \Phi$ of the variance matrix is not unique. If we transform the 'loading' matrix A by an orthogonal matrix T, then $(AT)(AT)' = AA'$. In this way, we can always rotate A by an orthogonal matrix

T, so that $A^* = AT$ yields the same Ω. Several approaches have been suggested to use this ambiguity in a factor analysis solution in order to create maximum contrast between the columns of A. A well-known method, due to Kaiser, is to maximize the *raw varimax criterion.*

Kaiser defined the *simplicity of the k-th factor*, denoted s_k, as the sample variance of its squared factor loadings. Thus

$$s_k = \frac{1}{p}\sum_{i=1}^{p}\left(a_{ik}^2 - \frac{1}{p}\sum_{h=1}^{p} a_{hk}^2\right)^2 \qquad (k=1, \ldots, m). \tag{1}$$

The *total simplicity* is $s = s_1 + s_2 + \ldots + s_m$ and the raw varimax method selects an orthogonal matrix T such that s is maximized.

Theorem 12

Let A be a given $p \times m$ matrix of rank m. Let ϕ be a real-valued function defined by

$$\phi(T) = \sum_{j=1}^{m}\left[\left(\sum_{i=1}^{p} b_{ij}^4\right) - \frac{1}{p}\left(\sum_{i=1}^{p} b_{ij}^2\right)^2\right], \tag{2}$$

where $B = (b_{ij})$ satisfies $B = AT$ and $T \in \mathcal{O}_{m \times m}$. The function ϕ reaches a maximum when B satisfies

$$B = AA'Q(Q'AA'Q)^{-1/2}, \tag{3}$$

where $Q = (q_{ij})$ is the $p \times m$ matrix with typical element

$$q_{ij} = b_{ij}\left(b_{ij}^2 - \frac{1}{p}\sum_{h=1}^{p} b_{hj}^2\right). \tag{4}$$

Proof. Let $C = B \odot B$, so that $c_{ij} = b_{ij}^2$. Let $\imath = (1, 1, \ldots, 1)'$ be of order $p \times 1$ and $M = I_p - (1/p)\imath\imath'$. Let e_i denote the i-th column of I_p and u_j the j-th column of I_m. Then we can rewrite ϕ as

$$\begin{aligned}
\phi(T) &= \sum_j\sum_i c_{ij}^2 - (1/p)\sum_j\left(\sum_i c_{ij}\right)^2 \\
&= \operatorname{tr} C'C - (1/p)\sum_j\left(\sum_i e_i'Cu_j\right)^2 \\
&= \operatorname{tr} C'C - (1/p)\sum_j(\imath'Cu_j)^2 \\
&= \operatorname{tr} C'C - (1/p)\sum_j \imath'Cu_ju_j'C'\imath \\
&= \operatorname{tr} C'C - (1/p)\imath'CC'\imath = \operatorname{tr} C'MC.
\end{aligned} \tag{5}$$

We wish to maximize ϕ with respect to T subject to the orthogonality constraint $T'T = I_m$. Let ψ be the appropriate Lagrangian function

$$\psi(T) = \tfrac{1}{2}\operatorname{tr} C'MC - \operatorname{tr} L(T'T - I), \tag{6}$$

where L is a symmetric $m \times m$ matrix of Lagrange multipliers. Then the differential of ψ is

$$\begin{aligned} \mathrm{d}\psi &= \operatorname{tr} C'M\mathrm{d}C - 2\operatorname{tr} LT'\mathrm{d}T \\ &= 2\operatorname{tr} C'M(B \odot \mathrm{d}B) - 2\operatorname{tr} LT'\mathrm{d}T \\ &= 2\operatorname{tr}(C'M \odot B')\mathrm{d}B - 2\operatorname{tr} LT'\mathrm{d}T \\ &= 2\operatorname{tr}(C'M \odot B')A\mathrm{d}T - 2\operatorname{tr} LT'\mathrm{d}T, \end{aligned} \tag{7}$$

where the third equality follows from Theorem 3.7(a). Hence, the first-order conditions are

$$(C'M \odot B')A = LT' \tag{8}$$

and

$$T'T = I. \tag{9}$$

It is easy to verify that the $p \times m$ matrix Q given in (4) satisfies

$$Q = B \odot MC, \tag{10}$$

so that (8) becomes

$$Q'A = LT'. \tag{11}$$

Post-multiplying with T and using the symmetry of L we obtain the condition

$$Q'B = B'Q. \tag{12}$$

We see from (11) that $L = Q'B$. This is a symmetric matrix and

$$\begin{aligned} \operatorname{tr} L &= \operatorname{tr} B'Q = \operatorname{tr} B'(B \odot MC) \\ &= \operatorname{tr}(B' \odot B')MC = \operatorname{tr} C'MC, \end{aligned} \tag{13}$$

using Theorem 3.7(a). From (11) follows

$$L^2 = Q'AA'Q \tag{14}$$

so that

$$L = (Q'AA'Q)^{1/2}. \tag{15}$$

It is clear that L must be positive semidefinite. Assuming that L is, in fact, non-singular, we may write

$$L^{-1} = (Q'AA'Q)^{-1/2} \tag{16}$$

and we obtain from (11)

$$T' = L^{-1}Q'A = (Q'AA'Q)^{-1/2}Q'A. \tag{17}$$

The solution for B is then

$$B = AT = AA'Q(Q'AA'Q)^{-1/2}. \tag{18}$$

□

An iterative zigzag procedure can be based on (3) and (4). In (3) we have $B=B(Q)$ and in (4) we have $Q=Q(B)$. An obvious starting value for B is $B^{(0)}=A$. Then calculate $Q^{(1)}=Q(B^{(0)})$, $B^{(1)}=B(Q^{(1)})$, $Q^{(2)}=Q(B^{(1)})$, etc. If the procedure converges, which is not guaranteed, then a (local) maximum of (2) has been found.

16. CANONICAL CORRELATIONS AND VARIATES IN THE POPULATION

Let z be a random vector with zero expectation and positive definite variance matrix Σ. Let z and Σ be partitioned as

$$z=\begin{pmatrix} z^{(1)} \\ z^{(2)} \end{pmatrix}, \qquad \Sigma=\begin{pmatrix} \Sigma_{11} & \Sigma_{12} \\ \Sigma_{21} & \Sigma_{22} \end{pmatrix}, \tag{1}$$

so that Σ_{11} is the variance matrix of $z^{(1)}$, Σ_{22} the variance matrix of $z^{(2)}$ and $\Sigma_{12}=\Sigma_{21}'$ the covariance matrix between $z^{(1)}$ and $z^{(2)}$.

The pair of linear combinations $u'z^{(1)}$ and $v'z^{(2)}$, each of unit variance, with maximum correlation (in absolute value) is called the *first pair of canonical variates* and its correlation is called the *first canonical correlation* between $z^{(1)}$ and $z^{(2)}$.

The *k-th pair of canonical variates* is the pair $u'z^{(1)}$ and $v'z^{(2)}$, each of unit variance and uncorrelated with the first $k-1$ pairs of canonical variates, with maximum correlation (in absolute value). This correlation is the *k-th* canonical correlation.

Theorem 13

Let z be a random vector with zero expectation and positive definite variance matrix Σ. Let z and Σ be partitioned as in (1), and define

$$B=\Sigma_{11}^{-1}\Sigma_{12}\Sigma_{22}^{-1}\Sigma_{21}, \qquad C=\Sigma_{22}^{-1}\Sigma_{21}\Sigma_{11}^{-1}\Sigma_{12}. \tag{2}$$

(a) There are r non-zero canonical correlations between $z^{(1)}$ and $z^{(2)}$, where r is the rank of Σ_{12}.

(b) Let $\lambda_1 \geq \lambda_2 \geq \ldots \geq \lambda_r > 0$ denote the non-zero eigenvalues of B (and of C). Then the k-th canonical correlation between $z^{(1)}$ and $z^{(2)}$ is $\lambda_k^{1/2}$.

(c) The k-th pair of canonical variates is given by $u'z^{(1)}$ and $v'z^{(2)}$, where u and v are normalized eigenvectors of B and C, respectively, associated with the eigenvalue λ_k. Moreover, if λ_k is a simple (non-repeated) eigenvalue of B (and C), then u and v are unique (apart from sign).

(d) If the pair $u'z^{(1)}$ and $v'z^{(2)}$ is the k-th pair of canonical variates, then

$$\Sigma_{12}v=\lambda_k^{1/2}\Sigma_{11}u, \qquad \Sigma_{21}u=\lambda_k^{1/2}\Sigma_{22}v. \tag{3}$$

Proof. Let $A=\Sigma_{11}^{-1/2}\Sigma_{12}\Sigma_{22}^{-1/2}$ with rank $r(A)=r(\Sigma_{12})=r$, and notice that the r non-zero eigenvalues of AA', $A'A$, B and C are all the same, namely $\lambda_1 \geqslant \lambda_2 \geqslant \ldots \geqslant \lambda_r > 0$. Let $S=(s_1, s_2, \ldots, s_r)$ and $T=(t_1, t_2, \ldots, t_r)$ be semi-orthogonal matrices such that

$$AA'S=S\Lambda, \qquad A'AT=T\Lambda, \tag{4}$$

$$S'S=I_r, \qquad T'T=I_r, \qquad \Lambda=\operatorname{diag}(\lambda_1, \lambda_2, \ldots, \lambda_r). \tag{5}$$

We assume first that all λ_i $(i=1, 2, \ldots, r)$ are *distinct*.

The first pair of canonical variates is obtained from the maximization problem

$$\begin{array}{ll} \underset{u,v}{\text{maximize}} & (u'\Sigma_{12}v)^2 \\ \text{subject to} & u'\Sigma_{11}u=v'\Sigma_{22}v=1. \end{array} \tag{6}$$

Let $x=\Sigma_{11}^{1/2}u$, $y=\Sigma_{22}^{1/2}v$. Then (6) can be equivalently stated as

$$\begin{array}{ll} \underset{x,y}{\text{maximize}} & (x'Ay)^2 \\ \text{subject to} & x'x=y'y=1. \end{array} \tag{7}$$

According to Theorem 11.17, the maximum λ_1 is obtained for $x=s_1$, $y=t_1$ (apart from the sign, which is irrelevant). Hence $\lambda_1^{1/2}$ is the first canonical correlation, and the first pair of canonical variates is $u^{(1)'}z^{(1)}$ and $v^{(1)'}z^{(2)}$ with $u^{(1)}=\Sigma_{11}^{-1/2}s_1$, $v^{(1)}=\Sigma_{22}^{-1/2}t_1$. It follows that $Bu^{(1)}=\lambda_1 u^{(1)}$ (because $AA's_1=\lambda_1 s_1$) and $Cv^{(1)}=\lambda_1 v^{(1)}$ (because $A'At_1=\lambda_1 t_1$). Theorem 11.17 also gives $s_1=\lambda_1^{-1/2}At_1$, $t_1=\lambda_1^{-1/2}A's_1$ from which we obtain $\Sigma_{12}v^{(1)}=\lambda_1^{1/2}\Sigma_{11}u^{(1)}$, $\Sigma_{21}u^{(1)}=\lambda_1^{1/2}\Sigma_{22}v^{(1)}$.

Now assume that $\lambda_1^{1/2}, \lambda_2^{1/2}, \ldots, \lambda_{k-1}^{1/2}$ are the first $k-1$ canonical correlations, and that $s_i'\Sigma_{11}^{-1/2}z^{(1)}$ and $t_i'\Sigma_{22}^{-1/2}z^{(2)}$, $i=1, 2, \ldots, k-1$, are the corresponding pairs of canonical variates. In order to obtain the k-th pair of canonical variates we let $S_1=(s_1, s_2, \ldots, s_{k-1})$ and $T_1=(t_1, t_2, \ldots, t_{k-1})$, and consider the constrained maximization problem

$$\begin{array}{llll} \underset{u,v}{\text{maximize}} & (u'\Sigma_{12}v)^2 & & \\ \text{subject to} & u'\Sigma_{11}u=v'\Sigma_{22}v=1, & & \\ \text{and} & S_1'\Sigma_{11}^{1/2}u=0, & & S_1'\Sigma_{11}^{-1/2}\Sigma_{12}v=0, \\ \text{and} & T_1'\Sigma_{22}^{1/2}v=0, & & T_1'\Sigma_{22}^{-1/2}\Sigma_{21}u=0. \end{array} \tag{8}$$

Again letting $x=\Sigma_{11}^{1/2}u$, $y=\Sigma_{22}^{1/2}v$, we can rephrase (8) as

$$\begin{array}{ll} \underset{x,y}{\text{maximize}} & (x'Ay)^2 \\ \text{subject to} & x'x=y'y=1, \\ \text{and} & S_1'x=S_1'Ay=0, \\ \text{and} & T_1'y=T_1'A'x=0. \end{array} \tag{9}$$

It turns out, as we shall see shortly, that we can take any *one* of the four constraints $S_1'x=0$, $S_1'Ay=0$, $T_1'y=0$, $T_1'A'x=0$, because the solution will automatically satisfy the remaining three conditions. The reduced problem is

$$\begin{aligned} &\underset{x,y}{\text{maximize}} && (x'Ay)^2 \\ &\text{subject to} && x'x=y'y=1, \qquad S_1'x=0, \end{aligned} \tag{10}$$

and its solution follows from Theorem 11.17: the constrained maximum is λ_k and is achieved by $x_*=s_k$ and $y_*=t_k$.

We see that the three constraints that were dropped in the passage from (9) to (10) are indeed satisfied: $S_1'Ay_*=0$, because $Ay_*=\lambda_k^{1/2}x_*$; $T_1'y_*=0$; and $T_1'A'x_*=0$, because $A'x_*=\lambda_k^{1/2}y_*$. Hence we may conclude that $\lambda_k^{1/2}$ is the k-th canonical correlation; that $u^{(k)'}z^{(1)}$, $v^{(k)'}z^{(2)}$ with $u^{(k)}=\Sigma_{11}^{-1/2}s_k$ and $v^{(k)}=\Sigma_{22}^{-1/2}t_k$ is the k-th pair of canonical variates; that $u^{(k)}$ and $v^{(k)}$ are the (unique) normalized eigenvectors of B and C, respectively, associated with the eigenvalue λ_k; and that $\Sigma_{12}v^{(k)}=\lambda_k^{1/2}\Sigma_{11}u^{(k)}$ and $\Sigma_{21}u^{(k)}=\lambda_k^{1/2}\Sigma_{22}v^{(k)}$.

The theorem (still assuming distinct eigenvalues) now follows by simple mathematical induction. It is clear that only r pairs of canonical variates can be found yielding non-zero canonical correlations. (The $(r+1)$-th pair would yield zero canonical correlations, since AA' possesses only r positive eigenvalues.)

In the case of *multiple* eigenvalues, the proof remains unchanged, except that the eigenvectors associated with multiple eigenvalues are not unique, and therefore the pairs of canonical variates corresponding to these eigenvectors are not unique either. □

BIBLIOGRAPHICAL NOTES

§1. There are some excellent texts on multivariate statistics and psychometrics, of which we mention in particular Anderson (1984) and Morrison (1976).
§2–§3. See Lawley and Maxwell (1971), Muirhead (1982) and Anderson (1984).
§5–§6. See Morrison (1976) and Muirhead (1982). Theorem 4 is proved in Anderson (1984). For asymptotic distributional results concerning l_i and q_i, see Kollo and Neudecker (1993). For asymptotic distributional results concerning q_i in Hotelling's (1933) model where $t_i't_i=\lambda_i$, see Kollo and Neudecker (1997).
§8–§10. See Eckart and Young (1936), Theil (1971), Ten Berge (1993), Greene (1993) and Chipman (1996). We are grateful to Jos Ten Berge for pointing out a redundancy in Theorem 17.8.
§11. For three-mode component analysis see Tucker (1966). An extension to four modes is given in Lastovička (1981), and to an arbitrary number of modes in Kapteyn *et al.* (1986).
§12–§13. See Rao (1955), Morrison (1976), and Mardia, Kent and Bibby (1992).
§14. See Lawley and Maxwell (1971), Clarke (1970) and Neudecker (1975).
§15. See Kaiser (1958, 1959), Sherin (1966), Lawley and Maxwell (1971) and Neudecker (1981).
§16. See Anderson (1984) and Muirhead (1982).

Bibliography

Abadir, K. M. and K. Hadri (1996). Problem 96.5.3, *Econometric Theory*, **12**, 868. (Solutions by H. P. Boswijk and M. Lu in *Econometric Theory*, **13**, 894–896.)

Abdullah, J., H. Neudecker and S. Liu (1992). Problem 92.4.6, *Econometric Theory*, **8**, 584. (Solution in *Econometric Theory*, **9**, 703.)

Abrahamse, A. P. J. and J. Koerts (1971). New estimators of disturbances in regression analysis, *Journal of the American Statistical Association*, **66**, 71–4.

Aitken, A. C. (1935). On least squares and linear combinations of observations, *Proceedings of the Royal Society of Edinburgh, A*, **55**, 42–8.

Aitken, A. C. (1939). *Determinants and Matrices*, Oliver and Boyd, Edinburgh and London.

Albert, A. (1973). The Gauss–Markov theorem for regression models with possibly singular covariances, *SIAM Journal of Applied Mathematics*, **24**, 182–7.

Amir-Moéz, A. R. and A. L. Fass (1962). *Elements of Linear Spaces*, Pergamon Press, Oxford.

Anderson, T. W. (1984). *An Introduction to Multivariate Statistical Analysis*, 2nd edn, John Wiley, New York.

Ando, T. (1979). Concavity of certain maps on positive definite matrices and applications to Hadamard products, *Linear Algebra and Its Applications*, **26**, 203–41.

Ando, T. (1983). On the arithmetic–geometric–harmonic–mean inequalities for positive definite matrices, *Linear Algebra and Its Applications*, **52/53**, 31–7.

Apostol, T. M. (1974). *Mathematical Analysis*, 2nd edn, Addison-Wesley, Reading.

Balestra, P. (1973). Best quadratic unbiased estimators of the variance–covariance matrix in normal regression, *Journal of Econometrics*, **1**, 17–28.

Balestra, P. (1976). *La Dérivation Matricielle*, Collection de l'Institut de Mathématiques Economiques, No. 12, Sirey, Paris.

Bargmann, R. E. and D. G. Nel (1974). On the matrix differentiation of the characteristic roots of matrices, *South African Statistical Journal*, **8**, 135–44.

Baron, M. E. (1969). *The Origins of the Infinitesimal Calculus*, Pergamon Press, Oxford.

Barten, A. P. (1969). Maximum likelihood estimation of a complete system of demand equations, *European Economic Review*, **1**, 7–73.

Beckenbach, E. F. and R. Bellman (1961). *Inequalities*, Springer-Verlag, Berlin.

Bellman, R. (1970). *Introduction to Matrix Analysis*, 2nd edn, McGraw-Hill, New York.

Ben-Israel, A. and T. N. Greville (1974). *Generalized Inverses: Theory and Applications*, John Wiley, New York.

Bentler, P. M. and S.-Y. Lee (1978). Matrix derivatives with chain rule and rules for simple, Hadamard, and Kronecker products, *Journal of Mathematical Psychology*, **17**, 255–62.

Binmore, K. G. (1982). *Mathematical Analysis: A Straightforward Approach*, 2nd edn, Cambridge University Press, Cambridge.

Black, J. and Y. Morimoto (1968). A note on quadratic forms positive definite under linear constraints, *Economica*, **35**, 205–6.

Bloomfield, P. and G. S. Watson (1975). The inefficiency of least squares, *Biometrika*, **62**, 121–128.

Bodewig, E. (1959). *Matrix Calculus*, 2nd edn, North-Holland, Amsterdam.

Boullion, T. L. and P. L. Odell (1971). *Generalized Inverse Matrices*, John Wiley, New York.

Bozdogan, H. (1990). On the information-based measure of covariance complexity and its application to the evaluation of multivariate linear models, *Communications in Statistics — Theory and Methods*, **19**, 221–278.

Bozdogan, H. (1994). Mixture-model cluster analysis using model selection criteria and a new informational measure of complexity, *Multivariate Statistical Modeling* (ed H. Bozdogan), vol. 2, Proceedings of the First US/Japan Conference on the Frontiers of Statistical Modeling: An Informational Approach, Kluwer, Dordrecht, 69–113.

Browne, M. W. (1974). Generalized least squares estimators in the analysis of covariance structures, *South African Statistical Journal*, **8**, 1–24. Reprinted in: *Latent Variables in Socioeconomic Models* (eds D. J. Aigner and A. S. Goldberger), North-Holland, Amsterdam, 205–226.

Chipman, J. S. (1964). On least squares with insufficient observations, *Journal of the American Statistical Association*, **59**, 1078–111.

Chipman, J. S. (1996). "Proofs" and proofs of the Eckart–Young theorem, *Stochastic Processes and Functional Analysis* (eds J. Goldstein, N. Gretsky and J. Uhl), Marcel Dekker, New York, 80–81.

Clarke, M. R. B. (1970). A rapidly convergent method for maximum-likelihood factor analysis, *British Journal of Mathematical and Statistical Psychology*, **23**, 43–52.

Courant, R. and D. Hilbert (1931). *Methoden der Mathematischen Physik*, reprinted by Interscience, New York.

Cramer, J. S. (1986) *Econometric Applications of Maximum Likelihood Methods*, Cambridge University Press, Cambridge.

Debreu, G. (1952). Definite and semidefinite quadratic forms, *Econometrica*, **20**, 295–300.

Dieudonné, J. (1969). *Foundations of Modern Analysis*, 2nd edn, Academic Press, New York.

Don, F. J. H. (1986). *Linear Methods in Non-Linear Models*, unpublished Ph.D. Thesis, University of Amsterdam.

Dubbelman, C., A. P. J. Abrahamse and J. Koerts (1972). A new class of disturbance estimators in the general linear model, *Statistica Neerlandica*, **26**, 127–42.

Dwyer, P. S. (1967). Some applications of matrix derivatives in multivariate analysis, *Journal of the American Statistical Association*, **62**, 607–25.

Dwyer, P. S. and M. S. MacPhail (1948). Symbolic matrix derivatives, *Annals of Mathematical Statistics*, **19**, 517–34.

Eckart, C. and G. Young (1936). The approximation of one matrix by another of lower rank, *Psychometrika*, **1**, 211–218.

Faliva, M. (1983). *Identificazione e Stima nel Modello Lineare ad Equazioni Simultanee*, Vita e Pensiero, Milan.

Fan, K. (1949). On a theorem of Weyl concerning eigenvalues of linear transformations, I, *Proceedings of the National Academy of Sciences of the USA*, **35**, 652–5.

Fan, K. (1950). On a theorem of Weyl concerning eigenvalues of linear transformations, II, *Proceedings of the National Academy of Sciences of the USA*, **36**, 31–5.

Fan, K. and A. J. Hoffman (1955). Some metric inequalities in the space of matrices, *Proceedings of the American Mathematical Society*, **6**, 111–16.

Farebrother, R. W. (1977). Necessary and sufficient conditions for a quadratic form to be positive whenever a set of homogeneous linear constraints is satisfied, *Linear Algebra and Its Applications*, **16**, 39–42.

Fischer, E. (1905). Ueber quadratische Formen mit reellen Koeffizienten, *Monatschrift für Mathematik und Physik*, **16**, 234–49.

Fisher, F. M. (1966). *The Identification Problem in Econometrics*, McGraw-Hill, New York.

Fleming, W. H. (1977). *Functions of Several Variables*, 2nd edn, Springer-Verlag, New York.

Gantmacher, F. R. (1959). *The Theory of Matrices*, vols I and II, Chelsea, New York.

Gauss, K. F. (1809). *Werke*, **4**, 1–93, Göttingen.

Godfrey, L. G. and M. R. Wickens (1982). A simple derivation of the limited information maximum likelihood estimator, *Economics Letters*, **10**, 277–83.

Golub, G. H. and V. Pereyra (1976). Differentiation of pseudoinverses, separable nonlinear least square problems and other tales, *Generalized Inverses and Applications* (ed. M. Z. Nashed), Academic Press, New York.

Graham, A. (1981). *Kronecker Products and Matrix Calculus*, Ellis Horwood, Chichester.

Greene, W. H. (1993). *Econometric Analysis*, Second Edition, Macmillan, New York.

Greub, W. and W. Rheinboldt (1959). On a generalization of an inequality of L. V. Kantorovich, *Proceedings of the American Mathematical Society*, **10**, 407–15.

Hadley, G. (1961) *Linear Algebra*, Addison-Wesley, Reading.

Hardy, G. H., J. E. Littlewood and G. Pólya (1952). *Inequalities*, 2nd edn, Cambridge University Press, Cambridge.

Hearon, J. Z. and J. W. Evans (1968). Differentiable generalized inverses, *Journal of Research of the National Bureau of Standards, B*, **72B**, 109–13.

Heijmans, R. D. H. and J. R. Magnus (1986). Asymptotic normality of maximum likelihood estimators obtained from normally distributed but dependent observations, *Econometric Theory*, **2**, 374–412.

Henderson, H. V. and S. R. Searle (1979). Vec and vech operators for matrices, with some uses in Jacobians and multivariate statistics, *Canadian Journal of Statistics*, **7**, 65–81.

Hogg, R. V. and A. T. Craig (1970). *Introduction to Mathematical Statistics*, 3rd edn, Collier-Macmillan, London.

Holly, A. (1985). Problem 85.1.2, *Econometric Theory*, **1**, 143–4; Solution in *Econometric Theory*, **2**, 297–300.

Holly, A. and J. R. Magnus (1988). A note on instrumental variables and maximum likelihood estimation procedures, *Annales d'Economie et de Statistique*, **10**, 121–138.

Hotelling, H. (1933). Analysis of a complex of statistical variables into principal components, *Journal of Educational Psychology*, **24**, 417–441 and 498–520.

Hsiao, C. (1983). Identification, *Handbook of Econometrics*, vol. I (eds Z. Griliches and M. D. Intriligator), North-Holland, Amsterdam, 223–283.

Kaiser, H. F. (1958). The varimax criterion for analytic rotation in factor analysis, *Psychometrika*, **23**, 187–200.

Kaiser, H. F. (1959). Computer program for varimax rotation in factor analysis, *Journal of Educational and Psychological Measurement*, **19**, 413–20.

Kalaba, R., K. Spingarn and L. Tesfatsion (1980). A new differential equation method for finding the Perron root of a positive matrix, *Applied Mathematics and Computation*, **7**, 187–93.

Kalaba, R., K. Spingarn and L. Tesfatsion (1981a). Individual tracking of an eigenvalue and eigenvector of a parameterized matrix, *Nonlinear Analysis, Theory, Methods and Applications*, **5**, 337–40.

Kalaba, R., K. Spingarn and L. Tesfatsion (1981b). Variational equations for the eigenvalues and eigenvectors of nonsymmetric matrices, *Journal of Optimization Theory and Applications*, **33**, 1–8.

Kantorovich, L. V. (1948). Functional analysis and applied mathematics, *Uspekhi Matematicheskikh Nauk*, **3**, 89–185. Translated from Russian by C. D. Benster, National Bureau of Standards, Report 1509, 7 March 1952.

Kapteyn, A., H. Neudecker and T. Wansbeek (1986). An approach to n-mode components analysis, *Psychometrika*, **51**, 269–75.

Koerts, J. and A. P. J. Abrahamse (1969). *On the Theory and Applications of the General Linear Model*, Rotterdam University Press, Rotterdam.

Kollo, T. (1991). *The Matrix Derivative in Multivariate Statistics*, Tartu University Press (in Russian).

Kollo, T. and H. Neudecker (1993). Asymptotics of eigenvalues and unit-length eigenvectors of sample variance and correlation matrices, *Journal of Multivariate Analysis*, **47**, 283–300. (Corrigendum, *Journal of Multivariate Analysis*, **51**, 210.)

Kollo, T. and H. Neudecker (1997). Asymptotics of Pearson–Hotelling principal–component vectors of sample variance and correlation matrices, *Behaviormetrika*, **24**, 51–69.

Koopmans, T. C., H. Rubin and R. B. Leipnik (1950). Measuring the equation systems of dynamic economics, *Statistical Inference in Dynamic Economic Models* (ed. T. C. Koopmans), Cowles Foundation for Research in Economics, Monograph 10, John Wiley, New York, chap. 2.

Kreijger, R. G. and H. Neudecker (1977). Exact linear restrictions on parameters in the general linear model with a singular covariance matrix, *Journal of the American Statistical Association*, **72**, 430–2.

Lancaster, P. (1964). On eigenvalues of matrices dependent on a parameter, *Numerische Mathematik*, **6**, 377–87.

Lancaster, T. (1984). The covariance matrix of the information matrix test, *Econometrica*, **52**, 1051–2.

Lastovička, J. L. (1981). The extension of component analysis to four-mode matrices, *Psychometrika*, **46**, 47–57.

Lawley, D. N. and A. E. Maxwell (1971). *Factor Analysis as a Statistical Method*, 2nd edn, Butterworths, London.

Leamer, E. E. (1978). *Specification Searches*, John Wiley, New York.

Liu, S. (1995). *Contributions to Matrix Calculus and Applications in Statistics*, Ph.D. Thesis, University of Amsterdam.

Luenberger, D. G. (1969). *Optimization by Vector Space Methods*, John Wiley, New York.

McCulloch, C. E. (1982). Symmetric matrix derivatives with applications, *Journal of the American Statistical Association*, **77**, 679–82.

McDonald, R. P. and H. Swaminathan (1973). A simple matrix calculus with applications to multivariate analysis, *General Systems*, **18**, 37–54.

MacDuffee, C. C. (1933). *The Theory of Matrices*, reprinted by Chelsea, New York.

MacRae, E. C. (1974). Matrix derivatives with an application to an adaptive linear decision problem, *Annals of Statistics*, **2**, 337–46.

Madansky, A. (1976). *Foundations of Econometrics*, North-Holland, Amsterdam.

Magnus, J. R. (1985). On differentiating eigenvalues and eigenvectors, *Econometric Theory*, **1**, 179–91.

Magnus, J. R. (1987). A representation theorem for $(\operatorname{tr} A^p)^{1/p}$, *Linear Algebra and Its Applications*, **95**, 127–34.

Magnus, J. R. (1988). *Linear Structures*, Griffin's Statistical Monographs, No. 42, Edward Arnold, London and Oxford University Press, New York.

Magnus, J. R. (1990). On the fundamental bordered matrix of linear estimation, *Advanced Lectures in Quantitative Economics* (ed. F. van der Ploeg), Academic Press, London, 583-604.

Magnus, J. R. and H. Neudecker (1979). The commutation matrix: some properties and applications, *Annals of Statistics*, **7**, 381–94.

Magnus, J. R. and H. Neudecker (1980). The elimination matrix: some lemmas and applications, *SIAM Journal on Algebraic and Discrete Methods*, **1**, 422–49.

Magnus, J. R. and H. Neudecker (1985). Matrix differential calculus with applications to simple, Hadamard, and Kronecker products, *Journal of Mathematical Psychology*, **29**, 474–92.

Magnus, J. R. and H. Neudecker (1986). Symmetry, 0–1 matrices and Jacobians: a review, *Econometric Theory*, **2**, 157–90.

Malinvaud, E. (1966). *Statistical Methods of Econometrics*, North-Holland, Amsterdam.

Marcus, M. and H. Minc (1964). *A Survey of Matrix Theory and Matrix Inequalities*, Allyn and Bacon, Boston.

Mardia, K. V., J. T. Kent and J. M. Bibby (1992). *Multivariate Analysis*, Academic Press, London.

Markov, A. A. (1900). *Wahrscheinlichkeitsrechnung*, Teubner, Leipzig.

Marshall, A. and I. Olkin (1979). *Inequalities: Theory of Majorization and Its Applications*, Academic Press, New York.

Milliken, G. A. and F. Akdeniz (1977). A theorem on the difference of the generalized inverses of two nonnegative matrices, *Communications in Statistics—Theory and Methods*, **A6**, 73–79.

Mirsky, L. (1961). *An Introduction to Linear Algebra*, Oxford University Press, Oxford.

Mood, A. M., F. A. Graybill and D. C. Boes (1974). *Introduction to the Theory of Statistics*, 3rd edn, McGraw-Hill, New York.

Moore, E. H. (1920). On the reciprocal of the general algebraic matrix (Abstract), *Bulletin of the American Mathematical Society*, **26**, 394–5.

Moore, E. H. (1935). *General Analysis*, Memoirs of the American Philosophical Society, Vol. I, American Philosophical Society, Philadelphia.

Moore, M. H. (1973). A convex matrix function, *American Mathematical Monthly*, **80**, 408–9.

Morrison, D. F. (1976). *Multivariate Statistical Methods*, 2nd edn, McGraw-Hill, New York.

Muirhead, R. J. (1982). *Aspects of Multivariate Statistical Theory*, John Wiley, New York.

Nel, D. G. (1980). On matrix differentiation in statistics, *South African Statistical Journal*, **14**, 137–93.

Neudecker, H. (1967). On matrix procedures for optimizing differentiable scalar functions of matrices, *Statistica Neerlandica*, **21**, 101–7.

Neudecker, H. (1969). Some theorems on matrix differentiation with special reference to Kronecker matrix products, *Journal of the American Statistical Association*, **64**, 953–63.

Neudecker, H. (1973). De BLUF-schatter: een rechtstreekse afleiding, *Statistica Neerlandica*, **27**, 127–30.

Neudecker, H. (1974). A representation theorem for $|A|^{1/n}$, *METU Journal of Pure and Applied Sciences*, **7**, 1–2.

Neudecker, H. (1975). A derivation of the Hessian of the (concentrated) likelihood function of the factor model employing the Schur product, *British Journal of Mathematical and Statistical Psychology*, **28**, 152–6.

Neudecker, H. (1977a). Abrahamse and Koerts' 'new estimator' of disturbances in regression analysis, *Journal of Econometrics*, **5**, 129–33.

Neudecker, H. (1977b). Bounds for the bias of the least squares estimator of σ^2 in the case of a first-order autoregressive process (positive autocorrelation), *Econometrica*, **45**, 1257–62.

Neudecker, H. (1978). Bounds for the bias of the least squares estimator of σ^2 in the case of a first-order (positive) autoregressive process when the regression contains a constant term, *Econometrica*, **46**, 1223–6.

Neudecker, H. (1980a). A comment on 'Minimization of functions of a positive semidefinite matrix A subject to $AX = 0$', *Journal of Multivariate Analysis*, **10**, 135–9.

Neudecker, H. (1980b). Best quadratic unbiased estimation of the variance matrix in normal regression, *Statistische Hefte*, **21**, 239–42.

Neudecker, H. (1981). On the matrix formulation of Kaiser's varimax criterion, *Psychometrika*, **46**, 343–5.

Neudecker, H. (1982). On two germane matrix derivatives, *The Matrix and Tensor Quarterly*, **33**, 3–12.

Neudecker, H. (1985a). Recent advances in statistical applications of commutation matrices, *Proceedings of the Fourth Pannonian Symposium on Mathematical Statistics* (eds W. Grossmann, G. Pflug, I. Vincze and W. Wertz), vol. B, Reidel, Dordrecht, 239–50.

Neudecker, H. (1985b). On the dispersion matrix of a matrix quadratic form connected with the noncentral normal distribution, *Linear Algebra and Its Applications*, **70**, 257–262.

Neudecker, H. (1989a). A matrix derivation of a representation theorem for $(\operatorname{tr} A^p)^{1/p}$, *Qüestiió*, **13**, 75–79.

Neudecker, H. (1989b). A new proof of the Milliken–Akdeniz theorem, *Qüestiió*, **13**, 81–82.

Neudecker, H. (1992). A matrix trace inequality, *Journal of Mathematical Analysis and Applications*, **166**, 302–303.
Neudecker, H. (1995). Mathematical properties of the variance of the multinomial distribution, *Journal of Mathematical Analysis and Applications*, **189**, 757–762.
Neudecker, H. and S. Liu (1993). Best quadratic and positive semidefinite unbiased estimation of the variance of the multivariate normal distribution, *Communications in Statistics—Theory and Methods*, **22**, 2723–2732.
Neudecker, H. and S. Liu (1995). Note on a matrix-concave function, *Journal of Mathematical Analysis and Applications*, **196**, 1139–1141.
Neudecker, H., S. Liu and W. Polasek (1995). The Hadamard product and some of its applications in statistics, *Statistics*, **26**, 365–373.
Neudecker, H., W. Polasek and S. Liu (1995). The heteroskedastic linear regression model and the Hadamard product: a note, *Journal of Econometrics*, **68**, 361–366.
Neudecker, H. and A. Satorra (1993). Problem 93.3.9, *Econometric Theory*, **9**, 524. (Solutions by H. Neudecker and A. Satorra, G. Trenkler, and H. Neudecker and S. Liu in *Econometric Theory*, **11**, 654–655.)
Neudecker, H. and T. Wansbeek (1983). Some results on commutation matrices, with statistical applications, *Canadian Journal of Statistics*, **11**, 221–31.
Norden, R. H. (1972). A survey of maximum likelihood estimation, *International Statistical Review*, **40**, 329–54.
Norden, R. H. (1973). A survey of maximum likelihood estimation, Part 2, *International Statistical Review*, **41**, 39–58.
Olkin, I. (1983). An inequality for a sum of forms, *Linear Algebra and Its Applications*, **52/53**, 529–32.
Penrose, R. (1955). A generalized inverse for matrices, *Proceedings of the Cambridge Philosophical Society*, **51**, 406–13.
Penrose, R. (1956). On best approximate solutions of linear matrix equations, *Proceedings of the Cambridge Philosophical Society*, **52**, 17–19.
Poincaré, H. (1890). Sur les équations aux dérivées partielles de la physique mathématique, *American Journal of Mathematics*, **12**, 211–94.
Polasek, W. (1986). Local sensitivity analysis and Bayesian regression diagnostics, *Bayesian Inference and Decision Techniques* (eds P. K. Goel and A. Zellner), North-Holland, Amsterdam, 375–87.
Pollock, D. S. G. (1979). *The Algebra of Econometrics*, John Wiley, New York.
Pollock, D. S. G. (1985). Tensor products and matrix differential calculus, *Linear Algebra and Its Applications*, **67**, 169–93.
Pringle, R. M. and A. A. Rayner (1971). *Generalized Inverse Matrices with Applications to Statistics*, Griffin's Statistical Monographs and Courses, No. 28, Charles Griffin, London.
Rao, A. R. and P. Bhimasankaram (1992). *Linear Algebra*, Tata McGraw-Hill, New Delhi.
Rao, C. R. (1945). Markoff's theorem with linear restrictions on parameters, *Sankhya*, **7**, 16–19.
Rao, C. R. (1955). Estimation and tests of significance in factor analysis, *Psychometrika*, **20**, 93–111.
Rao, C. R. (1971a). Unified theory of linear estimation, *Sankhya, A*, **33**, 371–477. Corrigenda in *Sankhya, A*, **34**, 477.
Rao, C. R. (1971b). Estimation of variance and covariance components—MINQUE theory, *Journal of Multivariate Analysis*, **1**, 257–75.
Rao, C. R. (1973). *Linear Statistical Inference and Its Applications*, 2nd edn, John Wiley, New York.
Rao, C. R. and S. K. Mitra (1971). *Generalized Inverse of Matrices and Its Applications*, John Wiley, New York.
Rogers, G. S. (1980). *Matrix Derivatives*, Marcel Dekker, New York.

Rolle, J.-D. (1994). Best nonnegative invariant partially orthogonal quadratic estimation in normal regression, *Journal of the American Statistical Association*, **89**, 1378–1385.

Rolle, J.-D. (1996). Optimization of functions of matrices with an application in statistics, *Linear Algebra and Its Applications*, **234**, 261–275.

Roth, W. E. (1934). On direct product matrices, *Bulletin of the American Mathematical Society*, **40**, 461–8.

Rothenberg, T. J. (1971). Identification in parametric models, *Econometrica*, **39**, 577–91.

Rothenberg, T. J. and C. T. Leenders (1964). Efficient estimation of simultaneous equation systems, *Econometrica*, **32**, 57–76.

Rudin, W. (1964). *Principles of Mathematical Analysis*, 2nd edn, McGraw-Hill, New York.

Schönemann, P. H. (1985). On the formal differentiation of traces and determinants, *Multivariate Behavioral Research*, **20**, 113–39.

Schönfeld, P. (1971). Best linear minimum bias estimation in linear regression, *Econometrica*, **39**, 531–44.

Sherin, R. J. (1966). A matrix formulation of Kaiser's varimax criterion, *Psychometrika*, **31**, 535–8.

Smith, R. J. (1985). Wald tests for the independence of stochastic variables and disturbance of a single linear stochastic simultaneous equation, *Economics Letters*, **17**, 87–90.

Stewart, G. W. (1969). On the continuity of the generalized inverse, *SIAM Journal of Applied Mathematics*, **17**, 33–45.

Styan, G. P. H. (1973). Hadamard products and multivariate statistical analysis, *Linear Algebra and Its Applications*, **6**, 217–40.

Sugiura, N. (1973). Derivatives of the characteristic root of a symmetric or a hermitian matrix with two applications in multivariate analysis, *Communications in Statistics*, **1**, 393–417.

Sydsaeter, K. (1974). Letter to the editor on some frequently occurring errors in the economic literature concerning problems of maxima and minima, *Journal of Economic Theory*, **9**, 464–6.

Sydsaeter, K. (1981). *Topics in Mathematical Analysis for Economists*, Academic Press, London.

Tanabe, K. and M. Sagae (1992). An exact Cholesky decomposition and the generalized inverse of the variance-covariance matrix of the multinomial distribution, with applications, *Journal of the Royal Statistical Society, Series B*, **54**, 211–219.

Ten Berge, J. M. F. (1993). *Least Squares Optimization in Multivariate Analysis*, DSWO Press, Leiden.

Theil, H. (1965). The analysis of disturbances in regression analysis, *Journal of the American Statistical Association*, **60**, 1067–79.

Theil, H. (1971). *Principles of Econometrics*, John Wiley, New York.

Theil, H. and A. L. M. Schweitzer (1961). The best quadratic estimator of the residual variance in regression analysis, *Statistica Neerlandica*, **15**, 19–23.

Tracy, D. S. and P. S. Dwyer (1969). Multivariate maxima and minima with matrix derivatives, *Journal of the American Statistical Association*, **64**, 1576–94.

Tracy, D. S. and R. P. Singh (1972). Some modifications of matrix differentiation for evaluating Jacobians of symmetric matrix transformations, *Symmetric Functions in Statistics* (ed. D. S. Tracy), University of Windsor, Canada.

Tucker, L. R. (1966). Some mathematical notes on three-mode factor analysis, *Psychometrika*, **31**, 279–311.

Von Rosen, D. (1985). *Multivariate Linear Normal Models with Special References to the Growth Curve Model*, Ph.D. Thesis, University of Stockholm.

Wang, S. G. and S. C. Chow (1994). *Advanced Linear Models*, Marcel Dekker, New York.

Wilkinson, J. H. (1965). *The Algebraic Eigenvalue Problem*, Clarendon Press, Oxford.

Wilks, S. S. (1962). *Mathematical Statistics*, 2nd edn, John Wiley, New York.

Wolkowicz, H. and G. P. H. Styan (1980). Bounds for eigenvalues using traces, *Linear Algebra and Its Applications*, **29**, 471–506.

Wong, C. S. (1980). Matrix derivatives and its applications in statistics, *Journal of Mathematical Psychology*, **22**, 70–81.

Wong, C. S. (1985). On the use of differentials in statistics, *Linear Algebra and Its Applications*, **70**, 285–99.

Wong, C. S. and K. S. Wong (1979). A first derivative test for the ML estimates, *Bulletin of the Institute of Mathematics, Academia Sinica*, **7**, 313–21.

Wong, C. S. and K. S. Wong (1980). Minima and maxima in multivariate analysis, *Canadian Journal of Statistics*, **8**, 103–13.

Yang, Y. (1988). A matrix trace inequality, *Journal of Mathematical Analysis and Applications*, **133**, 573–74.

Young, W. H. (1910). *The Fundamental Theorems of the Differential Calculus*, Cambridge Tracts in Mathematics and Mathematical Physics, No. 11, Cambridge University Press, Cambridge.

Zellner, A. (1962). An efficient method of estimating seemingly unrelated regressions and tests for aggregation bias, *Journal of the American Statistical Association*, **57**, 348–68.

Zyskind, G. and F. B. Martin (1969). On best linear estimation and a general Gauss–Markov theorem in linear models with arbitrary nonnegative covariance structure, *SIAM Journal of Applied Mathematics*, **17**, 1190–202.

Index of symbols

The symbols listed below are followed by a brief statement of their meaning and by the number of the page on which they are defined.

General symbols

$\equiv$	equals, by definition
$\Rightarrow$	implies
$\leftrightarrow$, iff	if and only if
$\square$	end of proof
min	minimum, minimize
max	maximum, maximize
sup	supremum
lim	limit, 70
i	imaginary unit, 13
e, exp	exponential
!	factorial
$\prec$	majorization, 215
$\lvert\xi\rvert$	absolute value of scalar ξ
$\bar{\xi}$	complex conjugate of scalar ξ, 13

Sets

$\in$, $\notin$	belongs to (does not belong to), 3
$\{x:x\in S,\ x$ satisfies $P\}$	set of all elements of S with property P, 3
$\subset$	is a subset of, 3
$\cup$	union, 3
$\cap$	intersection, 3–4
$\emptyset$	empty set, 3
$B-A$	complement of A relative to B, 4
A^c	complement of A, 4
$\mathbb{N}$	$\{1, 2, 3, \ldots\}$, 3
$\mathbb{R}$	set of real numbers, 4
$\mathbb{R}^n$, $\mathbb{R}^{m\times n}$	set of real $n\times 1$ vectors ($m\times n$ matrices), 4
$\mathbb{R}^n_+$	positive orthant of $\mathbb{R}^n$, 371
$\mathbb{C}^{n\times n}$	set of complex $n\times n$ matrices, 161
$\mathring{S}$	interior of S, 66
S'	derived set of S, 66
$\bar{S}$	closure of S, 66
∂S	boundary of S, 66
$B(c)$, $B(c;\ r)$, $B(C;\ r)$	ball with centre c (C), 65, 95
$N(c)$, $N(C)$	neighbourhood of c (C), 65

$\mathcal{M}(A)$	column space, 8
$\mathcal{O}$	Stiefel manifold, 359

Special matrices and vectors

I, I_n	identity matrix (of order $n \times n$), 6
0	null matrix, null vector, 5
K_{mn}	commutation matrix, 47
K_n	K_{nn}, 47
N_n	$\frac{1}{2}(I_{n^2}+K_n)$, 48
D_n	duplication matrix, 49
$J_k(\lambda)$	Jordan block, 18
$\imath$	sum vector $(1, 1, \ldots, 1)'$

Operations on matrix A and vector a

A'	transpose, 5
A^{-1}	inverse, 9
A^+	Moore–Penrose inverse, 32
A^-	generalized inverse, 38
$\operatorname{dg} A$, $\operatorname{dg}(A)$	diagonal matrix containing the diagonal elements of A, 6
$\operatorname{diag}(a_1, \ldots, a_n)$	diagonal matrix containing $a_1, \ldots, a_n$ on the diagonal, 6
A^2	AA, 7
$A^{1/2}$	square root, 7
A^p	p-th power, 217
$A^\#$	adjoint (matrix), 10
A^*	complex conjugate, 13
A_k	principal submatrix of order $k \times k$, 24
A_v	block-vec of A, 107, 190
(A, B), $(A:B)$	partitioned matrix
$\operatorname{vec} A$, $\operatorname{vec}(A)$	vec operator, 30
$\mathrm{v}(A)$	vector containing a_{ij} $(i \geqslant j)$, 48–49
$r(A)$	rank, 8
$\operatorname{tr} A$, $\operatorname{tr}(A)$	trace, 10
$\lvert A \rvert$	determinant, 9
$\lVert A \rVert$	norm of matrix, 10
$\lVert a \rVert$	norm of vector, 6
$M_p(x, a)$	weighted mean of order p, 227
$M_0(x, a)$	geometric mean, 229
$A \geqslant B$, $B \leqslant A$	$A-B$ is positive semidefinite, 22
$A > B$, $B < A$	$A-B$ is positive definite, 22

Matrix products

$\otimes$	Kronecker product, 27
$\odot$	Hadamard product, 45

Functions

$f: S \to T$	function defined on S with values in T, 70
ϕ, ψ	real-valued function, 170
f, g	vector function, 170
F, G	matrix function, 170
$g \circ f$, $G \circ F$	composite function, 91, 115

Derivatives

d	differential, 81, 82, 95
d^2	second differential, 104, 114–115
d^n	n-th order differential, 114
$\mathrm{D}_j\phi$, D_jf_i	partial derivative, 85
$\mathrm{D}^2_{kj}\phi$, $\mathrm{D}^2_{kj}f_i$	second-order partial derivative, 99
$\partial\phi(X)/\partial X$, $\partial F(X)/\partial X$, $\partial F(X)//\partial X$	matrices of partial derivatives, 171–172
$\phi'(\xi)$	derivative of $\phi(\xi)$, 80
$\mathrm{D}\phi(x)$, $\partial\phi(x)/\partial x'$	derivative of $\phi(x)$, 85, 87, 173
$\mathrm{D}f(x)$, $\partial f(x)/\partial x'$	derivative (Jacobian matrix) of $f(x)$, 87, 173
$\mathrm{D}\,F(X)$, $\partial \mathrm{vec}\, F(X)/\partial(\mathrm{vec}\, X)'$	derivative (Jacobian matrix) of $F(X)$, 95, 173
$\nabla\phi$, ∇f	gradient, 87
$\phi''(\xi)$	second derivative of $\phi(\xi)$, 110
$\mathrm{H}\phi(x)$, $\partial^2\phi(x)/\partial x\partial x'$	second derivative (Hessian matrix) of $\phi(x)$, 99, 100, 188
$\mathrm{H}f(x)$	second derivative (Hessian matrix) of $f(x)$, 101, 189
$\mathrm{H}F(X)$	second derivative (Hessian matrix) of $F(X)$, 114, 189

Statistical symbols

Pr	probability, 243
a.s.	almost surely, 247
$\mathscr{E}$	expectation, 244
$\mathscr{V}$	variance (matrix), 245
$\mathscr{V}_{as}$	asymptotic variance (matrix), 327
$\mathscr{C}$	covariance (matrix), 245
ML	maximum likelihood, 313
MSE	mean squared error, 252
$\mathscr{F}_n$	information matrix, 314
$\mathscr{F}$	asymptotic information matrix, 314
$\sim$	is distributed as, 250
$\mathscr{N}_m(\mu, \Omega)$	normal distribution, 250

Subject index

WILEY SERIES IN PROBABILITY AND STATISTICS
ESTABLISHED BY WALTER A. SHEWHART AND SAMUEL S. WILKS

Probability and Statistics Section

*ANDERSON · The Statistical Analysis of Time Series
ARNOLD, BALAKRISHNAN, and NAGARAJA · A First Course in Order Statistics
BACCELLI, COHEN, OLSDER, and QUADRAT · Synchronization and Linearity: An Algebra for Discrete Event Systems
BASILEVSKY · Statistical Factor Analysis and Related Methods: Theory and Applications
BERNARDO and SMITH · Bayesian Statistical Concepts and Theory
BILLINGSLEY · Convergence of Probability Measures
BOROVKOV · Asymptotic Methods in Queuing Theory
BOROVKOV · Ergodicity and Stability of Stochastic Processes
BRANDT, FRANKEN, and LISEK · Stationary Stochastic Models
CAINES · Linear Stochastic Systems
CAIROLI and DALANG · Sequential Stochastic Optimization
CONSTANTINE · Combinatorial Theory and Statistical Design
COOK · Regression Graphics
COVER and THOMAS · Elements of Information Theory
CSÖRGŐ and HORVÁTH · Weighted Approximations in Probability Statistics
CSÖRGŐ and HORVÁTH · Limit Theorems in Change Point Analysis
DETTE and STUDDEN · The Theory of Canonical Moments with Applications in Statistics, Probability, and Analysis
*DOOB · Stochastic Processes
DRYDEN and MARDIA · Statistical Analysis of Shape
DUPUIS and ELLIS · A Weak Convergence Approach to the Theory of Large Deviations
ETHIER and KURTZ · Markov Processes: Characterization and Convergence
FELLER · An Introduction to Probability Theory and Its Applications, Volume 1, *Third Edition*, Revised; Volume II, *Second Edition*
FULLER · Introduction to Statistical Time Series, *Second Edition*
FULLER · Measurement Error Models
GHOSH, MUKHOPADHYAY, and SEN · Sequential Estimation
GIFI · Nonlinear Multivariate Analysis
GUTTORP · Statistical Inference for Branching Processes
HALL · Introduction to the Theory of Coverage Processes
HAMPEL · Robust Statistics: The Approach Based on Influence Functions
HANNAN and DEISTLER · The Statistical Theory of Linear Systems
HUBER · Robust Statistics
HUŠKOVA, BERAN, and DUPAČ · Collected Works of Jaroslav Hájek – With Commentary

*Now available in a lower priced paperback edition in the Wiley Classics Library.

Probability and Statistics (Continued)

IMAN and CONOVER · A Modern Approach to Statistics
JUREK and MASON · Operator-Limit Distributions in Probability Theory
KASS and VOS · Geometrical Foundations of Asymptotic Inference
KAUFMAN and ROUSSEEUW · Finding Groups in Data: An Introduction to Cluster Analysis
KELLY · Probability, Statistics, and Optimization
LINDVALL · Lectures on the Coupling Method
McFADDEN · Management of Data in Clinical Trials
MANTON, WOODBURY, and TOLLEY · Statistical Applications Using Fuzzy Sets
MORGENTHALER and TUKEY · Configural Polysampling: A Route to Practical Robustness
MUIRHEAD · Aspects of Multivariate Statistical Theory
OLIVER and SMITH · Influence Diagrams, Belief Nets, and Decision Analysis
*PARZEN · Modern Probability Theory and Its Applications
PRESS · Bayesian Statistics: Principles, Models, and Applications
PUKELSHEIM · Optimal Experimental Design
RAO · Asymptotic Theory of Statistical Inference
RAO · Linear Statistical Inference and Its Applications, *Second Edition*
RAO and SHANBHAG · Choquet-Deny Type Functional Equations with Applications to Stochastic Models
ROBERTSON, WRIGHT, and DYKSTRA · Order Restricted Statistical Inference
ROGERS and WILLIAMS · Diffusions, Markov Processes, and Martingales, Volume I: Foundations, *Second Edition*; Volume II: Îto Calculus
RUBINSTEIN and SHAPIRO · Discrete Event Systems: Sensitivity Analysis and Stochastic Optimization by the Score Function Method
RUZSA and SZEKELEY · Algebraic Probability Theory
SCHEFFE · The Analysis of Variance
SEBER · Linear Regression Analysis
SEBER · Multivariate Observations
SEBER and WILD · Nonlinear Regression
SERFLING · Approximation Theorems of Mathematical Statistics
SHORACK and WELLNER · Empirical Processes with Applications to Statistics
SMALL and McLEISH · Hilbert Space Methods in Probability and Statistical Inference
STAPLETON · Linear Statistical Models
STAUDTE and SHEATHER · Robust Estimation and Testing
STOYANOV · Counterexamples in Probability
TANAKA · Time Series Analysis: Nonstationary and Noninvertible Distribution Theory
THOMPSON and SEBER · Adaptive Sampling
WELSH · Aspects of Statistical Inference
WHITTAKER · Graphical Models in Applied Multivariate Statistics
YANG · The Construction Theory of Denumerable Markov Processes

Applied Probability and Statistics Section

ABRAHAM and LEDOLTER · Statistical Methods for Forecasting
AGRESTI · Analysis of Ordinal Categorical Data
AGRESTI · Categorical Data Analysis

---

*Now available in a lower priced paperback edition in the Wiley Classics Library.

Applied Probability and Statistics (Continued)

ANDERSON, AUQUIER, HAUCK, OAKES, VANDAELE, and WEISBERG · Statistical Methods for Comparative Studies

ARMITAGE and DAVID (editors) · Advances in Biometry

*ARTHANARI and DODGE · Mathematical Programming in Statistics

ASMUSSEN · Applied Probability and Queues

*BAILEY · The Elements of Stochastic Processes with Applications to the Natural Sciences

BARNETT and LEWIS · Outliers in Statistical Data, *Third Edition*

BARTHOLOMEW, FORBES, and McLEAN · Statistical Techniques for Manpower Planning, *Second Edition*

BATES and WATTS · Nonlinear Regression Analysis and Its Applications

BECHHOFER, SANTNER, and GOLDSMAN · Design and Analysis of Experiments for Statistical Selection, Screening, and Multiple Comparisons

BELSLEY · Conditioning Diagnostics: Collinearity and Weak Data in Regression

BELSLEY, KUH, and WELSCH · Regression Diagnostics: Identifying Influential Data and Sources of Collinearity

BHAT · Elements of Applied Stochastic Processes, *Second Edition*

BHATTACHARYA and WAYMIRE · Stochastic Processes with Applications

BIRKES and DODGE · Alternative Methods of Regression

BLOOMFIELD · Fourier Analysis of Time Series: An Introduction

BOLLEN · Structural Equations with Latent Variables

BOULEAU · Numerical Methods for Stochastic Processes

BOX · Bayesian Inference in Statistical Analysis

BOX and DRAPER · Empirical Model-Building and Response Surfaces

BOX and DRAPER · Evolutionary Operation: A Statistical Method for Process Improvement

BUCKLEW · Large Deviation Techniques in Decision, Simulation, and Estimation

BUNKE and BUNKE · Nonlinear Regression, Functional Relations, and Robust Methods: Statistical Methods of Model Building

CHATTERJEE and HADI · Sensitivity Analysis in Linear Regression

CHOW and LIU · Design and Analysis of Clinical Trials

CLARKE and DISNEY · Probability and Random Processes: A First Course with Applications, *Second Edition*

*COCHRAN and COX · Experimental Designs, *Second Edition*

CONOVER · Practical Nonparametric Statistics, *Second Edition*

CORNELLL · Experiments with Mixtures, Designs, Models, and the Analysis of Mixture Data, *Second Edition*

*COX · Planning of Experiments

CRESSIE · Statistics for Spatial Data, *Revised Edition*

DANIEL · Applications of Statistics to Industrial Experimentation

DANIEL · Biostatistics: A Foundation for Analysis in the Health Sciences, *Sixth Edition*

DAVID · Order Statistics, *Second Edition*

*DEGROOT, FIENBERG, and KADANE · Statistics and the Law

DODGE · Alternative Methods of Regression

DOWDY and WEARDEN · Statistics for Research, *Second Edition*

DUNN and CLARK · Applied Statistics: Analysis of Variance and Regression, *Second Edition*

ELANDT-JOHNSON and JOHNSON · Survival Models and Data Analysis

EVANS, PEACOCK, and HASTINGS · Statistical Distributions, *Second Edition*

FLEISS · The Design and Analysis of Clinical Experiments

---

*Now available in a lower priced paperback edition in the Wiley Classics Library.

Applied Probability and Statistics (Continued)

FLEISS · Statistical Methods for Rates and Proportions, *Second Edition*
FLEMING and HARRINGTON · Counting Processes and Survival Analysis
GALLANT · Nonlinear Statistical Models
GLASSERMAN and YAO · Monotone Structure in Discrete-Event Systems
GNANADESIKAN · Methods for Statistical Data Analysis of Multivariate Observations, *Second Edition*
GOLDSTEIN and LEWIS · Assessment: Problems, Development, and Statistical Issues
GREENWOOD and NIKULIN · A Guide to Chi-Squared Testing
*HAHN · Statistical Models in Engineering
HAHN and MEEKER · Statistical Intervals: A Guide for Practitioners
HAND · Construction and Assessment of Classification Rules
HAND · Discrimination and Classification
HEIBERGER · Computation for the Analysis of Designed Experiments
HINKELMAN and KEMPTHORNE · Design and Analysis of Experiments, Volume 1: Introduction to Experimental Design
HOAGLIN, MOSTELLER, and TUKEY · Exploratory Approach to Analysis of Variance
HOAGLIN, MOSTELLER, and TUKEY · Exploring Data Tables, Trends, and Shapes
HOAGLIN, MOSTELLER, and TUKEY · Understanding Robust and Exploratory Data Analysis
HOCHBERG and TAMHANE · Multiple Comparison Procedures
HOCKING · Methods and Applications of Linear Models: Regression and the Analysis of Variables
HOGG and KLUGMAN · Loss Distributions
HOLLANDER and WOLFE · Nonparametric Statistical Methods
HOSMER and LEMESHOW · Applied Logistic Regression
HØYLAND and RAUSAND · System Reliability Theory: Models and Statistical Methods
HUBERTY · Applied Discriminant Analysis
JACKSON · A User's Guide to Principle Components
JOHN · Statistical Methods in Engineering and Quality Assurance
JOHNSON · Multivariate Statistical Simulation
JOHNSON and KOTZ · Distributions in Statistics

Continuous Multivariate Distributions

JOHNSON, KOTZ, and BALAKRISHNAN · Continuous Univariate Distributions, Volume 1, *Second Edition*
JOHNSON, KOTZ, and BALAKRISHNAN · Continuous Univariate Distributions Volume 2, *Second Edition*
JOHNSON, KOTZ, and BALAKRISHNAN · Discrete Multivariate Distributions
JOHNSON, KOTZ, and KEMP · Univariate Discrete Distributions, *Second Edition*
JUREČKOVÁ and SEN · Robust Statistical Procedures: Asymptotics and Interrelations
KADANE · Bayesian Methods and Ethics in a Clinical Trial Design
KADANE and SCHUM · A Probabilistic Analysis of the Sacco and Vanzetti Evidence
KALBFLEISCH and PRENTICE · The Statistical Analysis of Failure Time Data
KELLY · Reversibility and Stochastic Networks
KHURI, MATHEW, and SINHA · Statistical Tests for Mixed Linear Models
KLUGMAN, PANJER, and WILLMOT · Loss Models: From Data to Decisions
KLUGMAN, PANJER, and WILLMOT · Solutions Manual to Accompany Loss Models: From Data to Decisions
KOVALENKO, KUZNETZOV, and PEGG · Mathematical Theory of Reliability of Time-Dependent Systems with Practical Applications

*Now available in a lower priced paperback edition in the Wiley Classics Library.

Applied Probability and Statistics (Continued)

LAD · Operational Subjective Statistical Methods: A Mathematical, Philosophical, and Historical Introduction

LANGE, RYAN, BILLARD, BRILLINGER, CONQUEST, and GREENHOUSE · Case Studies in Biometry

LAWLESS · Statistical Models and Methods for Lifetime Data

LEE · Statistical Methods for Survival Data Analysis, *Second Edition*

LEPAGE and BILLARD · Exploring the Limits of Bootstrap

LINHART and ZUCCHINI · Model Selection

LITTLE and RUBIN · Statistical Analysis with Missing Data

MAGNUS and NEUDECKER · Matrix Differential Calculus with Applications in Statistics and Econometrics

MALLER and ZHOU · Survival Analysis with Long Term Survivors

MANN, SCHAFER, and SINGPURWALLA · Methods for Statistical Analysis of Reliability and Life Data

McLACHLAN and KRISHNAN · The EM Algorithm and Extensions

McLACHLAN · Discriminant Analysis and Statistical Pattern Recognition

McNEIL · Epidemiological Research Methods

MILLER · Survival Analysis

MONTGOMERY and PECK · Introduction to Linear Regression Analysis, *Second Edition*

MYERS and MONTGOMERY · Response Surface Methodology: Process and Product in Optimization Using Designed Experiments

NELSON · Accelerated Testing, Statistical Models, Test Plans, and Data Analyses

NELSON · Applied Life Data Analysis

OCHI · Applied Probability and Stochastic Processes in Engineering and Physical Sciences

OKABE, BOOTS, and SUGIHARA · Spatial Tesselations: Concepts and Applications of Voronoi Diagrams

PANKRATZ · Forecasting with Dynamic Regression Models

PANKRATZ · Forecasting with Univariate Box-Jenkins Models: Concepts and Cases

PIANTADOSI · Clinical Trials: A Methodologic Perspective

PORT · Theoretical Probability for Applications

PUTERMAN · Markov Decision Processes: Discrete Stochastic Dynamic Programming

RACHEV · Probability Metrics and the Stability of Stochastic Models

RÉNYI · A Diary on Information Theory

RIPLEY · Spatial Statistics

RIPLEY · Stochastic Simulation

ROUSSEEUW and LEROY · Robust Regression and Outlier Detection

RUBIN · Multiple Imputation for Nonresponse in Surveys

RUBINSTEIN · Simulation and the Monte Carlo Method

RUBINSTEIN and MELAMED · Modern Simulation and Modeling

RYAN · Statistical Methods for Quality Improvement

SCHUSS · Theory and Applications of Stochastic Differential Equations

SCOTT · Multivariate Density Estimation: Theory, Practice, and Visualization

*SEARLE · Linear Models

SEARLE · Linear Models for Unbalanced Data

SEARLE, CASELLLA, and McCULLOCH · Variance Components

STOYAN, KENDALL, and MECKE · Stochastic Geometry and Its Applications, *Second Edition*

STOYAN and STOYAN · Fractals, Random Shapes, and Point Fields: Methods of Geometrical Statistics

THOMPSON · Empirical Model Building

---

*Now available in a lower priced paperback edition in the Wiley Classics Library.

Applied Probability and Statistics (Continued)

THOMPSON · Sampling
TIJMS · Stochastic Modeling and Analysis: A Computational Approach
TIJMS · Stochastic Models: An Algorithmic Approach
TITTERINGTON, SMITH, and MAKOV · Statistical Analysis of Finite Mixture Distributions
UPTON and FINGLETON · Spatial Data Analysis by Example, Volume I: Point Pattern and Quantitative Data
UPTON and FINGLETON · Spatial Data Analysis by Example, Volume II: Categorical and Directional Data
VAN RIJCKEVORSEL and DE LEEUW · Component and Correspondence Analysis
WEISBERG · Applied Linear Regression, *Second Edition*
WESTFALL and YOUNG · Resampling-Based Multiple Testing: Examples and Methods for *p*-Value Adjustment
WHITTLE · Systems in Stochastic Equilibrium
WOODING · Planning Pharmaceutical Clinical Trials: Basic Statistical Principles
WOOLSON · Statistical Methods for the Analysis of Biomedical Data
*ZELLNER · An Introduction to Bayesian Inference in Econometrics

Texts and References Section

AGRESTI · An Introduction to Categorical Data Analysis
ANDERSON · An Introduction to Multivariate Statistical Analysis, *Second Edition*
ANDERSON and LOYNES · The Teaching of Practical Statistics
ARMITAGE and COLTON · Encyclopedia of Biostatistics. 6 Volume set
BARTOSZYNSKI and NIEWIADOMSKA-BUGAJ · Probability and Statistical Inference
BERRY, CHALONER, and GEWEKE · Bayesian Analysis in Statistics and Econometrics: Essays in Honor of Arnold Zellner
BHATTACHARYA and JOHNSON · Statistical Concepts and Methods
BILLINGSLEY · Probability and Measure, *Second Edition*
BOX · R. A. Fisher, the Life of a Scientist
BOX, HUNTER, and HUNTER · Statistics for Experimenters: An Introduction to Design, Data Analysis, and Model Building
BOX and LUCEÑO · Statistical Control by Monitoring and Feedback Adjustment
BROWN and HOLLANDER · Statistics: A Biomedical Introduction
CHATTERJEE and PRICE · Regression Analysis by Example, *Second Edition*
COOK and WEISBERG · An Introduction to Regression Graphics
COX · A Handbook of Introductory Statistical Methods
DILLON and GOLDSTEIN · Multivariate Analysis: Methods and Applications
DODGE and ROMIG · Sampling Inspection Tables, *Second Edition*
DRAPER and SMITH · Applied Regression Analysis, *Third Edition*
DUDEWICZ and MISHRA · Modern Mathematical Statistics
DUNN · Basic Statistics: A Primer for the Biomedical Sciences, *Second Edition*
FISHER and VAN BELLE · Biostatistics: A Methodology for the Health Sciences
FREEMAN and SMITH · Aspects of Uncertainty: A Tribute to D. V. Lindley
GROSS and HARRIS · Fundamentals of Queueing Theory, *Third Edition*
HALD · A History of Probability and Statistics and their Applications Before 1750
HALD · A History of Mathematical Statistics from 1750 to 1930
HELLER · MACSYMA for Statisticians
HOEL · Introduction to Mathematical Statistics, *Fifth Edition*

---

*Now available in a lower priced paperback edition in the Wiley Classics Library.

Texts and References Section (Continued)

JOHNSON and BALAKRISHNAN · Advances in the Theory and Practice of Statistics: A Volume in Honor of Samuel Kotz

JOHNSON and KOTZ (editors) · Leading Personalities in Statistical Sciences: From the Seventeenth Century to the Present

JUDGE, GRIFFITHS, HILL, LÜTKEPOHL, and LEE · The Theory and Practice of Econometrics, *Second Edition*

KHURI · Advanced Calculus with Applications in Statistics

KOTZ and JOHNSON (editors) · Encyclopedia of Statistical Sciences. Volumes 1 to 9 with Index

KOTZ and JOHNSON (editors) · Encyclopedia of Statistical Sciences: Supplement Volume

KOTZ, REED, and BANKS (editors) · Encyclopedia of Statistical Sciences: Update Volume 1

KOTZ, REED, and BANKS (editors) · Encyclopedia of Statistical Sciences: Update Volume 2

LAMPERTI · Probability: A Survey of the Mathematical Theory, *Second Edition*

LARSON · Introduction to Probability Theory and Statistical Inference, *Third Edition*

LE · Applied Survival Analysis

MALLOWS · Design, Data, and Analysis by Some Friends of Cuthbert Daniel

MARDIA · The Art of Statistical Science: A Tribute to G. S. Watson

MASON, GUNST, and HESS · Statistical Design and Analysis of Experiments with Applications to Engineering and Science

MURRAY · X-STAT 2.0 Statistical Experimentation, Design Data Analysis, and Nonlinear Optimization

PURI, VILAPLANA, and WERTZ · New Perspectives in Theoretical and Applied Statistics

RENCHER · Methods of Multivariate Analysis

RENCHER · Multivariate Statistical Inference with Applications

ROSS · Introduction to Probability and Statistics for Engineers and Scientists

ROHATGI · An Introduction to Probability Theory and Mathematical Statistics

RYAN · Modern Regression Methods

SCHOTT · Matrix Analysis for Statistics

SEARLE · Matrix Algebra Useful for Statistics

STYAN · The Collected Papers of T. W. Anderson: 1943–1985

TIERNEY · LISP-STAT: An Object-Oriented Environment for Statistical Computing and Dynamic Graphics

WONNACOTT and WONNACOTT · Econometrics, *Second Edition*

WILEY SERIES IN PROBABILITY AND STATISTICS

ESTABLISHED BY WALTER A. SHEWHART AND SAMUEL S. WILKS

---

*Now available in a lower priced paperback edition in the Wiley Classics Library.

Survey Methodology Section

BIEMER, GROVES, LYBERG, MATHIOWETZ, and SUDMAN · Measurement Errors in Surveys
COCHRAN · Sampling Techniques, *Third Edition*
COX, BINDER, CHINNAPPA, CHRISTIANSON, COLLEDGE, and KOTT (editors) · Business Survey Methods
*DEMING · Sample Design in Business Research
DILLMAN · Mail and Telephone Surveys: The Total Design Method
GROVES and COUPER · Nonresponse in Household Interview Surveys
GROVES · Survey Errors and Survey Costs
GROVES, BIEMER, LYBERG, MASSEY, NICHOLLS, and WAKSBERG · Telephone Survey Methodology
*HANSEN, HURWITZ, and MADOW · Sample Survey Methods and Theory, Volume I: Methods and Applications
*HANSEN, HURWITZ, and MADOW · Sample Survey Methods and Theory, Volume II: Theory
KASPRZYK, DUNCAN, KALTON, and SINGH · Panel Surveys
KISH · Statistical Design for Research
*KISH · Survey Sampling
LESSLER and KALSBEEK · Nonsampling Error in Surveys
LEVY and LEMESHOW · Sampling of Populations: Methods and Applications
LYBERG, BIEMER, COLLINS, de LEEUW, DIPPO, SCHWARZ, TREWIN (editors) · Survey Measurement and Process Quality
SKINNER, HOLT, and SMITH · Analysis of Complex Surveys

*Now available in a lower priced paperback edition in the Wiley Classics Library.

9 780471 986331